Beautiful SCENT

Joachim Mensing

Beautiful SCENT

The Magical Effect of Perfume on Well-Being

 Springer

Joachim Mensing
Miami Beach, FL, USA

ISBN 978-3-662-67258-7   ISBN 978-3-662-67259-4   (eBook)
https://doi.org/10.1007/978-3-662-67259-4

This Springer imprint is published by the registered company Springer-Verlag GmbH, DE, part of Springer Nature.
The registered company address is: Heidelberger Platz 3, 14197 Berlin, Germany

How nice that you have taken this book in your hand—and maybe even bought it. With 15 chapters, the book can already be described as a "thick tome"; And all those who do not count themselves among the absolute bookworms may still be a little skeptical, but you don't have to be. Because in front of you is a journey that can also be immersed in stages. I promise you: Our 15 destinations are anything but boring, and in the end you will belong to the insiders of a fascinating industry that fascinates others with its wealth of knowledge.

A journey into the world, trends and future, but also into the history of perfume and its industry with the latest findings from psychology, aromatherapy, brain research, and neuroperfumery, how scents can work for our well-being.

Preface

Dear reader,

Welcome to a journey into the world, trends, and future of perfumes.

This book—"Beautiful Scent", published Summer 2023—is a translation of the German edition—"Schöner Riechen", published 2021—but contains thrilling updates and captivating additions that will enhance our journey. On this journey I want to share with you current insider knowledge of a fascinating industry. In particular, I will introduce you to the latest findings from psychology, aromatherapy, brain research, and neuroperfumery and show you how scents can work even more targeted for our well-being.

Sometimes our journey also leads into the past, into the history of perfume, in which many new, surprising, but also amusing discoveries await us.

You may be wondering what motivated a psychologist like me to write a book called *Beautiful Scent*. Well, for more than 30 years, beautiful smells have determined my professional practice, both in therapy and, above all, in the fields of perfumery and the perfume industry. As a fragrance psychologist and trend coach with perfume training at one of the big fragrance manufacturers, I was involved in the creation of numerous, often even award-winning perfumes. For a long time now, I have been sharing my experiences, above all of how perfumes work on us humans and how they can work even better in the future, with an interested audience in lectures on the subject of "beautiful scent" and in the training course "perfume insider", which I offer you in this book. In all of this, I count myself among the lucky people who can say that they have made their hobby their profession. The world of scents and their effects has exerted an undiminished fascination on me for decades. And the most beautiful effect of a perfume is that it can specifically create more joie de vivre.

But why now a book about "beautiful scent" or the effect of scent and perfume? In the perfume industry, almost revolutionary developments can currently be observed. Groundbreaking discoveries and innovations in the most diverse areas of the perfume industry—for example in the field of neuroperfumery or in brain research relevant to the perfume industry, fragrance psychology and therapy, as well as fragrance selection and advice—are changing an entire industry. In addition, there are new approaches to the work of perfumers and fragrance development. I will report on all these topics in a well-understood, entertaining, and exciting way.

I will meet a scientific claim by means of literature references which will also help you to deepen the various aspects on your own.

Beforehand, I would like to explain something for a better terminological understanding. When I talk about perfumery, I mean two areas:

— The world in which perfumers create fragrances. This is about the world of the fragrance industry, but also about the world of scientific research. The latter, for example, investigates how we smell, how scents work, but also why we perfume ourselves at all.

— The world of stationary and online perfumeries, i.e. the so-called perfume specialist trade, which is about advising and selling perfumes, care and other beauty products to the end consumer.

Sometimes the two areas overlap. But I will always make it clear which one I am referring to when using the term "perfumery".

■ What do I want to achieve with this book?
The latest findings—in particular in the areas of neuroperfumery and fragrance psychology—are to be an inspiration for the practice of perfumery. This is primarily about perfume development and creation, but also about trade, marketing, and consulting, as well as the exciting field of fragrance therapy. In the first eight chapters of this book, I will virtually introduce you to these topics, refresh your knowledge about perfumery, and make you an insider of perfume and perfumery. But even if you are not working in this industry and are only interested in the topic of perfume, I would like to inspire you with this book. With the individual topics—in particular in the first eight chapters—I deliberately do not proceed in chronological order, because I want to create more tension and variety when reading. Nevertheless, everything fits into a big whole, and you will get to know all facets of the industry in a comprehensive and detailed manner.

More than that: I want to make you a beneficiary of a fascinating industry. For this, this book gives you many practical tips.

My goal is also to introduce you to the latest findings and innovations in such a way that you can use them for yourself today. I will show you that perfumes can do much more than just smell good. For example, fragrances are now being developed into real active perfumes that specifically increase well-being, health, and even joie de vivre. You can experience this for yourself in this book. One more thing: I want to make you your own perfumer—without the use of large tools.

■ Do you need any prior knowledge for this book?
Definitely not! Even if I report on the latest from research and practice as well as on the future of perfumery, you only need a little curiosity to discover a world that creates the most beautiful of all drugs: perfume.

■ What do you gain if you just take a sniff of the world of perfumes?
Based on my 30 years of professional experience, I make you an insider of a fascinating industry, including its trends and future prospects. With this book, you will become a witness of an industry that is currently experiencing a fragrant revolution, more precisely: the third revolution since the beginnings of perfumery about 6000 years ago. Probably the perfumery is even much older. The International Museum of Perfumery in the French Grasse dates its beginning back to 7000 BC. on the basis of objects found in the Near East, which can be regarded as perfume and cosmetic containers. This would make perfumery over 9000 years old.

The history of perfume and its effects alone are a fascinating topic in themselves—and the future prospects are just as exciting. So I can promise you that you will never be bored while reading this book. Often you will probably also smile. For example, you will learn which arguments the great philosophers put forward against the beautiful scents and how they fought against perfumes and perfumery; or which tricks and methods were used in the past to win the "influencers" of the old days for his perfume creation.

- **If you have little time, can you still read parts of the book and get well-informed?**
If you have little time and value quick information, this book offers you a "compact course for the perfume insider".

It provides comprehensive information on perfumes, the industry, current developments and trends, and gives all those who want to be active in our fascinating industry—who want to bring their own perfume to the market, for example, and perhaps position themselves as a brand or perfume influencer—a quick orientation. For this, I recommend you read the "smelling around" ▶ Chaps. 1, 2, 3, 4 and 5, 6, 7, 8 and from the ▶ Chap. 9, 12 and 13 the Sects. 9.6, 12.3, 12.4 and 13.3. These chapters and sections are the training content of my compact course for the perfume insider (contact: ▶ Online Perfume Academy). In particular, the ▶ Chap. 5, 6 and 7 give insider and practical knowledge of how trade is calculated for a new perfume, and provide many tips on how to develop and successfully position a perfume in the market.

In the following, I will give you a short overview of the essential contents of individual chapters that you might not even want to read in order to avoid tension and anticipation. Certainly, the overview is helpful in finding topics that might interest you.

But before I discuss the contents of individual chapters, allow me to take a small excursion into the world of perfume. Because the term perfume already demands a clarification due to the title of the book. It refers to two different areas:

— I use the term "perfume" on the one hand as a general term for scent. This includes everything that smells good or is supposed to smell good. In English, the word "fragrance" is used for "scent", with which something that smells good is also associated. This term does not only refer to perfume (English "perfume"), but also to other sources of scent. The understanding of perfume in the German-speaking world allows both associations. So the term "perfume" can refer to perfume products of the perfume industry as well as to other beautiful scents. As another differentiation, in addition to the word "fragrance" in English, the term "scent" is also used, which refers to a more characteristic smell. The German word "Geruch"—in American English "odor" or in British English "odour"—is used as a neutral term of the olfactory, but can also be associated with pleasant and unpleasant smells. In contrast, the term "Gestank" (English "stink") clearly refers to something that smells unpleasant. This shows the essential difference between perfume and stink: the former is experienced olfactorily positive, the latter negative. Of course, there are also perfumes that stink for some people, and scents whose smell is not experienced as good by everyone— but this would lead too far here.
Furthermore, in English, the terms "aroma" and "perfume" are also distinguished, which has also become established in German understanding. Both terms refer to pleasant smells. While "aroma" refers more to plants, spices and food, "perfume" is usually associated with scent and flowers.
— The term "perfume" is also used for a product or in the plural for various perfume types of perfumery. The core term "perfume" thus refers to an olfactory work of art with which one describes the most valuable—so to speak the crowning achievement of what perfumery and perfumers have to offer -. Created as a product for external use, perfume usually consists of a liquid mixture of various

so-called fragrance ingredients, or perfume components. In addition, there are also so-called "solid perfumes" that can be applied as a cream.

The fragrance ingredients of a perfume are dissolved as fragrance oil in alcohol and some distilled water. Depending on the concentration of the fragrance ingredients or the fragrance oil content, the following different perfume types are distinguished: "perfume", "eau de parfum", "eau de toilette", and "eau de cologne".

As an olfactory work of art, a perfume should not only serve the enjoyment of fragrance. Perfume lovers expect a perfume to surprise with a new and innovative fragrance impression at its launch. Furthermore, it must—and the perfume classics serve as the best example for this—at least subjectively increase the attractiveness and well-being of the wearer. To achieve this, selected and valuable fragrance ingredients—meaning great raw materials—must be used. But creating a perfume also requires perfume-making skills. So in addition to all artistic attributes, one can expect a good adhesive strength, that is, durability, during the fragrance from an excellent perfume. Perfumers often further differentiate, distinguishing between "sillage" and "longevity" of a perfume. Originally from France, the term sillage describes the lingering scent of the perfume a person wears, or the scent trail it leaves behind. For example, you enter an empty elevator and know exactly what the person inside smelled like. The term longevity describes the durability of a perfume on a fragrance strip or on the skin. It may even be the case that a perfume that lasts well on the skin is not necessarily associated with a high sillage impression. The crowning achievement of perfumery with usually the best adhesive strength is the pure perfume, the so-called "extrait de parfum", referred to in English as "perfume extract" or "pure perfume". The fragrance ingredient or fragrance oil concentration is between 15 and 30% and can even exceed 40%.

Sometimes the individual meanings for perfume also overlap. But I will always make it clear—as with the term perfume—which context I am referring to when using the word.

Joachim Mensing

Acknowledgments

Writing a book, even when you have the topic clearly in mind, is a journey into the unknown, at least as far as the time it takes to write is concerned. Without my understanding family, I would not have been able to make this journey. I would therefore like to thank my wife, Charmaine and our children Felicia, Shani, and Elaina for giving me the time, motivation, and support to write.

You also underestimate that new findings from research can change, rewrite, re-determine or weigh intermediate goals. Who would have thought, for example, that people apparently can smell even without parts of the olfactory bulb or that approaches to perfumery can already be found in the Stone Age. I am therefore grateful to Hanne-Lore Heilmann and Angela Müller, who helped me to edit the book and patiently gave me the opportunity to rewrite already finished text passages and adapt them to the current state of knowledge.

On a journey, an author quickly loses sight of the topics he or she wants to report on, which points are or could be of particular interest to others. I am therefore very grateful to Regina Spelman, who pointed out to me that I should bring in my experience as a psychologist to a greater extent.

It takes well-disposed people who support you on your journey and give you materials and tips on how to do it even better. I would like to thank in this context Kai Brüninghaus, Dr. Andreas Leistikow, and Dr. Alberto Peek, but especially Jacques Schumacher, who wonderfully enriched the book with his artistic pictures.

My very special thanks go to the participants of my perfume seminars, whose positive feedback also motivated me to write this book and especially to current findings from the Neuroparfümerie.

Finally, I would like to thank colleagues from the perfume and cosmetic industry, from trade and from associations, who inspired me in my work with many conversations.

About this Book

In the 1. Chapter I share with you a lot of basic knowledge about the topic "scent and history". You will learn a lot of fascinating things about the first scent rituals of people and gain new insights into the beginnings of perfume and perfumery. You will discover what was known in antiquity about the effects of scents, what importance scents already had for the quality of life and, above all,

— which fragrance effects one could already achieve in the Stone Age.

In the first chapter we also discuss the state of the art of current smell research, in particular what the influence of scents on our consciousness, our emotions and mood, and thus

— how scents work in the brain on our psyche;
— what influence smelling and smell as well as perfumes, fragrances, and aromatherapy have on us.

In doing so, we look at first examples from the neuroperfumery, how and where certain fragrances and essential oils stimulate individual brain regions and their networks.

Allow me to point out in this context that I use the terms brain area and brain region synonymously because, as I will show, it is difficult, if not impossible, to show a spatially bounded area of the cerebrum—as one associates it rather with the term brain area—for smelling.

In the 2. Chapter you can read what you need to know about the beautiful scents from the perspective of practice,

— how the current developments in perfumes look;
— what current trends and new expectations there are among fragrance users, such as the current development from perfume to active perfume.

But we also discuss questionable aspects of the topic of perfume—what one should know and consider when using fragrance and enjoying fragrance, especially as a perfume lover.

The ▶ Chap. 3, 4, 5, 6, 7 and 8 are primarily aimed at anyone who wants to quickly get a refreshingly different insight into the world of perfume and the latest perfumery. Here you will find more topics that I also discuss in my perfume insider workshops and training. For example, you will learn

— who are the people behind the scenes of perfumes today and how the calculation for a new fragrance works,
— how to market yourself and your perfume as an "influencer",
— which are the favorite ingredients of the perfumery and their current trends,
— how smelling works,
— who is the "maître des parfums" in our brain, creating sensory impressions and connecting scents with our self-image,
— who or what in the brain, so to speak as a fragrance manager, decides about our fragrance choice and favorite scents,

- how certain fragrance notes affect our personality, our emotions, mood and well-being,
- what the fragrance psychology knows about the use of perfumes and fragrance families and why people perfume themselves,
- why smelling is such a unique and fascinating sense that says a lot about our mental and physical well-being and our health,
- how and why our sense of smell can do so much good for us in the context of a scent therapy.

In addition, I will once again go into the exciting history of perfumery since its beginning in pre-Christian times. This includes two revolutions in the past that have changed the way we smell perfumes. I also discuss the role of women, without whom perfumery, as we know it, would not have taken place.

Chapters 9 and 10. have a focus on the future of smell and self-scent therapy. Here I go into the third revolution taking place in perfumery, which will completely change the entire industry. The latest discoveries from scent and brain research as well as from neuroperfumery, a very young science, are at the center. You will learn:

- what perfumes can do today—in addition to smelling good,
- how the future of perfume looks,
- which brain regions, i.e. networks, can be particularly well stimulated and with which perfumes,
- how the latest findings from neuroperfumery and brain research can be used in sales and scent consulting.

In addition, I present the latest findings from scent psychology and therapy and explain how perfumes can lead to more enjoyment of life.

Everything is described very application-oriented and understandable, with many suggestions for the practice of perfumery and the experience of perfumes. Especially in the 10. chapter I ask you to be creative and to create your own therapeutic perfume. My goal is that you experience the power of neuroperfumes, that is, of scents that are specifically designed to reprogram desired experience. With them, inner strength, joy and well-being can be increased as part of a fragrance-supported therapy.

In the 11. chapter I present to you current findings from brain research, neuro- and sales psychology for fragrance consulting, and in the 12. chapter you can discover your fragrance preferences and your associated desired experiences with the help of a neuropsychological self-test. I also give you many practical tips on how to make fragrance consulting even more fascinating in the stationary perfumery, because many trend researchers would sign the following: The new customer—and that's all of us—doesn't buy a product or a service anymore, but first and foremost a positive experience.

That's why I'm introducing you to a new, exciting, but also unexpected perfumery: the Experience Perfumery. So don't be surprised if you read about World Turtle Day or Scented Dancing in this context. Both serve the enjoyment of perfume and the increase of fragrance and self-experience, but also of health and health prevention.

Finally, in the 13, 14, and 15 chapter a number of well-kept insider information of the perfume industry, trade and marketing with you. You will learn about trends, innovations and developments with perfumers, how the perfume trade and its customers change, and who wins the race for the stationary perfumery of tomorrow. But I also show which strategic opportunities the stationary perfumery offer to benefit from a changing market with new consumer needs. I am also following these questions:

- How does the German perfume market differ from other markets?
- What are the perfume trends of the next years?
- Are there trends in fragrances that are similar to fashion trends globally?
- Are markets increasingly determined by national preferences?

In this context, I show that Germany is not in the middle of Europe on the map of fragrance preferences and market mentality, but a completely different country.

The last chapter of the book is a postscript and dedicated to the perfumers. I will investigate how and where the modern perfumer emerged, how perfumers can become their own luxury brand, and what opportunities, (today), do perfumers have.

Contents

The Research of the Fragrance Effect

... On Consciousness, Emotion and Mood for More Well-Being and Enjoyment—From the Stone Age to Current Brain Research

Contents

also wants to know when people first discovered the effect of scents and smells. For this, there are completely new and surprising findings from archaeology. They raise the questions of whether the art of perfumery is much older than previously assumed and when one can actually speak of perfumery per se.

Let us therefore first discuss what has been discovered about the effect of scents in the early days of man and in antiquity, and what is meant by perfumery, before turning to the first current findings of brain research on perfume, scent, and effect in this chapter.

1.1 History of the Fragrance Effect

1.1.1 Surprising Findings of Archaeology 77,000 Years Ago

This first and second chapter are mainly for those who are new to perfumery or in need of a compact refresher of basic knowledge. If you are already advanced, a quick read through should suffice. To make orientation easier and to facilitate quick reference later on, you will find a summary with literature references at the end of each chapter, in case you wish to explore the topics covered in more depth.

Perfume lovers will confirm that the right scent can be used to deliberately improve wellbeing and mood, as well as reducing stress. In this way, consciousness and mood can be improved and positive, joyful emotions can be triggered.

This raises the question: are these simply soft effects or subjective illusions, like the placebo effect known as the "effect of nothing", or is it actually possible to modulate consciousness, mood, and emotion through perfume and smell? If the latter is the case, one wants to see this effect confirmed by scientifically sound evidence. One

So far it was assumed that the perfume had arisen about 9000 years ago. This was based on the findings of perfume and cosmetic containers in the Middle East. Even today it is still assumed that the development of the perfume had taken a rapid upturn with the beginning of the first cities and the needs of their increasingly cramped living population. So Jericho, which was already bursting at the seams, had already erected a city wall in 8000 BC. Perfumes had to work as insecticides in addition for hygiene reasons alone. The high population density of all cities of antiquity—and even in modern times—caused odors and thus accompanying pests through waste and excrement. So Paris was said to smell like a large sewer in the 18th century. Residents and visitors were trapped in a fog of feces and urine.

When fragrances were first used to cover up the smell in the cities is not known. Fragrance applications or the attempts to at least make the pests more bearable with plants are very old. Whether perfume was first used as a fragrance to honor the gods, or to improve living conditions and health, I will discuss at a later point. Much speaks in favor of the latter due to current research. Here is an interesting note on the etymology of the term "perfume", which comes from Latin: The word consists of the preposition "per" (= through) and the noun "fumum" (= smoke, steam).

However, recent archaeological discoveries led to surprising findings with regard to the history of perfume. So plants were already used in the Stone Age. In South Africa, for example, 77,000 year old beds made of plants were found. They were strewn with the aromatic leaves of the quince from the family of the laurel plants. These leaves had a poisonous effect on insects and thus kept even the smallest pests such as fleas and lice away.

There is also much to suggest that perfume and perfuming go back to the fragrance to honor the gods. Fire has always had a magical fascination. Its discovery over a million years ago was a giant step in human development. The mastery and use of fire were the most essential difference between man and animal. Fire served the early humans not only for heating, but also for prolonging daylight. In addition, food could be heated with fire and thus chewed more easily and its shelf life extended. In addition, fire destroyed, for example, by smoking pathogens and offered protection against wild animals.

In ancient myths, gods brought fire to humans. In Greek mythology, fire belonged to the god Zeus, who watched over it. Therefore, it is likely that the early humans thanked their gods for fire and its benefits with fragrant incense.

Before we discuss the effects of perfumes or deliberately created scents and odors in different periods of cultural history from the perspective of research, let's first look at what is meant by perfume.

1.1.2 On the Difference between Perfume and Smell Application

Perfume can generally be referred to as a subjectively pleasant or attractive smell result. This arises from the conscious preparation of one or more chemicals that stimulate the sense of smell or the olfactory cells, which are exclusively or predominantly intended to olfactorily act on humans externally on the body and/or for their interests in their living space. Of course there is also the perfume of food, which is not relevant in this context. Those who use perfume are primarily interested in the good smell, without necessarily being aware of the preparation process and the effects.

In order to serve the different "scenting interests", perfume knows a wide range of application areas. So you can also perfume something that does not belong to your own body, for example other people or objects, animals and rooms. Even the invisible can be perfumed. The intention is always to create a pleasant or attractive smell result or something that is associated with a corresponding expectation. The question arises here whether people of the Stone Age already knew something like "perfume" and could "perfume" themselves or something in order to use the scent deliberately for a more pleasant experience, such as fewer mosquito bites. Probably, but not certainly, they used plants such as the Cape quince untreated. They may also have rubbed the leaves of the Cape quince on the skin to increase the effect, with the additional effect of a good smell. According to our current understanding, this would already have been a form of perfume.

The perfume of objects as a modern area of functional perfumery would also be conceivable for Stone Age people. Because if they had also expressed the leaves over the beds for better effect—which may have been the case—one would have to speak of a perfume of their objects. At least a precursor of perfume took place 77,000 years ago. Even if the scent was possibly not obtained from a substance, but only the pure, untreated plant was used in parts, the smell of the leaves was already used specifically. In contrast to this precursor, real perfume requires the pleasant smell to be obtained by a process such as at least expressing a plant, which is then used in a targeted manner. This means that certain incenses, such as those known from ancient Egypt and Mesopotamia, meet the criteria of perfume. Because the substances most commonly used for incense, such as resins, were obtained by the process of tree felling, or cutting the tree bark. Incense by simple burning of one or more plants is therefore not considered perfume.

It can be objected to this view that when mixing and burning various untreated substances, a pleasant or attractive smell can indeed be produced and this should be considered perfumery. Because strictly speaking, a conscious process takes place here, namely a mixing of ingredients in order to increase the scent experience and scent effect.

> "Perfuming" in this sense is a consciously induced process in which a pleasant and attractive smell is produced by preparing a substance or substances or by mixing two or more substances, which is then used externally to serve people.

Back to our example. If, in combination with the quince, other plants were used 77,000 years ago to produce a pleasant smell of the beds, then the Stone Age people would already have perfumed their beds.

This raises another, very subjective question: Must perfume necessarily go hand in hand with a pleasant smell? The French word "parfum", which we have adopted into the German language as "parfüm", expresses a pleasant smell of fragrances. I therefore suggest distinguishing between perfume and odour application.

tion. Odour applications can also serve or act on people. They may smell less good or even repulsive and do not require any prior processing—with the exception of collecting the plants or substances. In this sense, Stone Age people already knew targeted odour applications and their effects.

In this context, another question arises: Must perfume be based on a deliberately induced process? Depending on how this question is answered, rolling in the dirt would also be a form of perfume—but only from the dog's point of view, not from the human's. Not only dogs find dirt, mud, and other things that stink to us humans irresistible. They have different triggers for their scenting. One is certainly the search for more enjoyment, further for more well-being. This includes the masking of one's own smell by a camouflage smell that helps when hunting. These are occasions of perfume that are also known to humans—especially the scent-supported hunt. It does not have to take on Old Testament proportions like in the case of Judith, who saved the city of Betulia from Holofernes, an Assyrian general who was known for his looting, murder, and arson in the Middle East. Judith offered herself to him in pretense. In order to appear more attractive and inconspicuous in her actual intentions, she perfumed her body with the most tempting scents. Holofernes fell for it and was decapitated during the seduction of Judith that same night.

One could make it easy for oneself and explain that animals do not consciously perfume themselves, for example, for the hunt. But that would mean denying animals a consciousness. The number of researchers who agree with this at least in the case of mammals and some birds is decreasing. They rather assume that these have a form of consciousness. Then one would have to say that animals that deliberately perfume themselves with dirt or other substances—especially one after the other at different places—also perfume themselves in a certain way.

1

1.1.3 First Mentions of the Use and Effect of Fragrance in Ancient Times and Antiquity

The boundaries between perfume, medicine, pharmacy, and religion were fluid in antiquity. This is shown in the development and application of fragrances and the first perfumes. At the beginning of perfume, perfumes were mostly a mixture of plants or plant parts that were either dissolved in oil, processed into perfume creams, or used as fragrant fragrances such as resins. In modern perfume production, alcohol is usually used as a carrier medium, which alone influences the fragrance effect. Alcohol evaporates much faster than oil and distributes the fragrance faster and finer in the air. Perfumes based on oil or cream have development problems, especially with fresh citrus and flowery notes. Their top note takes a little longer to develop beautifully. Therefore, the discovery of alcoholic perfume must have seemed like a small olfactory firework, especially for the scent of perfume.

The alcoholic perfume was only known to the ancient Egyptians from about 400 BC. But they still held on to fragrant creams for a long time, which we today still know as so-called "solid perfumes". They offered the dual function of longer lasting fragrance and additional care, which the Egyptians who were conscious in this area particularly appreciated. For perfume ointments, they usually used ox fat as a carrier substance, for perfume oils the oil obtained from the kernels of the desert date, since the olive oil used in other cultures was not available due to climatic conditions bad for the olive tree. Only from the 10th century AD was alcohol used almost exclusively in perfumery in the Orient and Africa, and only from the 14th century in the Western cultural area, when the Crusaders brought the knowledge from the Orient back to their homeland.

Much of the beginnings of perfumery are shrouded in darkness, as there are no or only very few written records of it. I would like to introduce four authors here who also reported on the use of fragrances as well as perfume trends and the effects of their time: Theophrastus, Dioscorides, Pliny the Elder, and Claudius Galen.

» **Theophrastus** (Theophrastos) was born on the Greek island of Lesbos and lived around 300 BC. He was a pupil of Plato and Aristotle. With his small work *De odoribus* ("On the smells"), which also reports on the production of perfumes, he gave a first insight from a Western point of view into perfumery. The majority of the text deals with the then more popular fragrant "salve oils" for rubbing or anointing people and for the consecration of objects for their use in temples. Theophrastus also indicates the positive effect of some perfume oils on tumors, abscesses, headaches, and exhaustion.

Whether Theophrastus knew of earlier reports specifically on plants and their use is not handed down. But one can assume that Plato and Aristotle had a very extensive library. So what we today know as **Papyrus Ebers** could at least be known to him in content, even if this Egyptian scroll was written more than a thousand years before Theophrastus. It contains the texts on medical topics known from the reign of Amenophis I. These include instructions for the preparation of remedies, for example against toothache, injuries, and parasites. Furthermore, there is information on perfumes and (incense-) fumigations. The approximately 20 meter long scroll was dated to the middle of the 16th century BC.

In the even older **Gilgamesh Epic**, written around 2100 BC, the legendary king Ur (Mesopotamia) is told. He used incense to probably set the gods in a pleasant mood with frankincense and myrrh.

The Egyptians also achieved true mastery in the field of olfactory stimulation. They already started in the morning with time-related incense. The most famous Egyptian scents include the perfume consisting of many ingredients **Kyphi** ("Welcome to the gods"), which was used at the end of the day during the evening incense. The perfume, which is said to have even caused hypnotic states, was created from frankincense,myrrh,sandalwood, and other natural raw materials. There were also variants of Kyphi for other occasions. Scents or incense were applied throughout the day in ancient Egypt. In Heliopolis, one of the oldest Egyptian cities and the main seat of worship of the sun god Re, incense was probably burned in the morning with resins from the frankincense tree, at noon with resins of the myrrh, and in the evening at sunset with Kyphi, to honor Re, but also to do good for people. Kyphi was supposed to calm and relieve fears, provide a sleeping aid for more beautiful dreams, relieve asthma sufferers and—especially important in the Nile region—keep insects and pests away. Perfumes that contribute to the intensification of the mood and even to the modulation of the mood were therefore a matter of course even in the time of Theophrastus.

Perhaps Theophrastus had already gained insight into Indian and Chinese thought. Probably between the 15th and 8th centuries BC.—possibly even earlier—the Indian **Vedas** were written. They contain information on sandalwood,cinnamon, coriander, and myrrh. This knowledge went into the ayurvedic medicine and treatment, which developed around 500 BC and is still widespread today in India, Nepal, and Sri Lanka. The Vedas also contain first hints of healing fire or the Agnihotra ritual, which probably took place first in Nepal and which I will come back to later. The ability to distill, that is, to produce desired fragrances with water vapor, may

have been invented in the Indian subcontinent long before Theophrastus. In the 1970s, Italian archaeologist Paolo Rovesti found a terracotta object in a Pakistani research site that was dated 3000 BC and was probably used for distilling fragrances. It is generally assumed that the Sumerians already used a very simple vaporization process around 1300 BC to produce essential oils, which they could use to increase the fragrance effect. The time was ripe for this because perfumes were already being produced on a large scale by 2000 BC. Almost every large temple had a "perfume factory" to meet the great demand for scents for the effect on the gods and the population. The oldest perfume factory discovered to date is in Cyprus. It is estimated to be 2000 years old.

There were already trading centers for the packaging of fragrances or for the supply of perfume factories in antiquity, where certainly techniques and information on perfume production were exchanged. The trading centers included Babylon and another location on the west coast of present-day Israel, where maritime shipping took place mainly to Greece. Fixed trade routes that served as caravan routes as early as 1700 BC ran through the entire Middle and Near East to the trading centers. Perhaps they even led to the Far East, and Theophrastus had access to traditional Chinese medicine as it was set out in the *Book of the Yellow Emperorfor internal medicine*. This work was probably published around the time of Theophrastus. But there are also assumptions that the publication of individual texts could date back 2000 years.

Theophrastus certainly knew about the great Persian culture of scent. The elegant Damascus roses were cultivated in **Persia**, from where the much-praised rose scents came at that time. By the way, the intensely smelling Damascus rose was the medicinal plant of the year 2013. It has an

1

anti-inflammatory, antispasmodic, and antipyretic effect.

Perhaps Theophrastus also knew the **Second Book of Moses** (Exodus, ca. 1200 BC), which explains to the Hebrews the production of anointing oil for the consecration of priests, among other things. It consisted of myrrh,cinnamon, and calamus, mixed with olive oil.

There is ample evidence for the central role of scent in antiquity. It can also be assumed that people have wanted to know more about perfumes and their effects for a long time. Therefore, one can assume that Theophrastus' little book *De odoribus* was a bestseller in his time.

In the oils of antiquity, iris and the highly prized myrrh, one of the oldest perfume ingredients, were used preferentially. The myrrh fragrance consists of the gum resin hardened in the air, which is obtained from various myrrh trees by cutting the trunks and branches. Depending on the processing, myrrh smells as a fragrance slightly spicy-balsamic-sweet. Myrrh resin was used in three different ways: as perfume, for example in creams, for incense, and as medicine. Especially mixed with wine, myrrh was both a gustatory and an olfactory pleasure in ancient cultures.

Myrrh is also mentioned in the Old and New Testaments. Thus, in the Old Testament, there is a recipe for the production of anointing oil. In the New Testament, myrrh is given to the infant Jesus by the Three Wise Men together with gold and frankincense. Myrrh was considered a very valuable miracle drug with its diverse effects, which was not easy to obtain for many cultures—including the Egyptians.

The myrrh resin was used in the treatment of wounds. As a myrrh tincture, it is still recommended today for mild inflammations of the gums and mucous membranes. When taken orally, the plant works particularly well in combination with other substances such as chamomile and coffee charcoal, which are obtained from the seeds, that is, the beans, of coffee plants. This is also confirmed by modern medicine (Langhorst et al. 2013). Currently, the healing and protective effects of myrrh in combination with frankincense ("frankincense") are reported for many symptoms. In particular, anti-inflammatory and antibacterial, even cancer-fighting effects are mentioned (Cao et al. 2019).

In antiquity, the promised effects of myrrh as a fragrance were so great that the attraction of the sexes would even be increased to incest. In Greek mythology, Smyrna, the Greek word for myrrh, was the daughter of the king of Cyprus, who fell in love with her father and became pregnant by him. As punishment, Aphrodite turned her into a myrrh tree at the birth of her child Adonis, according to Greek legend. This gave people something: the original effect of perfume from a tree that everyone could use from then on to make themselves and others more attractive. The resin of the tree goes back to a fall from grace, just like the apple, only that it is an olfactory fall from grace.

» **Dioscurides** from the Roman province of Cilicia, a region in Asia Minor, was a Greek doctor who lived in the 1st century AD. He is considered a pioneer of pharmacology. Little is known about him except for the text collection *De materia medica*. Among other things, it reports on the fragrance effect of plants that he got to know on his travels to Egypt.*De materia medica* was considered the most important reference work on pharmacology for more than 1500 years and is still a historical source of information on plants, minerals and animal substances that were used as medicines in antiquity. Dioscurides was one of the first to call for the separation of pharmacy from perfumery—probably also because the weak effects of the scents of perfumes and incense could not be reconciled with the standards of pharmacy and medicine.

Pliny the Elder, also called Gaius Plinius Secundus, also lived in the 1st century AD and came from the region around Naples. He died there during the great eruption of Vesuvius at the age of 56. For his Roman employers he was active in many functions, such as scholar, officer, and civil servant. In doing so, he found time to write about natural history, the use and trends of perfume, and body care products. Through him we know today that the Romans liked to spray rose water with a fragrant scent in their theaters. In this context, he wrote that the Persians had invented perfume. Around 800 BC, they already were the main supplier of rose oil. He also explained Pliny the Elder, when and at what distance certain plants show the best scent effect and in which region they grow. He also reported on ancient perfume trends. These included the iris perfume from Corinth (Greece) and the rose perfume of the ancient Greek city of Phaselis, which was replaced by a quince blossom perfume from the Greek island of Kos. The perfume trends of the Roman Empire were mainly, but not exclusively, from Greece.

One of the most famous Egyptian perfumes was created in the city of Mendes in the eastern Nile delta and exported from there to Rome. The success of the Egyptian perfumers was in the apparently longer effect and durability of their perfumes. They were long superior to Greek perfume production. This was due to the better sense of touch of the perfumers for the right processing temperature and the order of the ingredients, essential factors in the effect of their perfumes.

Claudius Galen lived in the 2nd century AD and is considered one of the most important physicians of antiquity. He came from Pergamon, an ancient Greek city in today's Turkey. Galen loved the number four, from which various concepts such as the doctrine of the elements with fire, earth, air, and water as the basic elements of all being were derived mainly from a philosophical and medical point of view. He also developed the ancient doctrine of the four body fluids blood, mucus, yellow bile, and black bile further. Galen was skeptical of the effect of perfumes. He criticized above all the insufficient knowledge of the perfumers, which was responsible for the ineffectiveness of their creations. His criticism is unlikely to have referred to the plants themselves. For the Romans even referred to their loved ones as "my myrrh" or "my cinnamon".

Even before Galen, Hippocrates, and Dioscorides doubted the therapeutic effect of the fragrances mixed together by perfumers. Galen wanted to achieve his criticism essentially by separating the professions of perfumer and physician or pharmacist. But that took a long time. It was not until the World's Fair of 1867 in Paris that it finally happened. Perfumes were given their own place. The perfume department, which also included soap, was separated from the pharmacies. A separate trade part with now really independent profession emerged. The wonderful scent of soap and other products such as leather gloves had long since detached itself from the medical image. The customer wanted to offer pure pleasure, improvement of quality of life, and a more attractive experience of everyday things. Indirectly, Galen even contributed to the fact that the perfume split into the following three sectors:

- **Fine perfumery,** in which enjoyment of fragrance, attractiveness, and well-being are in the foreground,
- **functional perfumery,** which makes things more attractive and therefore also makes them, and
- **aromatherapy,** which adheres to the border area of medicine and pharmacy and wants to contribute to the alleviation of diseases and the improvement of well-being.

1

Later, I will discuss how these three areas are increasingly overlapping. This is also due to groundbreaking findings from new methods of brain and smell research as well as changed expectations of consumers of perfumes.

1.1.4 The Scent of the Blue Lotus Or: The Divine Perfume from the Primeval Waters

Today, the most is known about the use of fragrances in ancient Egypt. They are usually based on archaeological finds and knowledge. However, these only allow partly and mostly indirect, cautious statements about the use and effect of fragrances. Names of fragrances that were found on pictorial and written tablets or drawings are still unsecured, as they may also refer to another fragrance. Also, the assumed frequent use of individual fragrances could not always be confirmed by corresponding finds.

An example of a possible case of mistaken identity is Myrrh (Commiphora). Today—4000 years later—it is assumed that this anti-inflammatory and sensual stimulating plant or its resin used for incense in Egypt was called " ntyw". However, egyptologists are not quite sure about this. The term could also refer to Frankincense (Boswellia sacra), which was used more often in ancient times to worship the gods and is now said to have tumor-dissolving properties in cancer cells (Suhail et al. 2011). The term "ntyw" could have referred to pistachio (Pistacia lentiscus) in ancient Egypt, which is often used as a fragrance ingredient and for incense. Its essential oil has antibacterial, anti-inflammatory, and antioxidant effects. The term "ntyw" could also generally refer to the most luxurious and fragrant as well as the hardest to obtain imported resins or fragrances, of which frankincense and myrrh were the most im-

portant. The pistachio, which was probably called "snfr", grew on the east and west banks of the Nile.

For example, fragrances of Frankincense and Myrrh had to be imported from the mythical land of Punt with great effort. To this day, it is not certain whether it was located on the Horn of Africa, in Zimbabwe or in the Orient, today's Yemen. The most precious fragrance was frankincense— not least because the tree refused to grow in Egypt despite the pharaoh's efforts. According to sources from other cultures of antiquity, the gods were said to be particularly fond of the smell of frankincense smoke.

Depending on the type and preparation, Frankincense smells quite complex: as an essential oil balsamic-spicy and even slightly lemon-like with a coniferous undertone; as an incense balsamic-woody with a subtle lemon note.

The myrrh bush from the family of balsam plants was the most attractive for the people of that time. From it one obtained an aphrodisiac, which was used in different fragrance types. The effective resin was so popular that as early as 1500 BC. in Zimbabwe Myrrh terraces, today known as Nyanga terrace complex, were created. The large-scale cultivation of the trees obviously served the export of their resins (Duffey 2005).

The fascination and need for perfume seemed to be present from the beginning in the general population. In addition to Kyphi, frankincense, myrrh, and pistachio, the favorite scent of the Egyptians was the scent of the blue lotus, which smells the most intense of all lotus species and symbolized rebirth. Already the perfumers of that time tried to capture the scent in numerous creations. It smells fresh and green on the plant, somewhat spicy, with a sweet-balsamic undertone. Blue Lotus also burned well and diffused a unique green-spicy smell when smoked. The flower was often served and decorated at parties and banquets to create a special olfactory experience (Byl 2012).

In the morning, Blue Lotus smells particularly good. It was also the favorite plant of Nefertem, the protective god of perfume, to which it was dedicated. According to Egyptian mythology, the flower of the blue lotus emerged from the primeval waters with the perfume god sitting on it.

So even long before the "blue flower" of Romanticism was known in our cultural circle, the birth of the blue lotus flower contributed to the romanticization and mystification of the olfactory.

In general, people in antiquity were fascinated by the mystical-romantic effect and aura of scents. They offered a counterweight to the assessment of scents oriented purely on medical points of view. This metaphysical consideration gave perfumery something supernatural and shrouded in mystery from the very beginning, which still surrounds it today. The production of scent and its effects were also far more than just an extraction process in many other cultures. For example, Chinese Taoists believed as early as the 4th century BC that the extraction of the scent of a plant represents the liberation of its soul and that this effect is expressed in six moods: noble, luxurious, beautiful, quiet, withdrawn, and sophisticated.

Sensual scent was also associated with function. An example of this is the Chinese incense clock. Time could take on the form of a scent with this clock, which almost spatialized it. A cartridge filled with ground incense released scent at certain times. Scent effect thus also became part of the experience of time.

1.1.5 The Art of Smoking—On the Meditative and Artistic Effect of Scent

In Japan, the most sensual and artistic use of fragrance was developed as part of smoking rituals. With the introduction of Buddhism between 500 and 600 AD, the smoking of incense came to the island—a custom that had been adopted from the Chinese. In old Japan, special smoking vessels were created and their own ceremony developed. To this day, the ceremonial smelling from a smoking cup is practiced with eight to ten guests. The fragrance vessel is placed on the left palm at chest height, lightly covered with the right hand and passed on to the next person in a clockwise direction after two to three inhalations. In total, four to six different smokes are smelled. During each pass, the participants note what they smell and how they experience the effect of the fragrance. In the 16th century AD, a Zen monk named ten inspiring virtues that should guide the smoking. They are still valid today and will be described by me later.

The Bronze Age Indus culture in the northwest of the Indian subcontinent used specially made vessels for the burning of fragrance around 2000 BC. These were called burners, which were probably known from Mesopotamia, the Near East and perhaps also from Egypt. Before the development of their own devices for smoking incense, natural burners such as the mineral meerschaum were used, which tobacco lovers still know from meerschaum pipes.

Early on, national and regional fragrance mentality developed. This is also evident in techniques and equipment for smoking such as the pans and plates common in Egypt. They were initially made of stone, later, between the 5th and 4th millennia BC, of clay. From these pans, the production of metal vessels made of gold and bronze later developed, later also of iron, on which the smoking mixture was placed. In some cases, especially in ancient Egypt, these were partly small works of art with which the fragrance effect of the smoking became a very private experience and enjoyment. In the reports of the German Archaeological Institute, Department Cairo,

1

the discovery of a particularly beautiful find from 1914 in a temple was reported in 1978. In the bronze bowl at the end of a carved smoking arm, small smoking pastilles were burned on charcoal. Similarly beautifully designed smoking arms from the period 1500–1200 BC were also found.

In ancient Arabia, the smoking bowls were preferably made of gold or silver, in Africa of meerschaum, while large, noble shells were used in Central and South America.

Back to the Japanese art of incense burning. It took some time for the incense burning culture to come to Japan and for incense burners to be developed for smelling. But then an equally fascinating and extremely artistic incense trend took place. While incense was initially used exclusively for sacred purposes, it was discovered by Japanese nobles at the imperial court in Kyoto for regular scent games around the 8th century AD. On the one hand, it was about recognizing individual incense scents by their smell, on the other hand it was also about creating pleasant smelling fragrance mixtures. These were assigned to seasons, but also to themes from literature, painting, and architecture.

Incense was also burned with various other ingredients for coordinated dance rituals—in the 16th century AD as a sophisticated incense ceremony known as "Koh-Dō", which means "the way to listen to the scent" (Koh = scent, Dō = way). During the following centuries, the art of fragrance and the handling of incense burners were perfected. New fragrances were also added. Even today, this ritual is carried out by Koh-Dō masters according to strict rules. An incense burner is passed around in the circle. The guests "listen to the scent" and try to identify the respective scents as mixtures of resins, herbs, and woods in traditional fragrance variants with agarwood or oud while burning.

The art of incense included the slow and even burning of the fragrance ingredients to

increase their effect. This required special knowledge of how individual fragrance ingredients release their delicious smell under heat and with which materials this process could be optimally influenced. It was important that the incense work was initially heated up controlled at low temperature, so that the aroma of each incense could develop fully. For this purpose, incense ash is still filled as a substrate into the incense burner and a piece of incense coal is lit on it. The glowing coal is pressed into the ash and covered like a cone. With the help of a metal rod, an air hole about one centimeter deep is pushed into the center of the cone to the coal. The content of the incense burner then looks optically like a mini volcano. A mother-of-pearl or ceramic disc of about two centimeters in diameter with valuable aromatics is placed on it, which are protected in the incense burner from burning too quickly and thus have a longer effect.

After the ritual is finished, an incense burner can still be used. Sticks of incense placed inside stand upright and firm. They are popular again today because of their ability to cleanse a room and smell protective. In contrast to incense burners, the smell of incense sticks after burning in a room actually gets better, complementing the scent presentation and prolonging the scent effect.

Many incense stick lovers report that they stimulate the senses and make one feel more grounded and receptive. Similar to the incense burning that dates back to ancient times and was also used as a means against evil, incense sticks offer a special effect to modern city dwellers: one feels good about what one smells and sees. Incense sticks become a sort of countermeasure to environmental pollution, which comes with living in the city and inevitably enters homes from the outside. For this reason, incense sticks have also evolved in recent years. They not only smell like powerful patchouli, but also like green notes that remind one of the scent of a cedar forest, and like delicate, high-quality frankincense notes. It is no co-incidence that there are now Koh-Dō rituals as incense sticks, such as those of Koh-Dō master Keijirou Hayashi, who today leads "Hayashi Ryushodo", a 186-year-old business in Kyoto that packages the incense sticks in a minimalist white box.

An unknown Zen monk described the "Ten Virtues of Koh" in the 16th/17th century AD, to which incense burning and its effects still orient themselves today. Koh, the incense,

▬ connects with the transcendent and thus provides access to a world that lies beyond normal sensory perception;
▬ supports the cleansing of body, mind, and soul;
▬ eliminates negativity and has a cleansing effect on the environment;
▬ promotes and strengthens mindfulness;
▬ is a companion in times of loneliness or loneliness;
▬ brings peace and tranquility to the everyday hustle and bustle;
▬ never gets tired and bored, even if it is used a lot;
▬ also promotes satisfaction in small quantities;
▬ even after long storage does not lose its effect;
▬ even with daily use does not damage.

So the art of Koh-Dō is mainly about consciously and meditatively immersing oneself in the present, in the here and now, through the sense of smell. As a form of Zens, it trains mental vigilance that should lead to enlightenment.

However, this type of incense should not be abused. Not only smoking, but also excessive inhalation of incense can damage health. Nevertheless, the Japanese enjoy the highest life expectancy in the world—even though their country is still the main market for incense.

As already mentioned, the Vedas also contain first indications of healing fire or of the Agnihotra ritual, which was probably first practiced in Nepal. In this incense burning, ghee, a product similar to clarified butter, is burned together with other fragrance ingredients such as cow dung and rice at every sunrise and sunset in the fire of a pyramidal copper vessel. Over the centuries, more and more ayurvedic incense mixtures have been created.

Traditionally, the fragrance effect is explained with the fact that nature is in a state of complete balance of forces at sunrise and sunset. These times were considered the best for cleansing and building energies. Traditional main goals of the fragrance effects are still today psychological and spiritual regeneration, combined with a harmonization of body, soul and spirit. The fragrance effect is supported by chanted mantras, sacred syllables, words or verses that are tuned to the vibration ratios of sunrise and sunset.

1.1.6 Healing Fragrance Magic—From Medical and Physiotherapeutic Fragrance Applications to Divination and Spiritual Seances

Already in early times, fragrance was used in connection with medical treatment and prevention. From this developed a real herbal medicine with fragrances. So in ancient times fragrance enjoyed great popularity as medicine. Already thousands of years before the current movement of aromatherapy came about, essential oils and fragrances were used to treat various physical symptoms and diseases, especially for pain. In numerous ancient cultures there are predecessors of aromatherapy, as we know it today. They were based on the conscious and deliberately used medical effect of fragrances. In China, numerous writings were published on this topic. In the 15/16th century the Chinese *Materia Medica* ap-

1

peared. It lists—like the work of Dioscurides—among other things all plants from which one assumed medical properties. In a separate section on essential oils, the chamomile is mentioned, which is said to relieve headaches. Jasmine is described as a general tonic and ginger-scented oil is praised as a remedy for malaria.

The African climate promoted the use of fragrant oils early on as protection against sun and dehydration as well as for the faster healing of grazes and small skin injuries. Although, for example, in Egypt people liked to cleanse themselves, sufficient hygiene was often difficult.

In the first strongholds of the perfume industry, one was particularly exposed to the often hot and dry climate, to fine sand and dust, which could also lead to problems with the respiratory tract. For this reason, people in Africa, the Orient and Asia have long appreciated the aromatic and cleansing effect of fragrances—if they could afford them. This is also confirmed by excavations in Tayma, an oasis inhabited as early as 3000 BC at the old frankincense route in today's Saudi Arabia. Areas such as temples, houses, public buildings and graves were perfumed in different ways for cleansing and, of course, for spiritual effect.

The frankincense, also called "scent of the gods", was considered an absolute luxury perfume. There are different variants in terms of smell or more than 25 types. These include

— the black variant, the Borena frankincense or black frankincense from Ethiopia, Zimbabwe, or Kenya with an earthy and sweet smell,
— the Olibanum Eritrea from—as the name suggests—from Eritrea and Sudan with a sweet, honey-like and spicy smell,
— the green-white variant from Oman with a slightly aromatic-lemon and balsamic-minty smell.

These variants were among the most expensive for the Egyptians in antiquity—just be-

cause of the long trade routes. They were used for enjoyable incense, especially in temples, while other types served the needs of wealthy people. As the most effective frankincense for medical purposes, Boswellia serrata is still one of the oldest and most respected medicinal plants in Ayurveda. According to a legend, an elephant made an Indian prince suffering from arthritis and rheumatism aware of this frankincense—which is not too surprising, after all, elephants represent wisdom and longevity in India. Boswellia serrata seems to have helped the prince—and apparently also the pachyderm.

Another reason for the popular smoking was the medical or health concerns of a wide population in relation to the bad air, which was considered a carrier of disease. Also, bad air was seen as unworthy of the gods and therefore tried to disguise stench with fragrance. Not only in Mesopotamia, one of the birthplaces of perfumery, but also the olfactory accompaniments of diseases were interpreted as signs of the possession of demons and evil spirits. They tried to exorcise them with counter-smell or with smell-supported conjurations and exorcistic rituals.

In Europe, smoking was used for disinfection, especially of the air in sickrooms, especially during the plague. For this purpose, one also used frankincense and myrrh, but mostly cheaper fragrances such as mugwort, pine resin, linden, camphor, pine, lavender, rosemary, thyme and juniper.

Fragrance was also used for spiritual stimulation and the support of visions, for example when divining—and apparently not entirely unsuccessful. In order to fall into a trance, for example, the oracle priestesses of Delphi sat on smoldering laurel leaf fumes. The fragrant smoke was fed to them through holes in the ground and surrounded them magically.

Fragrance could also become a medium for extrasensory perception, for example when contacting the dead—for example when, with the help of perfume, the aura was amplified and the effect and success of the sessions with their extrasensory perceptions were increased. Spiritualist seances reached their peak in the 19th century. This also included olfactory neuroses and olfactory hallucinations or phantom smells, which I will discuss later. But also fragrances such as scented candles, pillows or incense sticks as well as complete smoking rituals were popular not only during seances themselves, but also at all kinds of spiritualist meetings—and not only during the Victorian era. They have always been part of spiritual technique and served the decoration and equipment at the respective meetings. In addition, there were purely spiritual fragrance seances, in which the effect of the seance was to be increased by the use of perfume. Also, the discoverers of America reported on smoking rituals of the indigenous population, which aimed at conjurations with psychophysical, but also purely medical effects. Fragrance oils were used by the natives mainly for therapeutic purposes. For this purpose, one relied on the combination of fragrance and heat, as one knows it from the sauna infusion with eucalyptus fragrance. For example, the Aztecs treated injuries with wraps and massages with fragrant ointments in sweat lodges, called Temazcalli.

The Inkas preferred a type of gel with valerian scent for massages, which was thickened with other herbs and seaweed. We would today call this a kind of thalassotherapy with soothing detoxification by algae. Apparently, the Inkas already knew that seaweed has a very positive effect on the body, especially on the immune system. With their many valuable vitamins, minerals and trace elements, the algae acted like a kind of fountain of youth, which the Spanish conquerors hoped to find in the New World, of course. The effect was strengthened by a therapy with valerian, which was supported by the scent. The sedative and

calming effect of the plant additionally provided deep relaxation.

Also in Central America, the Mayas knew the sweat lodge and its applications. Archaeological finds of the University of Boston under the direction of Normann Hammond have already proven these facilities for the time from 900 BC—that is, before the founding of Rome and thus before the beginning of the Roman bathing culture. As later in the artful Roman baths, Maya groups with up to twelve sauna guests could enjoy the treatments at the same time.

On the entire American continent, sweetgrass—also called fragrant meadow grass or vanilla grass (Hierochloe odorata), which actually smells like the spice of the same name—was the favorite therapy scent of the natives. To achieve an optimal effect, the plant was pressed and smeared in bundles on the body. Woven into a wreath, it was worn during smoking. Its enjoyable scent served as a mood enhancer for all age groups in case of depressive moods. In addition, vanilla grass was the accompanying scent of medical and religious rituals.

For example, the smoke treatment with coneflower (Echinacea) had proven to be effective for headaches. The aromatic plant was also rubbed to promote wound healing. The slightly sour and, for many, somewhat unpleasant smell probably supported the assumption that medicine must taste bitter—and also smell.

When looking at European antiquity, one notices that scents were not limited to a single effect. Apparently, a special perfume handed down was able to combine different effects particularly well. This was the legendary Megaleion, created by the perfumer Megallus from Sicily or Greece. The scent consisted of, among other things, burned resin, cinnamon, and myrrh and scored with a real double benefit: On the one hand, it smelled very good, on the other hand, it relieved tension and was popular in the treatment of wounds and inflammation.

1

All in all a small, fragrant all-round miracle cure.

Much of the scent effects in antiquity are still shrouded in darkness. Nevertheless, I join the opinion of the many who have published on this subject with all caution:

> ❯ Fragrances in antiquity knew various medically oriented application areas, which revolved around the following areas and symptoms, among others: headaches, pain, asthma, and other respiratory problems, stress, restlessness, lack of drive, eczema, insomnia, depressive moods, rheumatism, infections, and particularly disinfectant effects.

Early on, there were also areas of application for oils and fragrances. They were mainly focused on psychological, socio-cultural, and aesthetic reasons and motives, but partly overlapped with medical ones. They all still drive modern perfumery today. The desired fragrance effects have hardly changed over the millennia—even if they are now sometimes referred to and weighted differently.

I recommend the following book by Jonathan Reinarz to anyone interested in perfume history—especially its development from the Middle Ages and the Renaissance, that is, from the 15th century to the recent present: *Past Scents, Historical Perspectives on Smell* (Reinarz 2014). Reinarz is Director of the Department of the History of Medicine at the University of Birmingham in the UK.

Here is a list of fragrance effects or reasons for fragrance, as we typically know them from the past and present of perfumery:

- for intoxication, for homage, for sacrificial gifts, for thanksgiving, for religious ceremonies, and rituals,
- to express personality in terms of status and power, to emphasize, control, and also to self-elevation,
- for memory and nostalgia,
- for flavoring and seasoning,
- for mixing with other substances with a view to creation,
- for the fragrance of objects, areas, plants, animals, and humans,
- for intensifying and changing consciousness, self-experience, emotions, and mood,
- for the acquisition of aura, attention, attractiveness, attraction, and beauty,
- for occasions and festivals, seasons, weekdays, days, and times,
- to influence and acquire others or to enchant someone and to be loved more,
- for pleasure, love, sensual pleasure, to bewitch, pamper, and seduce,
- for luck, as a bringer of salvation,
- out of habit and tradition,
- as a gift, reward, and surprise,
- as a luxury, wealth and gain, for individual enrichment, for waste, and fascination,
- to be fashionable and trendy,
- for the creation and design of art, aesthetics, and beauty,
- for feeling good, for satisfaction, for relaxation, for sensuality, inner peace, harmony, mindfulness, spirituality, inner strength, transformation, self-discovery, and imagination,
- for the acquisition of creativity, inspiration, concentration, and intelligence,
- for motivation, invigoration, stimulation, acquisition of energy, for productivity, and new beginning,
- for delimitation, affiliation, and identification,
- for generating envy and curiosity,
- for disinfection, protection, defense, and expulsion,
- to pacify needs and addiction,
- for mourning and funeral care,
- for the care and wellness of mind, body, and soul,
- to cover up stink and body odor,
- as an aphrodisiac to invigorate and increase libido, desire, and appetite.

— increasing sensuality, e.g. by sandal-wood notes.

What particularly furthered the research into the effects of scent and smell, or their potential possibilities, was the beginning of systematic interdisciplinary cooperation. In this context, in particular, findings from the fields of perfume, medicine, pharmacy, chemistry, neurobiology, physics, sociology, psychology, economics, neuromarketing, and consumer neuroscience were shared. They were all fertilized by discoveries in olfactory brain research.

This type of research was based mainly on EEG studies in its early stages. In general, voltage fluctuations, i.e. potential differences, so-called evoked potentials, were measured at the surface of the head. They provide information about the activity of nerve cells in a certain area of the brain, for example in response to olfactory stimuli. In other words, the observed olfactory evoked potentials were evaluated as electrical responses of the cortex to olfactory stimuli. For this purpose, a person must be repeatedly exposed to olfactory stimuli under controlled conditions, such as a certain way of breathing.

As early as 1875, the English doctor and physiologist Richard Caton discovered that electrical activity could be seen on the cortex. The first studies on humans took place at the University of Jena from 1924. It took almost another 40 years before the field of olfactory research was able to develop an appropriate derivation technique for olfactory evoked potentials using EEG in 1966. In addition, a special device had to be developed for the systematic presentation of olfactory stimuli, which made it possible to measure exactly by means of a controlled presentation of odorous air.

At the beginning of the 1970s, the first measuring systems of this type, later called olfactometers, were used in smell research.

1.2 Scent Effect in Modern Times

1.2.1 Beginning of the Research of the Scent Effect

There is no agreement on when exactly the research of the scent effect began in modern times. It may coincide with the first smell studies in the second half of the 20th century under the use of the electroencephalogram (EEG). Four areas were of particular interest to the scientific confirmation of the scent effect also in view of the practice of perfumery:

— more relaxation, i.e. relaxation, e.g. by lavender,
— gaining energy and productivity, e.g. by citrus notes,
— experiencing reward and pleasure, e.g. by chocolate notes,

1

With the olfactometer it was now possible to controlledly give smelling test persons fragrance substances in different concentrations. In the next 40 years, the olfactometer was then further developed into a flow olfactometer which is controlled by a PC. Fragrance stimuli are given in a certain concentration and duration with constant humidity and temperature in a constant air stream (Moessnang and Freiherr 2013). In this way it can be investigated whether, for example, a fragrance stimulus that is no longer consciously perceived nevertheless shows brain activation. So the question is whether and how unconsciously smelled can influence. The first registration of potentials caused by smell in the 1960s was based on work from the years 1935 and 1949 (Fischer 2015).

In the 1980s, EEG studies led to the first really exciting smell findings. It was found, for example, that right-handers have a higher sensitivity to olfactory stimuli—provided they are presented on the right side of the nose. Left-handers, on the other hand, are more receptive to stimuli on the left side of the nose. Also—as one would expect—fragrance concentration and fragrance stimulus duration have a significant effect on EEG activity. Therefore, it is not yet certain whether fragrance substances in different concentrations show the same or at least similar effects during longer EEG recordings.

In the years after the first interesting smell findings, there was a veritable explosion of studies dealing with the effects of various fragrance substances on the brain, despite open application questions. Although these studies differ in their methodology and, from today's point of view, have deficiencies in their design, they nevertheless came to a common conclusion:

> Fragrance substances can act directly and/or indirectly and influence the psychological and physiological conditions of humans (Sowndhararajan and Kim 2016).

This corresponds to the current state of research, which has crystallized over the years. Later I will also present various examples of the effect of scent. Just so much: Thanks to the available studies, ancient applications can be very well understood. In this context, for example, the originally from India used remedy sandalwood can be mentioned.

1.2.2 Neuroparfümerie Causes a Breakthrough in Fragrance Effect Research

The real breakthrough to insights that hint at the potential and future of fragrance and odor effect took place during the past years through methods of neurobiology or the so-called young neuroperfumery—brain research in relation to perfumery. This allows for contrast-rich insights into the brain or its individual regions and networks through non-invasive imaging or neuroimaging methods using so-called functional magnetic resonance imaging, abbreviated fMRT or fMRI (English "functional magnetic resonance imaging"). It shows how, for example, brain areas and networks of the limbic system, the center for emotions, can be specifically stimulated by different scents and fragrance directions. In addition, indirect conclusions can be drawn about the emotional and cognitive effects of individual odors or perfumes—provided that the smelling people share them with or show them in their behavior.

Although numerous current research results are not or not yet released by industry, there are now neuroimaging maps. They show olfactory networks in connection with the scent effect and are constantly being further developed. They not only allow the enjoyable effect of scents such as chocolate and vanilla to be recognized. They also show in which brain regions they are expected to be perceived.

And it gets even better: These neuroimaging maps even allow olfactory functions to be compared in diseases and thus to expand the possibilities of early detection. Therefore, current research and medical practice are interested in developing examination techniques that allow for a "brain fingerprint" comparison and assessment of health and disease based on refined olfactory (Fjaeldstad et al. 2017). This can control whether a fragrance works or does not work in certain brain regions and leaves no traces. This could give an indication of how far, for example, Parkinson's, Alzheimer's, or other diseases may have progressed in a person. The following finding is increasingly gaining ground:

> Scents do not only address specific regions in the brain, they stimulate whole networks that often run through very different and widely separated areas of the brain. This also means: one must analyze a fragrance effect in its entire neuronal entanglement.

In the following, I will explain using a first example what has been discovered in terms of hedonic effect in the brain using neuroimaging, which network is involved and which possibilities for new fragrance creations arise from this.

The Hypothalamus plays a significant role in the experience and "wanting" of enjoying a scent. Its network not only controls hunger and thirst, as well as sexual desire, but also has reward, pleasure, and addiction centers. These are the nodes of the richly interconnected dopaminergic system. Dopamine—commonly known as the "happiness hormone"—is a messenger substance or neurotransmitter that transmits excitement from one nerve cell to another. It plays a central role in increasing motivation and motivation in the sense of "wanting something", even the enjoyment of a scent, even if the "wanting" does not necessarily have to be equated with pleasure yet,

also because the hypothalamus and its network, as I will show shortly, have their own preferences for scents.

The Dopamine transmission is involved in essential aspects of lust-related perception. This is how the individual reward, pleasure, and addiction centers are waiting for stimulation—also olfactory. The favorite scents of this system are chocolate scents, followed by Vanilla and cinnamon scents. Above all, it must smell sweet. But the system is not very choosy when it comes to craving sweets or craving sweets. The main thing is that it smells edible. In the perfume industry, it is the fragrance of the gourmand notes that smell like delicious desserts and fit exactly into the requirements profile of the dopaminergic system. They can even be addictive, as we will see later.

Before we go into further findings of olfactory pleasure perception, here are some background information on the research of smelling and scents.

▪ Centers of Excellence and the Mecca of Fragrance Research

At the Technical University Dresden, one of the strongholds of odor research, the topic of scent effect and gustatory perception has been pursued for decades. Here, various figures of the perfume industry, including the flavor chemist and researcher Günther Ohloff (1924–2005), studied. He led the research group of the Geneva company Firmenich, today one of the world's largest flavor manufacturers. Ohloff's scientific work comprises 228 publications, in addition to 111 patents. He also suggested theories about the effect of flavors on human emotions and social behavior. His book *Flavors and Olfactory Sense. The Molecular World of Scents*, published in 1990, is considered the standard work of flavor chemistry. It has since been revised by other scientists and is available under the title *Scent and Chemistry—The Molecular World of Odors* (Ohloff 1992). Ohloff also wrote a

1

much-noticed book on the cultural history of flavors (Ohloff et al. 2012).

Overall, the eastern part of Germany has long held a leading position in scent or flavor production. The Saxon chemical company Schimmel & Co. was at times the world market leader in flavor production. Not until almost half a century later, in 1874, did the flavor manufacturer Haarmann & Reimer emerge in western Germany, from which the world company Symrise later developed.

Scent or flavor manufacturer—the terms are used synonymously—usually deliver finished fragrance oils as ingredients for perfumes. These are marketed by perfume houses such as Guerlain,Chanel,Armani,-Gucci, or Hugo Boss under their name or as a license. One of the first perfume houses in Germany was founded by Johann Maria Farina (1685–1766). In 1709 he brought an Eau de Cologne of the same name to the market in Cologne. For the sake of completeness, however, it should be mentioned that it is not the most durable fragrance product ever launched. As early as the 14th century, the Queen of Hungary "brought Aqua Reginae Hungariae" or—as the fragrance was also called—"Hungarian Water" to the market. For 400 years—until the loss of its recipe—the fragrance dominated the market and is still the longest-selling perfume of all time. This record could be broken by Cologne by Farina, now called "Farina 1709", in 2109.

Another German hub of fragrance research is located at the Ruhr-Universität Bochum. Here, the fragrance researcher and cell physiologist Hanns Hatt, a graduate in biology, chemistry, and medicine, worked with an interdisciplinary team. Hatt was the first to discover that olfactory receptors also have an important function in cells outside the nose. They control cell growth, hormone regulation, and the release of messenger substances in the body. In addition, the popular science work *Das kleine Buch vom Riechen und Schmecken*

(Hatt and Dee 2012) by Hanns Hatt is now a classic for all those who are interested in smelling and how it works.

The Mecca of fragrance research, the Monell Chemical Senses Center, is located in the American city of Philadelphia. Here, scientists from numerous disciplines work together. The focus is on understanding the basic mechanisms and functions of taste and smell. The institute is rather medically oriented and investigates what the senses of humans say about health and illness. In addition, the center offers seminars that can be attended by anyone. In the spring of 2020, for example, events on the topic "The COVID-19 Nose" took place. Early on, researchers at the Monell Institute as well as the international fragrance research community observed that Covid-19 is accompanied by a partial loss of smell and taste. Other centers of fragrance research are, for example, the UFCST (University of Florida Center for Smell and Taste) or the CSGA (Centre des Sciences du Goût et de l'Alimentation) in the French city of Dijon.

Back to the smell & taste study conducted at the Smell & Taste Clinic, Department of Otorhinolaryngology at the Technical University of Dresden. The title of the study was: "Food-related smells activate dopaminergic brain regions" (Sorokowska et al. 2017). It was found that pleasure smells or the above-mentioned food smells like chocolate,vanilla, and cinnamon create a significantly higher activation of certain network regions than non-edible smells, such as those of flowers. An interesting phenomenon was observed in the exclusively right-handed test subjects, which had already been reported in other studies: The smell effect was particularly pronounced, but not exclusively, in network regions belonging to the right hemisphere responsible for face recognition. Pleasantly smelling odors have a corresponding effect on the evaluation of faces. They can gain attractiveness through scent—and vice versa.

In contrast, the left hemisphere with its dominance in speech production seems to react only partially to the effect of scents and to contribute relatively less to the experience of scent than the right hemisphere. This is also understandable for another reason: Smelling, as we often experience it, quickly makes us speechless, or we lack adequate expressions to describe the scent—probably also because we do not have our own language of scent in our culture that is not derived from other senses. This means that one could postulate that the left hemisphere, at least for many of us, was and is less challenged in olfactory perception or in olfactory socialization, where one learns the meaning of smells and scents already in early childhood.

1.2.3 A First Look at the Effect of Scents on Specific Regions of the Brain

A closer look at the brain hemispheres reveals specific regions and networks to which the sweet, edible pleasure scents mentioned in the study by Sorokowska (2017) above work. These include scents that work on …

- … the anterior zingular Kortex, also called "anterior cingulate cortex" or ACC. The effect could be observed on both sides of the hemispheres, left as well as right. The ACC is an area of the cerebral cortex and is attributed to the emotion center. This region and its network are, among other things, involved in the perception of pain. It is already known for a while that sweet smells increase the tolerance to pain. The pain, which can also be a soul pain, is not necessarily reduced by the sweet, edible fragrances, but it is easier to bear.
- … the Insula (right). As one of the five cerebral lobes, it is involved in the vegetative nervous system and initially processes unconscious body sensations. It is also involved in the perception of pain—especially with its right side, which assesses the perceived pain. In addition, it plays a major role as one of the addiction centers. Nature does not seem to be averse to creating a certain addiction to sweet, edible aromas on an unconscious level. This may be to prevent a deficiency of nutrients, or to ensure that babies always like the taste of breast milk. The insula is also the brain region most affected by negative external stimuli. This may also explain why it seeks sweet, edible fragrances to be pampered by them.
- … the Putamen located at the base of the brain (right). It belongs to the so-called Striatum and is part of the Basal Ganglia. The Putamen also has various functions, for example in the control of movement sequences and in learning. For olfactory research, it is interesting that the Putamen belongs to the dopaminergic system and plays a role in the reward or motivation for and striving for rewards. A low activation of the Putamen is therefore associated with a lower feeling of reward. Wanting less and achieving less leads to feeling less good. This could be demonstrated in children and adolescents suffering from chronic fatigue syndrome ("childhood chronic fatigue syndrome"—CCFS) (Mizuno et al. 2016). In adults, this disease is known as CFS.

With severe fatigue, which usually begins with a loss of drive, activities that require relatively little physical and mental effort, such as cooking or shopping, can become a torture. Sweet, edible fragrances stimulate the putamen, which is usually accompanied by a dopamine release. This could already be shown for other brain regions such as the amygdala. Something nice to smell is accompanied by a corresponding feeling. Or, as it is called in popular language: scents can make you happy.

1

The results of the study mentioned above suggest that scents can intervene in the reward and drive system. This does not mean that CCFS or CFS can be cured in this way. But a certain positive effect on the associated neuronal reward and drive system can be shown with this type of scent—especially when you are hungry or crave sweets.

1.2.4 A First Look at the Future of Fragrance

The results of the studies mentioned above should make perfume makers think. Thus, the information from imaging methods could be used as a guide for the creation of neuroscents or active perfumes that optimally address specific brain regions or their networks. Sweet, edible active scents could be used specifically against pain or evoke a feeling of reward and happiness in the brain. It is therefore not surprising that the often dessert-smelling Gourmand notes are among the fastest growing fragrance trends of recent years. This says a lot about the current wishes and needs of many fragrance users.

Since most brain regions have different functions, other possibilities for olfactory effects in a region can also be shown. The stimulation needs of the Putamen alone suggest the creation of motivation and drive scents for different situations, for example for learning.

Other brain regions also show a strong reaction to scents—but without being particularly choosy about the fragrance notes that activate them olfactorily. An example is the ventral tegmental area. This is a highly trainable group of nerve cells in the midbrain. So people who like to drink too much already have them activated by the smell of alcohol (Kareken et al. 2006).

Even if this book is about smelling better, we have to admit that unpleasant smells have a stronger and above all longer lasting effect on the brain. Nature is particularly interested in its preservation and wants to protect us from poisonous or harmful substances. The brain has a special interest in remembering everything that smells bad, harmful, stinking, or nauseating, but also forbidden, and reacting to it reflexively, because it endangers our well-being. The sense of smell is supported by the trigeminal nerve, which reacts to stinging, burning, biting, and sharp sensations in the nose and mouth, through which we also smell. Accordingly, there is brain activity with such substances, namely in such a way that they are not forgotten by the Hippocampus and its network, which also serves as a smell memory. According to a recent study, memories are stronger if the original experience is accompanied by unpleasant odors (Cohen et al. 2019). Stench and especially disgust-inducing odors have a greater effect on the brain and remain in memory longer than the beautiful scents.

1.2.5 Smelling Also Works the Other Way Around—The Fragrance Manager in the Brain

But the course of action also goes in the reverse direction—from the brain to the scent. The brain can in fact create its own scent atmosphere and smell emotion, even if a smell source does not justify this feeling. This way, the brain can also "smell good". It can even insist on a smell impression that—although it is not currently present for others—triggers psychophysical reactions. Most of the time, the sense of smell is influenced by a visually perceived stimulus that is experienced as attractive, for example the sight of a person.

It is above all the region of the brain's piriform cortex that, in conjunction with other regions, decides on the effect of perfume and smell. This region of the brain is closely linked to the sense of sight and has amazing abilities. It can not only amplify the effect of perfume and scent on consciousness, mood, and emotion, but also prescribe to the olfactory bulb, which is closely linked to the nose, what and how something smells. This can go so far that a certain substance is not smelled at all, is not smelled further or is only sniffed partially—depending on what the piriform cortex and its system perceive as important.

The piriform cortex also decides what, how, and where something is given to smell further processing in other brain regions. For this purpose it has a very emotional, often frightening smelling ally: the amygdala. It usually always smells with and exchanges information with the piriform cortex about the smell impression. The result can also be reported back to the olfactory bulb in a feedback loop. The effect of perfume and smell on consciousness, mood, and emotion is thus managed by an interesting scent manager.

With the piriform cortex we have a kind of little man in the brain who not only whispers gently how to deal with a smell stimulus. He can decide as a scent manager that a smell is forwarded directly to a region in the brain, for example to the orbitofrontal cortex. The latter controls personality traits such as extraversion and conscientiousness in the network and is particularly stimulated by certain perfumes and odors according to the findings of neuroperfumery. The piriform cortex thus also has an influence on our personality, which it motivates with olfactory impulses.

> Smelling is movement, movement is life, life is smelling—at least a lot of joie de vivre is lost without smelling.

Smelling creates movement, one could also say: almost reciprocal unrest in the brain. So it's not just about smelling from the nose in one direction, as one might think. Skin and other organs can also smell, as I will explain later. Before I continue the topic and go into the results of further current studies, I would like to briefly talk about life and movement, if only because smelling and smell, as I said, are also movements or originate from movements.

Life is movement. Without movement, a life as we know it is simply not possible. What we experience is therefore movement—at least it is movement in our brain. The experience that goes hand in hand with the movement, like the movement itself, can take place unconsciously, partially consciously or fully consciously. Experience and movement come in different shades, combinations, and dimensions. It can be experienced as emotional needs and desires, which in turn fuels experience and movement. Experienced movement can have a slow or fast effect as pleasure or displeasure, excitement or relaxation. We know this from the seven basic emotions that every human being knows: joy, surprise, fear, anger (rage), disgust, contempt, and sadness build up quickly, have an effect in a fraction of a second and can be read from our facial expressions. The same applies to our fourteen other learned basic emotions and basic feelings. They include compassion, disappointment, jealousy, relief, shame, envy, guilt, pride, attractiveness, enthusiasm, discomfort, trust, and love.

Psychologists distinguish between emotions and moods when it comes to these mental movements, even though the transition is fluid and both overlap. The difference between the two is mainly seen in the temporal development, even though more pronounced psychophysical characteristics such as heart racing and blushing are usually more common with emotions. Moods build up relatively slowly in comparison to emotions, but then stay in motion for a longer period of time. In the German language, it is easy to come up with over 200 adjectives that describe emotions, basic emotions and moods as well as the mental movement experienced.

1

Now back to the topic.

How do perfume, scent, and smell differ from other olfactory impressions? Smell is to be understood as a neutral term of the olfactory, whereas scent and perfume refer to a pleasant experience and stink to an unpleasant one.

Smelling is life and is therefore based on mental and physical movements in the most diverse forms—from intense to weak, from pleasant to unpleasant. It can range from contemptuous to loving, from disgusted to joyful and be experienced as rewarding and punishing. The particularly fascinating thing about smelling is that the movement starts unconsciously, can remain completely unconscious, but can also become conscious quickly and thus be experienced by us as smelling. In other words: pleasure or displeasure, excitement or relaxation, tension or solution or a combination of everything can be smelled both consciously and unconsciously. They are triggered by smell and stink as well as by scent and perfume as emotions, basic emotions, or moods and then have the corresponding effect on us.

Scent research confirms the effect of smells. Even if they are not conscious or only semi-conscious to us, or if we do not concentrate on them, they can influence our mood and emotions (Kadohisa 2013). Above all, smells can relax, invigorate, and soothe anxiety and stress (Kontaris et al. 2020). Later on, I will introduce various scent-supported therapies and exercises to achieve more joie de vivre, which can be implemented directly by oneself.

1.2.6 Research Areas of Scent Psychology

Questions such as how and why certain smells have psychological effects and how a particular perfume choice is made also belong to the research area of scent psychology. As a sub-area of perfumery and scent research, it investigates whether, where, how, and why scents and, as a result, perfumes and essential oils, when inhaled or applied, cause certain psychological, physical, or psychophysical effects. When determining scent effects and fragrance preferences, scent psychology is particularly interested in emotional and cognitive processing in the brain—but not exclusively, as the skin and other organs that have their own intelligence can also smell.

With the help of imaging methods, one can watch the brain smell, observe the effect of olfactory stimuli, but also recognize which fragrance preferences individual brain areas and their neuronal networks have. From this, the neuroperfumery is currently developing as its own research discipline. Through it, scent psychology is currently gaining numerous fascinating new insights. Based on these findings of neuronal fragrance preferences, the research area of scent psychology—and this interests everyone in practice who offers a perfume consultation—is which factors influence a fragrance choice, especially when deciding on a perfume.

In addition, scent psychology investigates psychosocial relationships. This area also speaks strongly to fragrance users, as it explains how a perfume affects the wearer's environment or what the scent evokes in oneself and in others. Here, findings from scent sociology, from the areas of personality psychology and self-concept research are in the foreground. Above all, the latter knows the connection between "how one would like to experience oneself more" and preferences for fragrance directions. The preference for certain perfumes can also be explained by olfactory socialization, i.e. by learning experiences. Here it plays a special role that certain scents or perfumes are linked to memories and the emotions and moods associated with them.

For over two decades, studies have empirically shown that smells can influence mood, emotion, physiology, and associated behavior (Herz 2022, 2009; Kontaris et al. 2020). For example, a study conducted by Canadian and US researchers reports that after just five minutes of a pleasant smell, a positive mood and relaxation can be induced, while after five minutes of an unpleasant smell like pyridine, which smells like decay, a negative mood and slight anxiety can arise (Villemure 2003). Accordingly, researchers are very interested in the effects of smells and the exploration of smell in the field of emotion, mood, but also pain research. It is increasingly recognized that odors not only change emotions and moods, but can also significantly influence the perception and memory of pain and suffering. Smells can thus change or even trigger emotional reactions and memories with considerable positive and certainly also negative emotional content (Herz and Engen 1996; Keogh et al. 2001). Even more: later I will show how smells influence evaluations of the attractiveness of people, but also of things like paintings (Rotton 1983; Ehrlichman and Bastone 1992).

1.2.7 Smelling, the Last Sign of Consciousness

Smelling is an expression and thus almost a sign of life. The importance of smell and its effect are even gaining importance in emergency medicine. The reaction to smell can even indicate consciousness in accidents, even if the patient only shows the smallest nonverbal reactions when smelling. The sniffing of odors, called "olfactory sniffing" in the jargon, signals consciousness in unresponsive patients with brain injuries and so chances of recovery. This is especially important in view of the fact that after a severe brain injury it is often very difficult to determine the condition of a patient and to predict his chances of survival or to make the right diagnosis in con-

nection with a therapeutic strategy such as pain treatment.

The medical error rate in determining a state of consciousness is 40%. In order to minimize this, the effects of fragrances are currently being researched as a diagnostic tool for the severity of loss of consciousness (Arzi et al. 2020). Furthermore, I will also describe how fragrances have been used for some time now to not only diagnose, but also treat diseases such as Parkinson, Alzheimer, and others. It should also be noted here that smell not only affects consciousness, but also encourages it to report circumstances that only allow for the smallest nonverbal reactions. The sense of smell is thus also an indication of life and sometimes even the last hope-giving sign of life.

Back to brain research.

The whole topic of how smell works and what the effect says is currently raising new questions with each new discovery. For example, the question arises through which neuronal circuits or processes individual brain regions can stage or influence consciousness, emotions, and moods by means of a stimulation caused by smell. There are surprises. For example, different brain regions work together, the involvement of which in the processing and effect of fragrances was previously unknown. It was also recognized that there is a struggle for competence between individual brain regions when it comes to who makes the final decision on an olfactory impression and the resulting behavior. It was also discovered that specific brain regions, better said: core areas of the brain, lead a double life and are responsible for different colored emotions and moods. For example, it was found that the Amygdala or in German (Mandelkern), which is located deep in our emotional center and is centrally responsible for the processing of smells, is involved in both negative emotions such as fear and anxiety as well as positive, pleasant, and relaxed experiences. For aromatherapy this offers the chance to reprogram core areas of the brain from fear and anxiety to positive experiences. However, for this the research on smells has to look at the entire process of smelling in much more detail in order to understand how, when, and where exactly these core areas stage individual activities.

One problem in researching the relationship between smell on the one hand and mood and emotion on the other hand lies in the structure of the brain, which can only partially distinguish between emotions and moods at the neuronal level, since different core areas are involved in each case. Another difficulty lies in the overlap of emotions and moods. So fear can be either a temporary emotional state or a persistent mood (Kontaris et al. 2020).

1.2.8 The Brain Can Create Its Own Smell

Smelling is based on different neuronal network processes. How smelling actually comes about when fragrance molecules come into contact with olfactory cells and thus generate neuronal impulses is still controversial in research. Sharma from the Indian Institute of Technology currently describes eleven theories (Sharma et al. 2019). The so-called Steric Theory is currently preferred most strongly. According to her, depolarizing molecules that are absorbed on the surfaces of olfactory cells generate a nerve impulse: "The smell quality is influenced by the timing of the nerve impulse, while the smell intensity is perceived by the total number of similarly excited cells as well as by the total number of impulses in the nerve" (Sharma et al. 2019).

In simple terms, the sense of smell begins with sensory neurons specialized in olfactory stimuli in the nose. They form, so to speak, the gateway to the world of smell and activate the olfactory bulb (Bulbus ol-

factorius"), the first smell station in the olfactory brain. This is followed by a complex process. From the olfactory bulb, neurons pass information on to a variety of anatomically and functionally different targets in the brain. A main target is the Amygdala, our phylogenetically oldest center for emotions and thus a central part of our emotional center in the limbic System. But before the amygdala smells, as already mentioned, the piriform Cortex is involved, which receives olfactory signals from the olfactory bulb first and forms the largest subregion of the olfactory brain. Because this part of the cortex does an enormous amount for smelling and also takes indirect influence on emotion and mood, the current research on smell focuses particularly on him.

In fact, research on the piriform cortex is constantly discovering new things. In analogy to the little man in the ear, it is, as already mentioned, increasingly seen as a kind of fragrance manager. It is both a fragrance relay and a fragrance memory as well as a very intelligent fragrance recognition station that decides in almost mysterious ways which other brain regions an olfactory information will be passed on to. In addition, the piriform cortex is involved in what we see and what we perceive with our other senses. So it influences what we perceive as a fragrance (Schulze et al. 2017). This way the piriform cortex can change or reshape the sense of smell in various ways through its network, usually intensify or weaken it. So it is also a fragrance interpretation center.

In research circles, it is currently being discussed how the piriform cortex can even give information to the olfactory bulb in a feedback loop and thus influence the sense of smell in advance (Wilson 2011). This in turn affects our consciousness, our emotions, and our mood and thus our behavioral reactions. For example, this is the case when the piriform cortex associates a smell with disgust or simply smells it out of the blue—regardless of whether a corresponding reaction is justified or whether an external olfactory stimulus is actually present. The latter is referred to as Phantosmie or in German Geruchshalluzination. Here you have a pleasant or unpleasant smell, even though there is no corresponding source of fragrance. Nevertheless, the sense of smell triggers corresponding moods and emotions. The brain is thus able to create its own smell and thus trigger psychological, psychophysical, and even psychosomatic reactions and respond to them.

In such a reaction, one does not immediately have to assume the worst and think that this form of sensory illusion is a symptom of a developing or already pronounced Schizophrenie,Epilepsie or Parkinson's disease. Persistent olfactory hallucinations occur more post-infectiously after a time-limited infection of the (upper) respiratory tract or as a post-traumatic smell disorder, for example caused by a head injury in an accident.

Olfactory hallucinations also play a role in classical psychoanalysis. The founder of psychoanalysis, Sigmund Freud, developed the theory of a "Geruchsneurose" as repression of a traumatic or forbidden event in 1892 at the case of Lucy R. The young Lucy came to Freud's treatment because she had largely lost her sense of smell and was additionally pursued by two subjective smell sensations. At first there was the smell of a burnt pastry and then the smell of cigars. Lucy said that she smelled these scents everywhere, even though she was assured the opposite. The young lady lived as a governess in the house of a factory director in Vienna. After a few therapy sessions, Freud came to the conclusion that these smells had once been objectively present. According to his analysis, they were related to her employer, whom Lucy had fallen in love

with unhappily, and his house. The experience of disappointed love developed into a trauma for her. The repressed memory manifested itself in the young woman in olfactory hallucinations.

There are also toxic-induced smell disorders, which, for example, are influenced by drugs—especially by the taking of psychotropic drugs such as antidepressants—but also by alcohol. In very rare cases, a tumor in the area of the olfactory nerve or an olfactory center for phantosmia can also be responsible for this sensory illusion, as is known from otolaryngology. Most of the time, this sensory illusion is due to too much stress, for which the olfactory bulb is very susceptible. Sufficient rest and sleep provide relief.

The phenomenon of a "stinky nose" is not a smell hallucination. Here, bacteria produce a slimy, foul-smelling coating. The affected people usually do not notice this themselves, but their fellow human beings do so all the more. If the affected person recognizes this, his self-experience and self-esteem suffer.

The sense of smell is basically divided as follows:
- normal sense of smell = normosmia,
- reduced sense of smell = hyposmia and
- absent sense of smell = anosmia.

There is also a hyperosmia, an increased olfactory sensitivity. This occurs, for example, in migraine. But you can also dream of smells and fine fragrances—regardless of whether they are present in the room as olfactory stimuli. A deviation from the average perception of a fragrance by the population is also called parosmia. These perception changes can be pleasant or unpleasant as well as more or less intense and are often influenced by additional visual stimuli.

This leads back to the piriform cortex and its network. As I said, through it flowing visual and other sensory as well as emotional information, it can whisper to the olfactory bulb what it smells like. This information seems to come mainly from the back part of the piriform cortex (Eng. "posterior piriform cortex", abbreviated: pPCX). But the front part of the piriform cortex (Eng. "anterior piriform cortex", abbreviated: aPCX) is also involved in smelling. According to the current state of research, a specialization of the parts of the piriform cortex can be seen. Nature has developed something brilliant for this. This way we are able to smell scents, for example from flowers, as a whole, but also the individual fragrance molecules they contain. The latter certainly requires some practice. Even trained noses quickly reach their limits here, as floral scents can contain more than 100 relevant fragrance molecules.

Natural odors are mixtures of many fragrance molecules. Plants in particular often emit not one, but several fragrance molecules at the same time (Pannunzi and Nowotny 2019). The art of perfumery is based on these two ways of smelling—as a whole and in individual parts. The front part of the piriform cortex probably smells like a trained perfumer. Here the mixture of the individual fragrance molecules is decoded. A categorization of the smells takes place. The back part of the piriform cortex smells—as is probably the case with most people—the smell impression as a whole (Kadohisa 2013; Wang et al. 2020).

The ability of humans to smell individual parts from complex scents has psychological and physical effects. So the brain, that is the piriform cortex and its network, can smell individual ingredients of a perfume creation that are not or only half conscious to us at first, but can influence our mood. Accordingly, then individual substances can make us feel more or less pleasant, more or less intense, which of course can also be seen in our behavior.

1.2.9 How Fragrances Unconsciously Influence Our Behavior—Two Examples

There are several examples in the literature of how individual fragrances can work on us unconsciously. The induction of odor-induced emotional reactions by copulins is often cited. These are short-chain fatty acids in female vaginal secretions. Copulins were thus diluted by the Vienna Ludwig Boltzmann Institute to such an extent that their smell was no longer perceived by male test subjects. Nevertheless, the following effect occurred: After the unconscious smelling of the copulins, the male test participants rated portrait photos of women as more attractive than before smelling. Even photos previously rated as less attractive became more attractive after unconscious smelling (Fröböse and Fröböse 2012).

I will discuss the results of the Ludwig Boltzmann Institute, which have been questioned by some researchers, in more detail in connection with pheromones.

Female attraction to men, which is also associated with an improvement in mood olfactorically, is also possible in a simpler way, namely without unconsciously smelling sexual attractants. According to a study, two of eight heterosexual men showed activation in certain brain regions when smelling a woman's fragrance. This reaction was particularly noticeable in the hypothalamus and the insula, two brain regions also responsible for reward and pleasure (Huh et al. 2008).

Currently, research is focusing on kisspeptin. This is a hormone that influences both puberty and brain processes in adult men. Kisspeptin also affects olfactory perception and makes men smell female perfumes like Chanel No 5 more beautiful (Yang et al. 2020). The research of smell is discovering more and more that nature knows different ways to help with olfactory perception according to gender and thus put them in a good mood. We will discuss some of their tools later.

1.2.10 Aromatherapy—How Essential Oils Work

That smells can have a positive effect on mood and health is not a discovery of modern times. Already the ancient cultures appreciated and knew their use. Even the perfumes of the first recorded perfumer, a woman named Tapputi, offered a pleasant smell with additional benefits. Around 1200 BC, she worked in Babylon with plants that also had a psychologically cleansing and thus therapeutic effect.

In the last century, the French chemist René-Maurice Gattefossé coined the term " aromatherapy", which we associate with plant-based essential oils for inhalation. They can also be partially applied to the skin or ingested. Originally, the essential oils were obtained by pressing plants or plant parts, especially leaves. Only later were methods such as steam distillation or other extraction methods used.

A precise definition of the current meaning of aromatherapy can be found on the website of the US National Cancer Institute (NIH). Aromatherapy is therefore "the use of essential oils from plants (flowers, herbs, or trees) as a complementary health approach. The essential oils are usually used by inhalation or by applying a diluted form to the skin." Essential oils from fruits such as orange oil, which are certainly also classified as plants, are not explicitly mentioned.

The aromatherapy has given rise to the aromachology. It pursues a more scientific approach and investigates why and how certain scents trigger a psychophysiological reaction. But aromatherapy itself makes a scientific claim—especially with regard to

the effects when inhaling the plant oils. Empirical studies confirmed this claim for several oils. The American psychiatrist Rachel Herz already proved this years ago with her study "Facts and Fiction about Aromatherapy: A Scientific Analysis of the Olfactory Effects on Mood, Physiology and Behavior", which was published as a professional article, that the oils and their smells used in aromatherapy can affect mood, physiology and behavior (Herz 2009).

It is now generally accepted that inhaling essential oils can, for example, affect blood pressure and heart rate. Endocrinology, the study of hormones, metabolism and the diseases associated with them, also confirms the effect of aromatherapeutic applications. A study showed that women who inhaled lavender aroma while watching a stressful video had lower CgA levels than the control group receiving a placebo (Toda and Matsue 2020). CgA stands for chromogranin A, a protein that is also considered an indicator of stress when values are high. More and more studies are therefore coming to the conclusion that aromatherapy with essential oils offers clinical advantages. It represents an alternative and an additional medical treatment for high blood pressure, fatigue, psychological stress, and various other diseases and symptoms (Kawai et al. 2020).

Further studies examined the influence of essential oils on moods and cognitive effects. Thus, the US biologist Sachiko Koyama and his colleague Thomas Heinbockel (Koyama and Heinbockel 2020) as well as the English psychologist Mark Moos with his team (Moos et al. 2003, 2008) went to the question of how plants act when inhaled as essential oils on cognitive performance and mood. The effect of the plants depends on the place of their growth and their age. There are also seasonal influences. The purity of the oils, the

1

way they are obtained and their administration are also decisive.

Here are examples from empirical studies on the effect of some plants as essential oils:

- Lavender has a calming effect. When inhaled, reaction time slowed down for tasks that require memory and attention (Moos et al. 2003).
- Rosemary increases performance. The aroma makes you more alert (Moos et al. 2003). Test subjects feel fresher and more active emotionally. Blood pressure, heart rate and breathing rate increase with the inhalation of rosemary oil (Sayorwan et al. 2013).
- Peppermint improves vigilance and linguistic skills (Moos et al. 2008). In combination with rosemary, it increases memory functions. Peppermint is one of the oldest and most popular medicinal plants with a wide variety of effects that were already appreciated by the ancient Egyptians. Fragrance research has long confirmed that the smell of peppermint also has a very positive effect on athletic performance and fitness. For example, it was found that adults did more push-ups in the presence of peppermint flavors and improved their running performance (Herz et al. 2022).
- Ylang-Ylang has a similar effect to lavender. It promotes (inner) peace and provides deceleration. Test subjects reacted relatively slowly when inhaling (Moos et al. 2008).
- Sage (garden sage) and Spanish sage increase cognitive performance, specifically memory. This effect was even observed in Alzheimer's patients, as reported by researchers at the University of Messina in Italy (Miroddi et al. 2014). In particular, specific chemical components of sage and 1,8-cineole are discussed as being particularly promoting for cognitive performance,

which occur in larger quantities in eucalyptus and other plants. Which components finally work in which combination and in what quantities is often not yet researched. Lavender and rosemary each have 505 and 450 components. Most plants have at least 100 to 250, only a few plants are content with a small number—like Guaica wood with only 25 (Koyama and Heinbockel 2020).

More and more, the statements of Valerie Ann Worwood, one of the more well-known advocates of aromatherapy, are being confirmed that essential oils, among other things, promote relaxation and concentration. In her book published in 1997, *The Fragrant Mind: Aromatherapy for Personality, Mind, Mood, and Emotion* (Worwood 1997), she takes it one step further: she specifically recommends essential oils for the enhancement of experience for different personality types as well as generally for mood change.

The confirmation of the effects of essential oils, of smells in general, and of individual plant ingredients has led to a boom in olfactory brain research. It has been found, for example, that lavender and eugenol (an organic compound with a strong clove-like smell that is also found in cloves) or chamomile have a particularly strong effect on certain areas of the brain and networks of the temporal lobe or the frontal lobe (one of four lobes of the cerebrum), and can make the state of wakefulness more relaxed. For rosemary, an effect was found on the front part of the brain (the frontal lobe) that is typically associated with increased vigilance or attention, depending on the situation experienced.

The empirical studies of how certain fragrance ingredients work in and for the brain has prompted further research into the search for specific neuroscents, or as some call active scents. For Céline Manetta

(Senior Manager of Human and Consumer Insights at IFF—a leading fragrance manufacturer), the results pointed towards "Aromatherapy 2.0".

Neuroscents are in demand: According to IFF, the company "surveying more than 3,000 women and men in 14 different countries and found that nine out of 10 wanted to improve their well-being. Eighty-seven percent desired a fragrance with emotional and physical benefits, while 45 percent expressed that they expect fine fragrance to better their mood and comfort levels... Following coronavirus-related lockdowns, there is a swell of demand for perfume with wellness attributes" BeautyInc newsletter (2023). Therefore, while we might have mixed feelings, neuroscents are seen as the next generation of fragrances, created on the basis of advanced science and backed by facts. Studies in neuroperfumery, for example, have found that lavadin (part of the "Lavandula family" but a naturally occurring hybrid plant) is even more relaxing than lavender. These types of research methods generally combine people's verbal and neurological responses (fMRI) to smells.

Each of the world's leading fragrance manufacturers and suppliers, (we present them in Chapter 6) have now developed their own tools to identify beneficial neuroscents:

- Firmenich uses tools like EmotiWAVES to identify natural ingredients in essential oils that have been neuroscientifically proven with an fMRI to stimulate a specific emotion.
- In November 2022, Givaudan released MoodScentz+ to improve the effect of fragrances
- IFF partners with SleepScore Labs to find active scents that improve sleep.

Current techniques and discoveries even inspire to guide consumers in-store to fragrances that best suit their emotional desires and needs. L'Oréal and neurotechnology company Emotiv have teamed up to develop such devices. One of these devices – Scent Sation – was piloted at the Dubai Mall in November 2022 and is set to launch in the UK, Germany, Spain, and Switzerland by 2023/2024. We will discuss ideas for in-store perfume consultation using fMRI technologies in chapter 10.

Later on, I will discuss other smells and fragrances in their significance for mental experience, such as phenylethanol, which is a main component of rose and is said to have an antidepressant, or at least a mood-calming anti-stress effect. We will also discuss how and where individual smells activate the brain for specific experience. Furthermore, I will show that certain areas of the brain and networks are addicted to certain fragrances and want to smell them first.

Let me mention one more thing here to make you curious about possibilities for fragrance-supported therapies, which I will discuss later and suggest as a self-experiment: it is fascinating how quickly a smell can work. Studies confirm that odor-induced emotional reactions can occur with significant changes in mood in less than two minutes (Ehrlichman and Bastone 1992; Chen and Haviland-Jones 1999). The first recognition of a fragrance can already take place in the range of less than 500 milliseconds, and distinguishing between two fragrances takes no longer than one to two seconds (Junek et al. 2010; Draguhn 2010). Of course, there are other senses that are faster, such as sight, which perceives something in the range of 300 milliseconds. But we humans smell very quickly—and better than many think.

1

Summary

This chapter discussed the scent effect on consciousness, emotion, and mood. For this, it went back to antiquity. Furthermore, it was about the topic of smell. It starts unconsciously, can remain completely unconscious, but can also become conscious quickly and thus be experienced as smelling. I have also presented the effects of odors, fragrances, and perfumes, as well as essential oils, on moods, emotions, and cognitive experience on the example of some first newer studies. Our smell brain shows amazing abilities. This even goes so far that brain regions and networks around the piriform cortex, which is part of the smell brain, can change olfactory perception from the outside world so that it creates feelings and behaviors that are incomprehensible to outsiders. However, therapy with fragrances can also be used for a whole range of diseases and symptoms.

References

Arzi A et al. (2020) Olfactory sniffing signals consciousness in unresponsive patients with brain injuries. In: Nature (2020). 581(7809):428–433

Byl SA (2012) The essence and use of perfume in ancient Egypt. University of South Africa

Cao B et al (2019) Seeing the unseen of the combination of two natural resins, frankincense and myrrh: changes in chemical constituents and pharmacological activities. Molecules 24(17):3076

Chen D, Haviland-Jones J (1999) Rapid mood change and human odors. Physiol Behav 68:241–250

Cohen AO et al (2019) Aversive learning strengthens episodic memory in both adolescents and adults. Learn Mem 26:272–279

Draguhn A (2010) Geschmack und Geruch. In: Klinke R, Pape HC, Kurtz A, Silbernagl S (eds) Physiologie, 6. ed. Thieme, Stuttgart, pp 742–757

Duffey AE (2005) Hatshepsut's expedition to Pwn.t. In: Byl SA (2012) The essence and use of perfume in ancient Egypt. University of South Africa

Ehrlichman H, Bastone L (1992) The use of odour in the study of emotion. In: van Toller S, Dodd GH (eds) Fragrance: the psychology and biology of perfume. Elsevier, London, pp 143–159

Fischer J (2015) EEG-Ableitung der olfaktorisch evozierten Potenziale bei streng einseitiger Stimulation des Riechepithels mit dem Olfaktometer. Dissertation der Medizinischen Fakultät der Friedrich-Schiller-Universität Jena

Fjaeldstad A et al (2017) Brain fingerprints of olfaction: a novel structural method for assessing olfactory cortical networks in health and disease. In: Sci Rep. 7:42534

Fröböse G, Fröböse R (2012) Lust und Liebe – alles nur Chemie? Wiley-VCH, Hoboken

Hatt H, Dee R (2012) Das kleine Buch vom Riechen und Schmecken. Albrecht Klaus, München

Herz RS et al (2022) A three-factor benefits framework for understanding consumer preference for scented household products: psychological interactions and implications for future development. Cognitive Research: Principles and Implications 7:28

Herz RS (2009) Aromatherapy facts and fictions: a scientific analysis of olfactory effects on mood, physiology and behavior. Int J Neurosci 119:263–290

Herz RS, Engen T (1996) Odor memory: review and analysis. Psychon Bull Rev 3:300–313

Huh J et al (2008) Brain activation areas of sexual arousal with olfactory stimulation in men: a preliminary study using functional MRI. J Sex Med 5(3):619–625

Junek S, Kludt E, Wolf F, Schild D (2010) Olfactory coding with patterns of response latencies. Neuron 67(5):872–884

Kadohisa M (2013) Effects of odor on emotion, with implications. Front Syst Neurosci 7:66

Kareken DA et al.(2006) Alcohol-related olfactory cues activate the nucleus accumbens and ventral tegmental area in high-risk drinkers: preliminary findings. Wiley Online Library

Kawai E et al (2020) Increase in diastolic blood pressure induced by fragrance inhalation of grapefruit essential oil is positively correlated with muscle sympathetic nerve activity. J Physiol Sci 70(1):2

Keogh E, Ellery D, Hunt C, Hannent I (2001) Selective attentional bias for pain-related stimuli amongst pain fearful individuals. Pain 91:91–100

Kontaris I et al (2020) Behavioral and neurobiological convergence of odor, mood and emotion: a review. Front Behav Neurosci 14:35

Koyama S, Heinbockel T (2020) The effects of essential oils and terpenes in relation to their routes of intake and application. Int J Mol Sci 21(5):1558

Langhorst J et al (2013) Randomised clinical trial: a herbal preparation of myrrh, chamomile and coffee charcoal compared with mesalazine in maintaining remission in ulcerative colitis – a double-blind, double-dummy study. Alimentary Pharmacol Therap 38(5):490–500

Miroddi M et al (2014) Systematic review of clinical trials assessing pharmacological properties of Salvia species on memory, cognitive impairment and Alzheimer's disease. CNS Neurosci Ther 20(6):485–495

Mizuno K et al (2016) Low putamen activity associated with poor reward sensitivity in childhood chronic fatigue syndrome. Neuroimage Clinical 12:600–606

Moessnang C, Freiherr J (2013) Olfaktorik. In: Schneider F, Fink GR (eds) Funktionelle MRT in Psychiatrie und Neurologie. Springer, Berlin

Moos M et al (2003) Aromas of rosemary and lavender essential oils differentially affect cognition and mood in healthy adults. Int J Neurosci 113(1):15–38

Moos M et al (2008) Modulation of cognitive performance and mood by aromas of peppermint and ylang-ylang. Int J Neurosci 118(1):59–77

Ohloff G (2012) Irdische Düfte – himmlische Lust: Eine Kulturgeschichte der Duftstoffe. Springer Basel AG, Frankfurt am Main

Ohloff G, Pickenhagen W, Kraft P (1992) Scent and chemistry: the molecular world of odors (Riechstoffe und Geruchssinn: Die molekulare Welt der Düfte 1. ed.). Wiley, Weinheim

Pannuzi M, Nowotny T (2019) Odor stimuli: not just chemical identity. Front Physiol 10:1428

Reinarz J (2014) Past scents: historical perspectives on smell. University of Illinois Press, Urbana

Rotton J (1983) Affective and cognitive consequences of malodorous pollution. Basic Appl Soc Psychol 4:171–191

Savorwan W et al (2013) Effects of inhaled rosemary oil on subjective feelings and activities of the nervous system. Sci Pharm 81(2):531–542

Schulze P, Bestgen AK, Lech RK, Kuchinke L, Suchan B (2017) Preprocessing of emotional visual information in the human piriform cortex. Scientific Reports vol 7, Article number: 9191

Sharma A et al (2019) Sense of smell: structural, functional, mechanistic advancements and challenges in human olfactory research. Curr Neuropharmacol 17(9):891–911

Sorokowska A et al (2017) Food-related odors activate dopaminergic brain areas. Front Hum Neurosci 11:625

Sowndhararajan K, Kim S (2016) Influence of fragrances on human psychophysiological activity: with special reference to human electroencephalographic response. School of Natural Resources and Environmental Sciences, Kangwon National University, Chuncheon 24341, Korea; in Scientia Pharmaceutica

Suhail MM et al (2011) Boswellia sacra essential oil induces tumor cell-specific apoptosis and suppresses tumor aggressiveness in cultured human breast cancer cells. BMC Complement Altern Med 11:129

Toda M, Matsue R (2020) Endocrinological effect of lavender aromatherapy on stressful visual stimuli. Contemp Clin Trials Commun 17:100547

Villemure C (2003) Effects of odors on pain perception: deciphering the roles of emotion and attention. Pain 106:101–108

Wang L et al. (2020) Cell-type-specific whole-brain direct inputs to the anterior and posterior piriform cortex. Front Neural Circuit 14(4)

Wilson DA (2011) Cortical processing of odor objects. Neuron 72(4):506–519

Worwood VA (1997) The fragrant mind: aromatherapy for personality, mind, mood and emotion. Bantam Books, London

Yang L et al (2020) Kisspeptin enhances brain responses to olfactory and visual cues of attraction in men. JCI Insight 5(3):e133633

Perfumes in Change

The Practice Of Perfume Effects—From Fragrance Counselling To Scent Assisted Therapy

Contents

© The Author(s), under exclusive license to Springer-Verlag GmbH, DE, part of Springer Nature 2023
J. Mensing, *Beautiful SCENT*,
https://doi.org/10.1007/978-3-662-67259-4_2

2

Trailer

At my university, over 30 years ago, people were quite surprised at the passion I, as a psychologist and sociologist, had developed for perfumes and, in particular, for the psychology behind fragrance choice. Of course, some of my colleagues were also perfume users and usually wore French perfume classics as eau de parfum or eau de toilette, but there was no academic interest in exploring the world of perfumes.

At that time, the perfume offer was quite limited compared to today. Most perfumes were launched on the market with beautiful stories, targeting target groups such as "the young sporty" through sociodemographic and lifestyle characteristics. Only a few suspected that psychological and neuropsychological factors would play such a big role in fragrance choice.

In recent years, perfumes have changed. New creation techniques, ingredients, but above all a new awareness have led to a change in the perfume industry. Consumers are now offered completely different fragrance options.

There are also completely new types of perfume that are increasingly able to do more than just smell good. This is the subject of this chapter. I also want to give newcomers and career changer to this fascinating industry, background information on perfume practice and aromatherapy and their claims.

2.1 Characteristics of Perfumes Today

High-quality essential oils, which are obtained from lavender or sage, for example, are not only a gift from nature, but must also be considered a cultural asset. Because their careful extraction requires great experience. Plant ingredients can directly trigger a psychophysical effect. This gives them the potential—especially in the context of aromatherapy—to act directly on personality, consciousness, mood, and emotions (Worwood 1997).

But what about perfumes? Can they also be used for aromatherapy? And did they show an effect?

First, it must be defined what is considered perfume. In this context, it is better to speak of fine perfume to distinguish it from the scent of a household product. Later on, I will go into the different areas of perfumery and explain the difference between fine perfume and functional perfume as well as aromatherapy. I will also go into detail about the perfume market and the world of beautiful scents.

Perfume, as we most commonly know it today, is a liquid mixture of different fragrance ingredients (in German Riechstoffe)—a term that is often used synonymously with fragrance- or volatile compounds (Duftstoffe) with a molecular weights of less than $300\,\mathrm{g\,mol^{-1}}$ (grams per mole/molecule of substance), that humans perceive through the olfactory system. These mixtures of fragrance ingredients are usually used as fragrance oils dissolved in alcohol and some distilled water to create a pleasant smell. This fragrance experience is applied to the body or clothing as part of an external application. Perfume lovers also expect, more or less explicitly, that the scent impression of a new perfume is also new and innovative. In addition to liquid perfumes, there are of course also so-called "solid perfumes", which can be applied as a cream. They originated from the first perfumes of antiquity.

Depending on the concentration of the fragrances or the fragrance oil content in the alcohol, different perfume variants are distinguished, such as: "perfume", "eau de parfum", "eau de toilette", or "eau de cologne". There are also other perfume specialties, which are often attributed to functional perfumery or overlap with fine perfumery. These include hair perfumes, body perfumes, or perfume deodorants.

Traditionally, a perfume is also characterized by its scent profile, in which one smells the top note first, then the heart and finally the base or the bottom which, based on its impression, gives indication of its olfactory family, by breaking down the scent and classifying its smell. This is also often visualized as a scent pyramid with the respective characteristic ingredients or main accords for the scent profile. Today, this scent profile is no longer considered a typical perfume characteristic. Because in perfumery there is also a trend in which linear or almost linear is created on purpose, (the top note extends more or less into the base), and a single perfume can fall into several different olfactory families. Nevertheless, most perfumes are still created with a typical olfactory family in mind, (as we will discuss later), with a scent impression that unfolds harmoniously from the initial smell of the top note (typically smelled

during the first 15 min after application) via the heart note (developed after the first quarter of an hour and usually lasts up to two hours) to the base (often begins after two hours and can even last for several days).

In addition to a good fragrance, one also expects it to be long-lasting i. H longevity. This was already the case in antiquity. Thus, Theophrastus testified in relation to the Greeks: "A long-lasting perfume is what women are looking for" (De odoribus, "On odors", paragraph 42). For this, today's perfumers often use fragrance retardants, so-called Fixateure. They hold lightly volatile components such as citrus notes by heavier volatile or longer lasting ingredients such as resins. In the preface to this book, we pointed out that perfumers like to distinguish between the "sillage" and "longevity" of a perfume. The term "sillage" describes the lingering scent of the perfume a person wears, or the scent trail it leaves behind. The term longevity describes the performance, or durability (the time a fragrance lasts) of a perfume on a fragrance strip or on the skin. The performance of a perfume can last anywhere from 60 minutes to 12 hours and more (an eau de parfum generally lasts around 8 hours and tends to stay longer on fabrics than on the skin). A perfume can also last up to 18 hours and, as we said, even days, and is then called a "Beast Mode" perfume, a current fragrance trend that we will discuss later (Chapter 14). In addition, perfumers like to distinguish between "sillage" and "projection". Projection is about how far a perfume travels from the skin (or scent strip). If a perfume is projected well, you can smell it very well from an arm's length away. In contrast, a fragrance with poor projection can only be smelled from a short distance of a few centimeters, while, as mentioned, silage is the scent trail left by someone wearing perfume after walking past.

2

The division into feminine and masculine perfumes is also increasingly seen as outdated today. Since the beginning of the first emancipation movements in the late 19th century, women have also discovered men's perfumes for themselves and vice versa.

Although the idea of therapy is not usually in the foreground when a perfume is created, but rather more the feeling of well-being and the acquisition of attractiveness of the wearer and his representation in front of others, it can also influence personality, consciousness, mood, and emotions. However, according to the EU Cosmetic Regulation, application may only be external. Here, cosmetic products, to which fragrance and perfuming also belongs, are defined by their purpose as follows: "Substances or preparations of substances which are exclusively or predominantly intended to be applied externally to the human body or in the mouth for cleansing, protecting, maintaining a good condition, perfuming, changing the appearance, or influencing the body odor."

2.2 Types and Use of Perfumes— Areas of Interest of Perfume Lovers

The market for beautiful scents or perfumes ("fragrances") in different variants is very large and worldwide reaches the 51-billion-dollar mark (2022). Already in 2016, consumers worldwide spent 46.7 billion dollars on scents. However, perfume, which is most often offered in the concentration of an Eau de Parfum, followed by Eau de Toilette, is the smallest segment in the entire beauty industry.

For decorative cosmetics, i.e. make-up products, 72 billion dollars were spent in 2019, for facial and skin care 140 billion dollars and for hair and body care 236 billion dollars. The latter category includes shower and bath products, soap, oral and dental care products, deodorants, shav-

ing care, pre- and aftershaves, hair, hair removal, and foot care products as well as sun protection, and baby care. In 2019, the market for all beauty products—including women's and men's fragrances—reached 14 billion euros in Germany alone. In 2022, after two COVID-19 years, the market reached the size of 2019 again. In Germany, buyers spent 14.3 billion euros on personal care and cosmetic products. Of this, women's and men's fragrances generated 2019 around 1.5 billion euros and is currently growing, (2022 to 1.7 billion euros with a + 3% increase to 2021) in Germany. In comparison, the fragrance segment in the United States was around $5.1bn in 2022. The use of fragrance still shows a gender-specific tendency today. In Germany, the women's fragrance market is almost twice as large as the men's fragrance market at 500 million euros. However, the comparison is not quite correct, as many men use shaving care and, in this context, aftershaves in particular, as perfume. In Germany alone, this represents a market of 200 million euros.

Every year, over 2,000 new perfume launches in various variants flood the market. In the case of fine fragrances, six classifications are typical, with classifications based on the concentration of fragrance compounds or fragrance oils being relatively relative. There are no legal regulations as to what is to be considered as Eau de Parfum or Eau de Toilette. A regulation is opposed to the fact that individual fragrance oils smell differently when they are dissolved in alcohol and some distilled water for traditional perfume production.

In the following, I would like to introduce the most important perfume variants.

■ **Everything you need to know about the precious perfume**

The word "perfume" is associated with the crowning glory of perfumery. The fragrance oil concentration is between 15 and 40%. The rest—and thus the majority—is alcohol with distilled water added. From a 20%

concentration, one speaks of pure perfume or Extrait de Parfum.

» *Why do we use other perfume variants in addition to perfume?*

It is certainly a question of money, but it also depends on how you like to be perfumed. Pure perfumes are less suitable for generous application. As much as you love long-lasting, more intense perfumes, there are still some things to consider. Especially with fragrance concentrations of 15 to 40%, a careful approach is required so that you can smell the perfume longer. Perfumes can trigger a smell signal cascade in the olfactory cells and their receptors, which then react hyperactively—a process that is interrupted after a while. Although the fragrance components of a perfume are still present, they are no longer or only barely perceived. There is an adaptation. Everyone knows the situation that you have applied too much perfume and at the same time you smell it less and less.

» *Where is the best place to apply perfume?*

Ideally, a perfume needs skin contact. It is best applied to warm body parts with a noticeable pulse. But beware: larger quantities should not come too close to the nose. Earlobes, neck, chest, temple or neck are suitable for perfume. For pure perfumes, you should limit yourself to light dabbing. Wrist, on the other hand, can receive a little more of the delicious elixir. The back of the knees are the perfect place for pure perfumes if you want to apply them more generously and at the same time avoid a too rapid adaptation.

» *Does the way you apply perfume reveal as much or almost as much about you as the perfume you wear?*

There are a number of ways to apply perfume that can have different effects on yourself and others, or that can put you in different moods. Here are some examples:
- Many men love to give their scent, especially after shaving, on the neck and cheeks or to clap. The scent is then first invigorating, energizing, and disinfecting tonic.
- Spray the perfume into the air and run into the cloud. Perfuming is thus the principle of the fairy tale of the Brothers Grimm, the fragrance particles fall like a fairy tale from the sky. Maybe the most innocent and shy way of perfuming, in which you only get a hint of perfume and can be sure not to appear over-perfumed.
- Dab perfume into the knee hollows. This quickly becomes the most subtle and sophisticated perfume application, especially for women. Depending on the movement or position of the legs, one can surprise others with the fragrance effect. If women then set their legs in pose with the right clothes, this often confuses men at first, then curiosity and finally the desire to seek proximity.
- Luxuriate in the scent. In other words: enjoy your own over-perfuming. This has something majestic and reminds of the court of King Louis XV. Water was not an option as a cleaning agent at that time. With perfume, the own smells were covered. The etiquette at the court also prescribed to use a different scent every day. The court of Lous XV was therefore also called the "perfume court".

» *Does a perfume smell different on everyone?*

In the perception of fragrance, differences in temperature and humidity play a role, but also individual condition and skin type. In addition, everyone has an individual body odor, which is influenced by nutrition, lifestyle, age, gender, immune system, and health condition. Experience has shown that a perfume is still relatively easy to recognize in the top note/ headnote, that is, in the first 15 min, with different carriers—even if it is applied discreetly. If the perfume melts during the heart note on the skin, that is, after about two hours of wear,

2

it begins to smell more individual and its recognition becomes more difficult. It gets really difficult to recognize a perfume in the base or in the fond, that is, after several hours of wear—unless it has a very distinctive aftertaste and/or is known. Of course, the previous experience also plays a role in the recognition of a perfume, if one knows, for example, how a perfume develops. In general, one can say: The longer a perfume is applied discreetly to the skin, the more individual it smells.

» *How many individual fragrance, or ingredients, does a perfume consist of? When can you smell something?*

A fragrance is a chemical substance that stimulates the sense of smell. It is based on molecules composed of elementary units. Molecules are two- or more-atom particles held together by chemical bonds. Everything that smells to humans constantly emits small amounts of specific fragrance molecules that stimulate the olfactory cells in the upper part of the nose. You can calculate how many molecules of fragrance must be present in one breath per olfactory cell in order for there to be a sensation. The amount is different for each fragrance and can, for example, be eight molecules.

When it comes to the sense of smell, one distinguishes between the threshold of perception and thethreshold of recognition. The threshold of perception is the concentration of fragrance so low that one only suspects that one smells something. With the threshold of recognition, on the other hand, one can name the fragrance—if one knows it. There is a perfume on the market with a single fragrance ingredient, the synthetic "Iso E Super". Most perfumes consist of 30 to 70 fragrance ingredients, but can also be composed of several hundred. In theory, perfumers can choose from a palette of fragrances of about 3000 natural and synthetic fragrance ingredients. This is the case when the price of a perfume is not an issue, ingredients are accessible and the

perfume does not collapse when using too many fragrance ingredients. This is only a theoretical size, because a lot of fragrance does not automatically mean a lot of nice smell.

» *Can you mix a completely individual perfume yourself?*

Yes, you can, from two perfumes—which you must not tell any perfumer. The method is called perfume layering. I will introduce it in detail later. For perfumers, the idea of mixing two artworks is horrifying. They see a perfume as a creation of usually many different coordinated ingredients that do not allow intervention. Nevertheless, one is often surprised how the olfactory artworks complement each other when layering. However, you have to stick to certain rules. For example, when layering, the heavier perfume must be sprayed over the lighter, fresher, more volatile fragrance, so that both perfumes mix well.

» *How are perfumes won?*

The raw materials for perfume oil are traditionally produced by five methods, which are further refined in further processing steps:

1. **Distillation**. A separation process in which a fragrance ingredient is separated or dissolved by evaporation or condensation.
2. **Maceration**. Most often, chopped substances are soaked in a liquid such as alcohol, oil and water, but also with other means to obtain soluble components.
3. **Enfleurage**. Heat-sensitive, freshly harvested flowers or plants are allowed to give off their volatile essential oils in fat, oil or other substances.
4. **Extraction**. Substances are separated or purified by various methods, including steam distillation.
5. **Expression**. Certainly one of the oldest methods of perfumery to obtain something that smells good and is beneficial. Originally, for example, plant leaves

were pressed by hand. Even today this is done by means of mechanical pressing processes.

» *How long does a perfume last, and where is the best place to store it?*

It may sound surprising, but it's true: perfume does not belong in the bathroom. In many bathrooms, for example, there are temperature fluctuations and high humidity due to hot showers, which can quickly change the smell of an olfactory work of art. Perfumes love cooler, drier, and darker rooms with constant temperatures. They especially don't like sunlight, heat, and oxygen. The bedroom might be the ideal place to store them. If treated well, the fine essences can last for more than three years. This is especially true for perfumes from the Amber-Oriental fragrance family. These fragrance notes can last for decades, even if the top note eventually disappears. For perfumes in the "fresh-green-citrus" fragrance family, such as Bergamot notes, it is even recommended to store them at refrigerator temperature. If a perfume has changed color significantly and become thicker, it means that it has exceeded its life expectancy.

» *What areas and areas of interest fascinate perfume lovers?*

Perfume and perfumery span a wide range of specialties. Twelve areas in particular are fascinating. Here are some German-language literary classics and newer treatises as examples and to get a taste. Some of them are also available as English or French editions Just from the titles alone, they give a good overview of the areas and interests in perfume and perfumery:

Perfume history:
- Corbin, A. (1984) Pesthauch und Blütenduft. Eine Geschichte des Geruchs. Wagenbach, Berlin
- Morris, E.T. (2006) Düfte – Die Kulturgeschichte des Parfums. Albatros, Düsseldorf

- Faure, P. (1990) Magie der Düfte. Eine Kulturgeschichte der Wohlgerüche von den Pharaonen zu den Römern. Artemis, München und Zürich
- Le Guérer, A. (1992) Die Macht der Gerüche. Eine Philosophie der Nase. Klett-Cotta, Stuttgart.
- Schlögel, K. (2020) Der Duft der Imperien: „Chanel No 5" und „Rotes Moskau". Hanser, München

Perfume ingredients:
- Hall, R., Klemme, D., Nienhaus, J. (1985) H&R Lexikon Duftbausteine – Die natürlichen und synthetischen Komponenten für die Kreation von Parfums. Glöss, Hamburg.
- Martinetz, D., Hartwig, R. (1998) Taschenbuch der Riechstoffe – Ein Lexikon von A–Z. Harri Deutsch, Thun und Frankfurt a./M.
- Legrum, W. (2012) Riechstoffe, zwischen Gestank und Duft: Vorkommen, Eigenschaften und Anwendung von Riechstoffen und deren Gemischen. Springer Spektrum, Wiesbaden
- Nagel, B. (2020) PARFUM. PUR. Düfte, Farben, Kulinarik: und eine Prise Poesie. Art Parfum, Oy-Mittelberg

Perfume making:
- Stead, C. (1996) Parfum aus ätherischen Ölen selbst herstellen—Komponieren Sie Ihren ganz persönlichen Duft. ECON Taschenbuch, Düsseldorf,
- Aftel, M. (2004) Die Kunst der Alchimisten—Alles über Parfum. Rütten & Loening, Berlin
- Ellena, J. C. (2012) Der geträumte Duft—Aus dem Leben eines Parfümeurs. Insel-Verlag, Berlin

Perfume artworks:
- Turin, L., Sanchez, T. (2013) Das kleine Buch der großen Parfums: Die hundert Klassiker. Dörlemann, Zürich
- Girard-Lagorce, S. (2001) 100 legendäre Parfums. Tosa, Wien

- Mayer Lefkowith, C. (2000) Glanzstücke der Parfümindustrie. Brandstätter, Wien

Smelling:
- Hatt, H., Dee, R. (2012) Das kleine Buch vom Riechen und Schmecken. Albrecht Klaus, München
- Burdach, K., J. (1991) Geschmack und Geruch. Gustatorische, olfaktorische und trigeminale Wahrnehmung. Hans Huber, Bern, Stuttgart, Toronto
- Pause, B. (2020) Alles Geruchssache: Wie unsere Nase steuert, was wir wollen und wen wir lieben. Pieper, München

Fragrance molecules:
- Ohloff, G. (1992) Riechstoffe und Geruchssinn: Die molekulare Welt der Düfte. Wiley, Weinheim
- Peter, K., Vollhardt, C., Schore, N.E. (2011) Organische Chemie. Wiley, Weinheim
- Breitmaier, E. (2005) Terpene: Aromen, Düfte, Pharmaka, Pheromone. Wiley, Weinheim

Fragrance psychology:
- Gschwind, J. (1998) Repräsentation von Düften. Wißner, Augsburg
- Jellinek, P. (1973) Die psychologischen Grundlagen der Parfümerie. Hüthig, Heidelberg
- Mensing, J. (2005) Duft-Guide. Der schnelle Führer zu Ihren Ideal-Düften. In: Roller, U., Spelman, R. Parfums – Edition 2005. Ebner, Ulm. S. 206–215

Olfactory brain research/Neuroperfumery:
- Pause, B. (2004) Über den Zusammenhang von Geruch und Emotion und deren Bedeutung für klinisch-psychologische Störungen des Affektes. Pabst Science Publ, Lengerich
- Moessnang, C., Freiherr, J. (2013) Olfaktorik. In Schneider, F., Fink, G., R. (Hrsg.). Funktionelle MRT in Psychiatrie und Neurologie. Springer, Berlin Heidelberg
- Spitzer, M., Bertram, W. (2009). Hirnforschung für Neu(ro)gierige: Braintertainment 2.0. Schattauer, Stuttgart

Sociology of smell:
- Raab, J. (1998) Die soziale Konstruktion olfaktorischer Wahrnehmung. Eine Soziologie des Geruchs. Dissertation, Universität Konstanz
- Ehrensperger, A. (2015) Parfümgeschichten: über die Sprachlosigkeit sinnlicher Erfahrungen. University of Zurich
- Bandura, J. (2005) Der Geruch und der Geruchssinn—eine soziologische Betrachtung über die soziale Konstruktion der olfakto- rischen Wahrnehmung. Studienarbeit des Fachbereichs Soziologie—Kultur, Technik und Völker. Universität Duisburg-Essen (Institut für Soziologie)

Scent marketing:
- Rempel, J., E. (2006) Olfaktorische Reize in der Markenkommunikation. Theore- tische Grundlagen und empirische Er- kenntnis zum Einsatz von Düften. Springer, Berlin Heidelberg
- Schiansky, M. (2011) Mit allen Sinnen: Duftmarketing. Diplomica, Hamburg
- Knoblich, H., Scharf, A., Schubert, B (2003) Marketing mit Duft. De Gruyter Oldenbourg, München

Aromatherapy:
- Worwood, V., A. (1992) Liebesdüfte—Die Sinnlichkeit ätherischer Öle, Goldmann Ratgeber. Goldmann, München
- Schnaubelt, K. (1995) Neue Aromatherapie—Gesundheit und Wohlbefinden durch ätherische Öle. vgs, Köln
- Lawless, J.(1996) Kleine Aroma-Apotheke, ECON Taschenbuch, Düsseldorf

Perfume-Belletristik:
- Süskind, P. (1985) Das Parfum. Die Geschichte eines Mörders. Diogenes, Zürich

- Janson, B. (2012) Der verbotene Duft. Ullstein, Berlin
- Rose, M., J. (2013) Das Haus der verlorenen Düfte. Aufbau, Berlin

» *Which perfume museums should one visit at least once from their location?*

There are wonderful perfume museums all over the world. I will only mention two. They are located in places where the heart of the perfume industry beats particularly strongly:

- Musée International de la Parfumerie
  2 Boulevard du Jeu de Ballon
  F-06130 Grasse
- Osmothèque
  36 Rue du Parc de Clagny
  F-78000 Versailles

The Osmothèque is part of the ISIPCA, one of the most famous perfume schools.

After this excursion, I will return to the other perfume variants.

▪ Eau de Parfum

This is not only in Europe, but worldwide, the most popular perfume variant. It has, after the perfume with 10 to 14%, the second highest concentration of fragrance oil or fragrance. Consumers associate with this perfume variant a certain value, also because one believes that a Eau de Parfum can be used more sparingly. This is usually also the case for the durability of the fragrance, especially when the concentration is between 12 and 14%. However, the fragrance oil concentration is not a guarantee for their durability or subjective impression. This is influenced by several factors such as climate, environment, nutrition, mood, skin condition and -quality, but also by medication and the accustoming to the perfume.

Perfumery is also not additive. A high fragrance concentration does not guarantee a better smell. Often this is even rather the case when a fragrance oil is diluted in more alcohol, thus lighter concentrated or set. Perfumers are familiar with the problem that their olfactory work of art can collapse or suddenly develop a different fragrance character at a higher concentration, because now certain ingredients smell more pronounced. Finding the right concentration for a fragrance oil is an art in itself.

Often one starts with the first mixing of an oil at 12% and then works step by step up or down. It can happen that the same mixture smells better with 12% than with 14%. This also depends on the alcohol used. So Germany's perfumers had to work for a long time with state-supported alcohol from sugar beets, which stung a little in the nose at first smell. French perfumers have always been able to fall back on better alcohol. I still remember my early days in perfumery, when in Germany the alcohol was made "rounder" with a hint of musk. So one hoped to be able to keep up with the alcohol used in France from sugar cane, wheat or fruits—even in Eau-de-Vie quality. Already the choice of the alcohol, which makes up the largest part of a perfume in the classical alcoholic perfumery, is therefore decisive for the fragrance perception.

▪ Eau de Toilette

This type of perfume has lost much of its former popularity in recent years. Many perfume lovers find Eau de Toilettes either too light, and they prefer a more valuable sounding Eau de Parfum—or they prefer an even lighter scent. In fact, there are two scent trends independent consumer trends in perfumery. On the one hand, lighter perfuming is in demand, on the other hand, scents with character and expressiveness are sought for certain occasions. As a fragrance trend in 2023, we could see the appearing of highly concentrated fragrances, some of the most frequently cited examples of scents labelled as "beast mode" fragrances include Sauvage Elixir by Dior and Maison Francis Kurkdjian's BaccaratRouge 540 (both the Extrait and the Eau de Parfum).

2

Most Eau de Toilettes have a fragrance concentration of about 6 to 8%. However, the concentration can also be higher—especially for men's fragrances. Many men associate the term "perfume" with something feminine, but still expect their own scent to be durable and personal. That's why there are also Eau de Toilettes for men with 12 to 14 percent fragrance oil concentration or more, which at least elevates them to the ranks of Eau de Parfums.

■ **Eau de Cologne**

This is the classic presentation of many scents, especially from the "fresh-green-citrus" direction. Even today, the invigorating bergamot notes with a three to five percent fragrance concentration dominate in this scent direction.

■ **Eau Fraîche**

The perfume variant Eau Fraîche, also called Splash Cologne, typically comes with a fragrance oil concentration of 1 to 3% as a refreshing summer scent. Most shaving water / after shave or balm products as well as soaps, bath and shower gels also have a fragrance concentration in this range. The fragrance content is usually less than 1% for creams and lotions, and not more than 0.5% for skin care products. The fragrance content of most lightly perfumed room sprays and air fresheners starts at about 0.5%. But there are also products with a fragrance content of up to 5%. Fragrance candles often have a fragrance concentration of about 1.5%. Not all fragrances are suitable for scenting the various products. For example, fragrances for candles must still smell good under heat influence.

■ **Body Mist**

This perfume variant, also called in German Körperduftspray, is particularly popular in the USA and has an oil concentration of 0.3 to 3%. Many body mists are scented skin or body care products. Some do not contain alcohol, which could dry out the skin. Those who cannot perfume themselves at work often use body mists as a substitute for perfume. They are also a good alternative for allergy sufferers, of which there are particularly many in the USA and in Europe. In addition, body mists are significantly cheaper than eau de parfums.

2.3 The Trend in the Perfume Industry Towards Nature

Some perfume products are already very close to essential oils in terms of claim and quality. These are fragrances based on plant essential oils. The range of scents of these quality perfumes is currently increasing more and more, as essential oils can be obtained from various parts of plants. These include flowers, flower buds, leaves, branches, fruits and their parts, bark, roots, seeds, needles, wood and the rhizome, which grows underground or close to the ground.

However, for many plants—even if they smell very good or unusually interesting—the proportion of fragrance content is so low that their processing is not worthwhile. You would like to use these unusually smelling plants or plant parts, because the art of perfume-making also consists in creating a new and innovative fragrance experience. In these cases, one often resorts to natural identical fragrance ingredients (in German naturidentische Duftstoffe), which are chemically identical to a plant or plant part, but are not or only partially obtained from naturally occurring substances. However, this can result in the therapeutic fragrance effect (in German Duftwirkung) being limited or not occurring at all under aromatherapeutic requirements. Nevertheless, there is also a clear trend towards more natural and thus towards natural essential oils in perfumes. This current trend can also be formulated as follows:

> The transitions between fragrance enjoyment, care, and therapy are becoming more fluid.

Even if natural identical fragrances are and must continue to be used in the creation of perfumes due to the limited availability of various plant parts, the industry is paying more and more attention to a positive or even health-promoting effect of the ingredients. They are increasingly being researched by the fragrance-producing industry.

Perfume research has focused in recent years particularly on the aspect of skin care through perfume ingredients. For example, in 2020 the perfume oil manufacturer Firmenich brought the biotechnologically produced fragrance "Dreamwood" to market, inspired by sandalwood in "Mysore quality". This is a quality designation named after an Indian place. This sandalwood is produced in Indian distilleries from wood chips of the sandalwood tree. The new fragrance smells warm and creamy and has an antimicrobial, that is, growth-inhibiting, and calming effect on the skin. Therapeutically, sandalwood is of particular importance as a fragrant massage oil. In 2014, research also came as a surprise with a surprising message: The skin can smell sandalwood.

This raises the question of perfume efficacy, which is increasingly difficult to answer: Are natural fragrances better than artificial or synthetic? There are natural, fully synthetic, and semi-synthetic fragrances. At the end of the 19th century, synthetic fragrances were celebrated as an innovative and revolutionary new movement in perfumery. They usually have the advantage that they are much more durable and can also be produced more cheaply. In many cases, although not always, synthetic fragrances also offer an advantage for allergy sufferers. Because the synthetic fragrances smell like the unadulterated image and often better or allow new olfactory impressions, but do not contain the allergens that can be contained in natural fragrances. Nevertheless: The tide has long since turned again, and consumers increasingly wish for natural ingredients in their perfumes—since synthetic fragrances have been criticized in recent years, mainly because of the parabens and phthalates they may contain—but not only because of this—the public rightly has a negative opinion about the effects on human health.

But what if research develops more and more synthetic and semi-synthetic fragrances that reduce the risk of allergies, work better in perfumes and offer additional health benefits such as optimization of skin condition? In this context, semi-synthetic fragrances are of particular interest. They are isolated from natural substances and then possibly further processed. Later on, I will go into the importance of plant peptides for the future of perfumery. This raises a completely different question: How strong can perfumes actually work? But more on that later.

2.4 Personal Requirements for Perfumes

Perfume lovers today have more and more opportunities to very precisely determine their personal requirements and criteria for their perfume and thus take into account different types of fragrance use. For example, perfumes are available or under development and can be selected
- as vegan, organic and as bio-perfume;
- as certified natural perfume, e.g. from BDIH, NaTrue or Ecocert;
- as alcoholic and alcohol-free perfume or with special alcohol such as bioethanol;
- as perfume made from natural, biotechnological, semi-synthetic, or pure synthetic ingredients or a combination of these;
- in a light (e.g. as an eau de toilette) to concentrated concentration (e.g. as an eau de parfum or extrait de parfum);

2

- with one molecule, with few ingredients, or with several;
- as perfume made from one plant species or from several plants or plant parts of a specific region;
- as liquid or solid (solid cream / balm) perfume;
- as perfume without questionable or unwanted effects to avoid personal allergy risks. In this context, it is interesting that more and more testing methods are discovered whether fragrances could potentially trigger skin reactions, without the need for animal testing or human clinical trials;
- as a fragrance to counteract bad odors such as cigarette smoke (e.g. with Vernova Pure® technology);
- as a fragrance with added benefits such as anti-aging, moisturizing, cleansing, and slimming properties (e.g. offered with Robertet ActiScents technology);
- as a bespoke fragrance, mixed with the ingredients of your choice at an in-store refill station (e.g. patented by Coty in 2023);
- as a sustainable fragrance with bio-based packaging for more environmental friendliness (e.g. offered by Sulapac since 2023);
- as fragrance with ingredients and formulas designed to be mood-boosting;
- as a fragrance created to enhance synergistic experiences;
- as a fragrance with neuroactive ingredients and benefits, one of the latest trends in the cosmetics and personal care industry (we'll get into that later);
- as a fragrance designed for your needs and desires with AI (we will talk about this in detail);
- as a vintage fragrance to indulge in classic simplicity.

2.5 Perfume in Conflict: Prohibited and Unwanted Effects

The legislator has identified a number of natural and synthetic fragrance ingredients that are considered potentially allergenic and/or harmful to health. Their use in fragrance products is prohibited or their concentration is very limited. The list of these ingredients—which are also to be warned of without explicit ban—is constantly getting longer.

According to EU law, 26 ingredients with allergy risks must be indicated in perfumes from 2019 if 0.001% of the respective fragrance remains on the skin or hair in the end product, for example in a perfume, and it is therefore a leave-on product. If the product can be washed off like a shampoo, the value is 0.01%.

The INCI (International Nomenclature of Cosmetic Ingredients) offers allergy sufferers in particular the opportunity to check a product for the 26 suspicious ingredients before purchase. Fragrances are then summarized under the collective terms "perfume", "fragrance", "aroma", or "flavour". If the suspicious fragrance ingredients are contained in a cosmetic product, i.e. also in a perfume, they are listed on the packaging with their INCI names in descending order of concentration and with the date of manufacture. The ingredients with a concentration of less than 1% are then listed in no particular order.

Various consumer organizations such as Öko-Test demand even further restrictions on ingredients. They are not satisfied that the responsibility for suspicious ingredients in products and thus their safety for consumers lies solely in the hands of manufacturers. As early as 2012, the scientific advisory board of the EU Commission published a paper in which 82 fragrance ingredients and essential oils were named as proven contact allergens. The concentration of many other fragrances should be limited. Many even demand that the individual ingredients be listed in all fragrance formulations.

So far, the EU law has only required the general indication "perfume" on the packaging for fragrances with approved substances in order to protect the formulations of the perfumers as a secret. Accordingly, many brand manufacturers do not disclose these formulations or describe them with olfactory impressions and some information on ingredients in the head, heart and base of the perfume. A perfume usually consists of 30 to 70 ingredients. Many consumers therefore fear that the constantly increasing allergic reactions are due to secret substances. These reactions are recorded two to three times more often in women. This is especially true for household products to which women are more exposed.

Approximately 3% of the European population are estimated to be allergic to some fragrance components. It is estimated that well over 2 million Americans suffer from significant fragrance allergies or sensitivities and the number is rising. These can have either natural or synthetic origins. A service for consumers could be that the ingredients of products are listed on official websites. In this way, one could search for substances with allergy risks and cross them off the shopping list immediately.

But the real problem is different. While consumer groups such as Women's Voices of the Earth fear secret chemicals, an inverse fear haunts industry: consumers could find out what is not in the fragrance. So wonderfully described fragrance components from exotic countries could turn out to be simple chemicals. Nevertheless, there is an increasingly trend towards more honesty, evidence, and fairness in the fragrance

and cosmetics industry. And that is also demanded by the legislator.

For example, since July 2019, the EU has banned certain marketing statements that advertise compliance with legal requirements. For example, the statement "This product complies with the requirements of the Cosmetic Products Regulation (Kosmetikverordnung)" is not allowed. More fairness is also taken into account. So the claim is prohibited: "In contrast to product X, the product does not contain the ingredient Y, which is known for its irritating effect." The legislator also requires a convincing, not exaggerated advertising message. For example, a statement such as "This product gives you wings" should no longer be communicated.

The regulations are also changing in the USA. Elena Knezevic (Editor-in-Chief of Fragrantica—a website for perfume lovers) outlines for the years 2023 and 2004: Perfume and cosmetics companies are currently preparing for the new regulations (MoCRA—The Modernization of Cosmetics Regulation Act), some of which will be mandatory in the next year (2024). This law gives the FDA (U.S. Food and Drug Administration) new powers over the industry. From December 2023, cosmetic companies must register with the FDA and report products and ingredients as well as fragrance allergens. In other words, consumers get more information about the products they use.

In order to make scents more attractive to consumers, the trend towards natural perfumes focuses on certain plants that have ripened only under very specific conditions. They carry seals of quality such as NaTrue or Demeter in accordance with the specifications of natural cosmetics manufacturers. The focus is on plants recognized by certified natural cosmetics, whose origin is known, which have been grown and processed in accordance with the controlled-organic guidelines. This of course excludes

2

the use of genetically modified plant material. This makes it possible to comply with the new COSMOS ORGANIC quality seal, which is defined by international environmental organizations as the minimum quality requirements for "organic" fragrance and cosmetic products. The new, harmonized COSMOS standard combines the previous standards of BDIH (Germany), COSMEBIO, ECOCERT (both France), ICEA (Italy), and SAO (England).

A fragrance user can of course associate positive and harmless, but also questionable and negative things with a perfume from his or her individual point of view. The numerous perfume variants make a generally valid statement difficult. Therefore, before smelling a perfume, you should first create a personal profile or a requirements profile for your scent. In this way, you can concentrate only on bio-perfumes with natural and unobjectionable ingredients for yourself when you are looking for or advising on a perfume. Vegan perfumes made from at least 90% natural and sustainably sourced ingredients have already become the new standard in fine perfumery. An example is Gaultier Divine—Eau de Parfum by Jean Paul Gaultier (2023).

Depending on the profile of requirements, a perfume can indeed also come close to an essential oil. In this case it is a pure natural product—with or without alcohol—and can even have been created from a specific plant—like most essential oils. But there are also essential oils from different plants or from a recommended mixture that gives an oil a certain perfume character. For example, a mixture of essential oils from calming lavender, mood-enhancing orange and concentration-enhancing lemon is associated with better learning. In other words, perfumes and essential oils can overlap and are doing so more and more, as the current trend shows.

2.6 Consumers Want More than Just Smelling Good— From Perfume to Active Perfume

From a fragrance psychology perspective, there are even more questions:
— What does olfactory offer the best effect to increase personal feeling and experience, such as specifically the acquisition of more self-attractiveness and self-confidence? Are they perfumes or essential oils?
— How must a fragrance smell for someone so that it is suitable for the "therapy of the self", which is more focused on the search, acquisition, and increase of mindfulness and identity?
— Or simply asked: What is best for personal self-optimization?

When answering these questions, it is certainly also about what promotes health through smell. But here the personal "self" is clearly in the foreground in its experience and feeling. Both will often overlap.

But the search, acquisition and increase of more self-attractiveness, mindfulness, and self-confidence can also be a separate therapeutic goal, especially when it comes to processing one's own identity. I will later propose fragrance-supported self-coachings for the "therapy of the self".

Of course, an experienced fragrance enjoyment is needed to achieve the respective optimization goals. But it is mainly about the possession of one's own olfactory tool that works for one and thus can do more than just smell good. Accordingly, ideally you need an "active fragrance" or an "active oil" or an "active perfume". Such perfumes are currently the focus of research in neuroperfumery. Many see the future of fragrances in neuroactive fragrances (neuroscents). We will discuss this topic later using the latest findings.

This olfactory "effect tool" can be a fragrance, a perfume, or an essential oil. How and whether it works can often only be decided by the user himself. It cannot be claimed that the same fragrance or the same tool will also have the same effect on others. Whether it is an "active fragrance" depends on the respective subjective experience.

This subjective experience characterizes the difference to classical aromatherapy, in which—as in medicine—one focuses on effects of plants that are as universal as possible, for example against stress. In contrast, the method I refer to as scent therapy claims—even if not exclusively—to contribute to "therapy of the self", without certain fragrance notes or oils having to claim general validity in their effects. Thus, in scent therapy, the individual himself must decide what works best for him olfactorily.

Often he has to experiment and become his own private perfumer. This can be achieved with the layering of perfumes method without much effort, which we will talk about later. Of course, one can fall back on the knowledge and application methods of aromatherapy. But here too it applies: Everyone has to decide for himself whether and with which olfactory tools a "therapy of the self" can be achieved. There are good reasons for this. The experience of scent arises in regions of the brain that are responsible for personal feelings, moods, emotions as well as for consciousness and personality and are colored by individual memories and associations. In order to achieve a targeted personal self-optimization through scent, one should not exclude olfactory tools that do one good.

For the "therapy of the self" it is therefore not a question of whether it is better to use a perfume, an essential oil, or another olfactory medium in the scent therapy. With the self-optimization intended here, perfumes are also a welcome tool, however

each person defines them for themselves. Of course, organic perfumes that are similar in quality to essential oils are a good first choice.

This also answers my question from the beginning: Can perfumes also be used for scent therapy, or is scent therapy only possible with essential oils?

However, the difference between aroma and scent therapy still needs to be clarified, which I will do in the following.

2.7 The Difference between Aromatherapy and Scent Therapy

The terms aromatherapy and scent or fragrance therapy are mostly used synonymously. However, on closer inspection, there are differences in the self-understanding of these methods.

In comparison to scent or fragrance therapy, **aromatherapy** has a more medicine-oriented self-claim. It is based on the use of essential oils from plants that are used in a therapy to improve physical and mental well-being. The focus is therefore on plants with person-spanning or general, objective effects. The goal of aromatherapy is to be used as an alternative or supplement to a number of diseases and symptoms. It thus acts as an additional medical treatment, e.g. for stress and depression.

Scent therapy is focused on the individual, subjective well-being of its users. All kinds of olfactory media are used, which in addition to essential oils can also include perfumes. The user decides on the type and form of administration. What is important is what is good for each individual. This means that there is no claim that the same scent or the same tool will work in the same way for others. This means that the subjective experience of the individual decides whether something is an "effective scent" for him and whether he can work therapeutically with it.

2

The main goal of fragrance therapy is the "therapy of the self" and thus the search for and increase of more self-attractiveness, mindfulness, and self-confidence in the context of processing one's own identity.

Of course, scent therapy and aromatherapy overlap in many therapy goals—even if not everyone working in these areas would agree with my distinction. Nevertheless: In general, scent therapy does not see itself as an independent treatment method in contrast to aromatherapy, but as an olfactory component of a fragrance-supported therapy. Later in this book I will go into this type of therapy with exercises and explanations of creativity with scent landscapes, mindfulness, as well as the "Scented Loving-Kindness Meditation" and self-coaching with affirmations.

» *So, now you have the most important basics about perfume and perfumery—the foundation for every insider. Maybe some of what you already know as a perfume lover was already known to you in Chapters 1 and 2, but it is important to me that everyone has the same level of knowledge for the further journey into the world of perfume.*

Summary

In this chapter we went into the types and use of perfumes as well as the wishes and interests of perfume lovers. Currently, a trend is emerging that the traditional perfume is changing or that new subgroups of perfumes are being created. In recent years, for example, a new generation of perfume has arisen, which is particularly suitable for aromatherapy, as are essential oils. More and more consumers are looking for perfumes as "active fragrances" that can do more than just smell good. These include, for example, bio-perfumes that are similar in quality to essential oils. In this context, there is an increasing awareness of more honesty, proof, and fairness in the

fragrance and cosmetics industry, from which allergy sufferers also benefit. The current trend can also be formulated as follows: The transitions between fragrance enjoyment, care and therapy are becoming more fluid. Perfume lovers today have the opportunity to very precisely determine their personal requirements and criteria for their fine perfume, for example by only selecting scents that are certified as natural perfume with a quality seal.

References

Aftel M (2004) Die Kunst der Alchimisten – Alles über Parfum. Rütten & Loening, Berlin

Bandura J (2005) Der Geruch und der Geruchssinn – eine soziologische Betrachtung über die soziale Konstruktion der olfaktorischen Wahrnehmung. Studienarbeit. Universität Duisburg-Essen (Institut für Soziologie)

Breitmaier E (2005) Terpene: Aromen, Düfte, Pharmaka, Pheromone. Wiley, Weinheim

Burdach KJ (1991) Geschmack und Geruch. Gustatorische, olfaktorische und trigeminale Wahrnehmung. Hans Huber, Bern/Stuttgart/Toronto

Corbin A (1984) Pesthauch und Blütenduft. Eine Geschichte des Geruchs. Wagenbach, Berlin

Ehrensperger A (2015) Parfümgeschichten: über die Sprachlosigkeit sinnlicher Erfahrungen. Schweiz Arch Volkskunde 111(2):167–186. University of Zurich

Ellena J, C. (2012) Der geträumte Duft – Aus dem Leben eines Parfümeurs. Insel-Verlag, Berlin

Faure P (1990) Magie der Düfte. Eine Kulturgeschichte der Wohlgerüche von den Pharaonen zu den Römern. Artemis, München/Zürich

Girard-Lagorce S (2001) 100 legendäre Parfums. Tosa, Wien

Gschwind J (1998) Repräsentation von Düften. Wißner, Augsburg

Hall R, Klemme D, Nienhaus J (1985) H&R Lexikon Duftbausteine – Die natürlichen und synthetischen Komponenten für die Kreation von Parfums. Glöss, Hamburg

Hatt H, Dee R (2012) Das kleine Buch vom Riechen und Schmecken. Albrecht Klaus, München

Janson B (2012) Der verbotene Duft: Historischer Roman. Ullstein, Berlin

Jellinek P (1973) Die psychologischen Grundlagen der Parfümerie. Hüthig, Heidelberg

Knoblich H, Scharf A, Schubert B (2003) Marketing mit Duft. De Gruyter, Oldenbourg, München

Lawless J (1996) Kleine Aroma-Apotheke. ECON Taschenbuch, Düsseldorf

Le Guérer A (1992) Die Macht der Gerüche. Eine Philosophie der Nase. Klett-Cotta, Stuttgart

Legrum W (2012) Riechstoffe, zwischen Gestank und Duft: Vorkommen, Eigenschaften und Anwendung von Riechstoffen und deren Gemischen. Springer Spektrum, Wiesbaden

Martinetz D, Hartwig R (1998) Taschenbuch der Riechstoffe – Ein Lexikon von A – Z. Harri Deutsch, Thun/Frankfurt a./M.

Mayer Lefkowith C (2000) Glanzstücke der Parfümindustrie. Brandstätter, Wien

Mensing J (2005) Duft-Guide. Der schnelle Führer zu Ihren Ideal-Düften. In: Roller U, Spelman R (eds) Parfums – Edition 2005, 10. ed. Ebner, Ulm, pp 206–215

Moessnang C, Freiherr J (2013) Olfaktorik. In: Schneider F, Fink GR (eds) Funktionelle MRT in Psychiatrie und Neurologie. Springer, Berlin/Heidelberg

Morris ET (2006) Düfte – Die Kulturgeschichte des Parfums. Albatros, Düsseldorf

Nagel B (2020) PARFUM. PUR. Düfte, Farben, Kulinarik: und eine Prise Poesie. Art Parfum, Oy-Mittelberg

Ohloff G (1992) Riechstoffe und Geruchssinn: Die molekulare Welt der Düfte. Wiley, Weinheim

Pause B (2004) Über den Zusammenhang von Geruch und Emotion und deren Bedeutung für klinisch-psychologische Störungen des Affektes. Pabst Science Publ, Lengerich

Pause B (2020) Alles Geruchssache: Wie unsere Nase steuert, was wir wollen und wen wir lieben. Piper, München

Peter K, Vollhardt C, Schore NE (2011) Organische Chemie. Wiley, Weinheim

Raab J (1998) Die soziale Konstruktion olfaktorischer Wahrnehmung. Eine Soziologie des Geruchs. Dissertation, Universität Konstanz

Rempel JE (2006) Olfaktorische Reize in der Markenkommunikation. Theoretische Grundlagen und empirische Erkenntnis zum Einsatz von Düften. Springer, Wiesbaden

Rose MJ (2013) Das Haus der verlorenen Düfte. Aufbau Taschenbuch, Berlin

Schiansky M (2011) Mit allen Sinnen: Duftmarketing. Diplomica, Hamburg

Schlögel K (2020) Der Duft der Imperien: „Chanel No 5" und „Rotes Moskau". Hanser, München

Schnaubelt K (1995) Neue Aromatherapie – Gesundheit und Wohlbefinden durch ätherische Öle. vgs Verlagsgesellschaft, Köln

Spitzer M, Bertram W (2009) Hirnforschung für Neu(ro)gierige: Braintertainment 2.0. Schattauer, Stuttgart

Stead C (1996) Parfum aus ätherischen Ölen selbst herstellen – Komponieren Sie Ihren ganz persönlichen Duft. ECON Taschenbuch, Düsseldorf

Süskind P (1985) Das Parfum. Die Geschichte eines Mörders. Diogenes, Zürich

Theophrast (2015) De odoribus, Edition, Übersetzung, Kommentar. Eigler U, Wöhrle G (eds). Walter de Gruyter GmbH & Co KG, Berlin

Turin L, Sanchez T (2013) Das kleine Buch der großen Parfums: Die hundert Klassiker. Dörlemann, Zürich

Worwood VA (1992) Liebesdüfte – Die Sinnlichkeit ätherischer Öle. Goldmann Ratgeber, Goldmann, München

Worwood VA (1997) The fragrant mind: aromatherapy for personality, mind, mood and emotion. Bantam Books, London

Psychology of Perfume Choice

How We Smell, Who or What in the Brain Decides on the Scent, and Why Perfumes Do So Much Good

Contents

3

Customers often give perfumers, the perfume industry and retailers puzzles with their reactions to perfumes. Often, the choice of perfume in the point of sale perfume store seems spontaneous and determined by chance, and not really predictable. Therefore, it is difficult for marketing, neuromarketing, and economics to provide explanations for the purchase of perfume. The perfume and brain research provides new results on possible influencing factors on the choice of perfume, which overload established explanatory models. Now, findings from the neuroperfumery shed light on the darkness. In the choice of perfume, there is quickly a struggle in the head of the consumer, which brain regions and networks can prevail in the decision. Two regions are particularly in the foreground, about which I want to report in connection with recent findings and developments in the field of olfactory research. This makes it clear what makes smelling and thus the choice of perfume so special.

3.1 Plea for the Sense of Smell: We Smell More than We Thought

The olfactory sense is one of the oldest of humans, but was long underestimated and even devalued. The latest scientific findings have completely changed this view. One can even speak of a "renaissance of smelling". A growing number of researchers are challenging the notion that the sense of smell has declined during human evolution. They even assume an opposite development (Shepherd 2013).

We may even be able to smell with our tongue, because in addition to taste receptors, there are also olfactory receptors (Malik et al. 2019). Therefore, the statement that our tongue only knows the gustatory perceptions of sweet, sour, salty, bitter, and umami, (meaty, savory, delicious) may have to be reconsidered.

The mere existence of smell receptors does not by itself allow the conclusion that we can smell with all of them—especially not in the way that the olfactory process takes place through the nose: We smell with more than 10 million olfactory cells and their smell receptors, which are located in the nasal mucosa of the two nasal cavities. So far it has been assumed that the smell impression only develops through the smell cells in our nose, which is closely connected to the mouth. But odorants are also released when chewing and then enter the nasopharyngeal space when swallowing and exhaling. I will go into this aspect in more detail later. Nevertheless, it is conceivable that an unconscious smell with the tongue also creates a smell impression. Thus it would be possible that the interplay of smell and taste begins already on the tongue.

The effect of sandalwood—an old anti-inflammatory remedy originally from India, used as incense sticks and massage oil—had long given rise to the suspicion that we can also smell with our skin. Hanns Hatt, professor of physiology at the Ruhr-Universität Bochum and probably Germany's best-known smell researcher, and his team have now confirmed the olfactory ability of the skin (Busse et al. 2014). Earlier, these scientists surprised with the statement that, for example, there are also smell receptors in the prostate, in the intestine, which is known as intelligent, and in the often overused kidney. We humans therefore do not only smell with our nose, but with our whole body—both internally and externally. Thus it is likely that we not only constantly sniff ourselves, but also our entire environment—partly consciously, partly unconsciously.

■ **Do we smell in stereo?**

Olfactory research is now based on further assumptions. Currently, it is being discussed whether humans as well as various animal species can smell in stereo to at least some extent. Most right-handers can smell better through the right nostril (left-handers through the left), but both nostrils work. A group led by Yuli Wu from the Institute of Psychology of the Chinese Academy of Sciences in Beijing has now concluded that humans can navigate by means of stereo smelling (Wu et al. 2020). In their experiment, the test subjects turned their heads or noses in the direction from which the higher concentration of fragrance was to be smelled. This is certainly no comparison to the smoothshark and its sense of smell. In contrast to humans, its nostrils are far apart, so it can smell in stereo and thus automatically get an orientation of where the scent comes from, e.g. when fragrance molecules first reach one nostril and then the other with a delay. Certainly more studies are needed to find out to what extent humans can actually navigate olfactorily. Maybe only 10% of us can really smell in stereo in everyday life, suggests Thomas Hummel, head of the "Smell and Taste" department at the University Hospital Dresden. This assumption, in his opinion, applies especially to younger, fragrance-in-

terested noses. But we humans are, when it comes to the sense of smell, definitely more capable than originally assumed.

■ **Olfactory early warning system**

The smell impression, which is perceived by nose, skin, and organs, is, inter alia, part of an early warning system of the immune system. Because our own smell reflects inner experience like emotions, feelings and moods, it is also an indicator of the respective health condition. This can be observed very well in dogs, which can smell much better than us, (up to 80 times better), in relation to human smells. They are almost perfect diagnosticians. For example, the so-called cancer dogs are able to smell lung, breast, bowel, and bladder cancer with amazing accuracy. In addition, they can also detect epileptic seizures and hypoglycaemia in people in their environment at an early stage, in addition to numerous other risks for humans. Dogs are therefore ideal smell watchers for us humans (Preuk 2013).

Even the medical doctors of old times, such as the most famous doctor of antiquity, the Greek Hippocrates (around 460 BC—around 370 BC), knew immediately when they smelled certain odors of their patients. Because: Disease smells. Dozens of diseases can lead to a characteristic smell. Above all, the different smells of urine serve the diagnosis. For example, a sweet-spicy, maple syrup-like urine indicates a metabolic disorder. But also certain smells of excretions help in the diagnosis and, for example, indicate intestinal diseases. Furthermore, certain breath and sweat smells are associated with specific diseases. For example, the breath of patients with liver disease smells of ammonia, other smells are associated with schizophrenia. More detailed studies exist for Parkinson's disease, which is also announced by the smell of the skin (Trivedi et al. 2019). Even a lost or reduced sense of smell, can show first symptoms for self-analysis of health. For example, the sense of smell is impaired

3

at the beginning of Alzheimer's disease, but also in infectious diseases such as Covid-19.

3.2 Scent Memory: Smelling Relativizes Space and Time

- **Olfactory déjà-vu**

Back to "healthy" smelling. Rather, a short film runs in us more or less conscious (Rolls 2004). How it comes about is complex and by no means completely comprehensible. The good news is that much more research is being done on smell than a few years ago. For example, the two American scientists Linda Buck and Richard Axel were even awarded the Nobel Prize in Medicine for their findings. As already mentioned, currently at least eleven theories compete with each other. It is, simply put, among other things about the question: Is smelling based on different vibration frequencies, or do the fragrance molecules fit in the transferred sense like a key in a lock, by docking specifically on receptors and thus triggering specific activations? Therefore, one would like to have a super microscope in research, with which one could observe the activities of the molecules.

But back to the smell film. It is first produced on the olfactory mucosa. This is located on the roof of the nasal cavity, where the fragrance molecules first meet. On this field there are—generously estimated—up to 30 million olfactory cells, others speak of only 10 million. There are olfactory hairs on it, whose surfaces are equipped with about 350 different receptor types. Via switches, the olfactory stimuli reach individual brain regions. On their way, they are converted from a chemical stimulus into electrical impulses. This alone makes the sense of smell so unique compared to other senses. How smelling exactly works, we will discuss later.

The conversion also opens up an especially exciting field for fragrance psychologists. Because we humans have an excellent sense of smell, which is located in the emotional center (Hatt 2006). Emotion in com-

bination with memory or vice versa makes the reproduction of complex films possible, which touch us with topics that are often very emotional and go back far into our childhood. So smell also contains links from past to present.

More: Since odors are only perceived unconsciously according to the latest brain research before they become conscious to us—if at all—there can be another effect when smelling: a rare but possible olfactory déjà-vu. You believe you have already smelled or experienced the smell of a current situation. This impression can be so strong that you have the vision to pre-experience or pre-smell the development of a situation because you believe you have already experienced it. Then the smell would so to speak point to the near future. This also has to do with the fact that we smell, among other things, with a certain brain region, the already mentioned amygdala. He is superior to all other regions in the speed of his sensory perception.

The Amygdala is our early warning system in the emotional center and reacts even when the cerebrum does not yet know anything. Smell can thus relativize our sense of time, also because we can immerse ourselves in different times quickly through smelling. Long-forgotten things from the past can be activated as a current, spontaneous impression or a feeling for what is to come.

- **Smelling in weightlessness**

Several regions and networks of our brain work together in the memory of scent, including the seahorse, called Hippocampus in Latin. The ability of our olfactory memory to store conscious and unconscious is almost fascinating. For example, baby research has shown that a fetus can smell the amniotic fluid of their mothers, that is, their first primal scent (in German "Ur-Duft" or "Ur-Parfüm"), unconsciously, and that a person can remember it years later. In addition, the smell of one's own amniotic fluid, which is usually perceived at a pleasant temperature in the protective weightlessness, still has a positive psychological effect even years later. This says a lot about the quality of the olfactory storage.

The same applies to the second "Ur-Parfüm" that we humans perceive as newborns when breastfeeding: the mix of own smell of the mother, her skin and mother's milk, which accompanies us more or less consciously in memory for a lifetime. The Anglo-Dutch perfume maker Quest, today Givaudan, found that sweet, vanilla-like, musky fragrance notes, as they occur in mother's milk and are used in the creation of many perfumes, have a noticeably relaxing feel-good effect even in adults. This can also be seen on a neuronal level in brain studies. Only slowly does one understand what a great role these "Ur-Parfüms" can play in a therapy supported by scent even in adults. Later on, I will show you some of these therapies that you can use yourself. In particular, I will also discuss milk or milk mousse fragrance notes and their effects.

3.3 Artificial Olfactory Intelligence: The Future Has Already Begun

The interest in diagnostics with the nose is currently increasing sharply. It is currently being developed into a computer-controlled diagnosis, that is, into a kind of medical "E-nose", based on questions of medicine, psychology, and therapy, using artificial intelligence. Simply put, a computer smells diseases and conditions. In this way, a research group at the Technion, the Technical University of Israel in Haifa, developed a so-called SniffPhone within an EU funding program (Cordis) together with a consortium of scientists from six European countries. The sensors of this smartphone evaluate the mouth odor of the user for the early detection of stomach cancer, the fifth most common

cause of cancer-related deaths in Europe. In 2018, the SniffPhone was awarded the Innovation Prize of the European Commission as the best project in its category.

The application of the portable device is extremely simple. The measurement results are sent from the smartphone to a cloud via Bluetooth. Medical staff have access to the data and notify the SniffPhone user. According to Cordis, this non-invasive method has the potential to revolutionize cancer prevention. Meanwhile, a whole industry is trying to win the future market for itself with new diagnostic devices, that is, electronic noses. The development is now taking an unimaginable step further. The Volkswagen Foundation is therefore supporting a research project at Friedrich-Alexander University Erlangen-Nuremberg that is to use artificial intelligence to predict which molecule structure produces certain odors. The molecule structure is to be determined even before the odor arises, which is particularly interesting for health prophylaxis. In this way, the probability of a certain odor occurring can be calculated, which in turn serves as an indicator of a possible upcoming disease. In other words, the probability of future health risks is calculated on a molecular level. In addition, the success of treatment courses could be recognized very early in this way.

If these analysis methods were further developed—especially in mobile form—they would be ideal for the early detection and evaluation of emerging psychosomatic complaints. So before a smell develops, psychological stress could be recognized and, using a smartphone, a targeted and situation-specific intervention in the form of exercises, for example for more resilience, relaxation, mindfulness, self-confidence or self-esteem, could be proposed. Such devices would be a further development of the nose of a dog, which is quite rightly referred to as a smell watcher and which is astonishingly capable.

However, the current focus of research on this analysis method is on the resource-saving creation of fragrances—which would take us back to our actual topic. Because perfume creators go through a lot of trial and error, that is, many futile attempts, to find new olfactory impressions for perfumes. This costs resources, especially time and money. It is no coincidence that more and more fragrance manufacturers are venturing into new territory with artificial intelligence (AI). For example, the Symrise AG, a publicly traded, second-ranked global provider of flavors, fragrances and ingredients, based in Holzminden, Lower Saxony, has developed "Philyra" together with IBM Research, the world's largest industrial research organization with twelve laboratories on six continents. "Philyra" has access to a huge database consisting of fragrance formulas, data on fragrance families and historical data. With the help of AI, for example, it creates a fragrance specifically for Brazilian men of the Millennial generation or suggests ingredients for new creations to perfume creators.

The term Wirkparfüm or active perfume is particularly interesting to psychologists in connection with the topic of fragrance molecules. These are, as already mentioned, fragrances that can do more than just smell good. With the knowledge of which molecules result in which olfactory impression, perfumes can be created specifically for special emotional needs and thus for individual brain regions. But more on that later. Back to smelling and what makes it so special.

3.4 Increasing the Enjoyment of Fragrance: Merging of the Senses

The sense of smell has traditionally been attributed a supporting function in connection with other senses (Knoblich et al. 2003). But that is also what makes the special fascination of smelling so special: it is often experienced together with other senses such as sight and hearing, which makes it even more intense. So fragrances are not only perceived by so-called synesthetes—that is, people who can merge the senses with each other—for example, in color and in shape. In fact, this applies to almost all people to a certain extent. You don't have to be a little Kandinsky to, for example, see fresh citrus notes as orange, light green and yellow flying triangles following the sound and rhythm of a saxophone. In other words, smell is ideally suited to being merged with other senses, so to speak multi-sensorially, in order to experience an increase (Tamura et al. 2018)—although there are individual and cultural differences that can make the merging of individual senses different.

The general relationships have long been made their own by perfumers and fragrance lovers in our cultural area. For example, they speak of a fragrance smelling green or of a perfume being harmonious and round or stinging in the nose. However, what smells rather blue, red, yellow, or green can quickly be associated differently with color blindness or deficient color vision. It is estimated that 8 to 9% of all men suffer from color blindness or deficient color vision in the colors red and green. There are also other visual impairments.

- **The sensory world of the Makú Indians— an example**

Cultures such as the Makú Indians on the Rio Uneiuxi in the Amazon region do not distinguish at all between green and blue in their language Nadëb and will therefore not be able to say what distinguishes both colors in smell. They also seem to have no abstract concepts of colors and to associate colors more with things like the brown manioc root (their main food, depending on the stage of utilization with different brown smells) or their river (which probably also smells dangerous during the often prolonged heavy rains, because it floods their settlement with brown mud). An observation of the non-existence of abstract color concepts was also made with Stone Age peoples on New Guinea.

The Makú Indians also have their own sense of time in their world, which was shaped by two gods (hostile brothers) who lived at the upper and lower reaches of the river before they were Christianized by an Irish missionary in the 1980s. Time periods and memories were measured until Christianization according to events such as before and after major floods, which were attributed to the conflict of the divine brothers (Mensing et al. 2017). It is possible that the new white god from the skin color of the missionary now saw their gods like themselves as brown. The assumption is therefore that contact with "whites" has changed their self-perception and that the color "brown" is now associated with further smell associations in their facets. Perhaps now a kind of brown also goes with tribe membership and -smell. Through these relationships and developments, the color brown may have further differentiated itself among the Makú Indians, and individual shades of brown are now associated with increasingly different smells. During my visit to the Makú Indians in the second half of the 1980s, the color brown pointed to a considerable number of smell things that were equally important in meaning as the color green (blue). However, it was not checked whether the Makú Indians might have red-green color blindness (protanopia). With this type of color blindness, red cannot be perceived. This leads to confusion, for example red with brown or green. It was also not examined whether the lack of differentiation between green and blue is due to green or blue blindness.

- ▪ **Intensification of the senses**

The fusion with other senses makes smelling, as already mentioned, more intense and attractive. It can also be expressed this way: it smells nicer through multisensory experience. In fragrance marketing, this connection has been used for a long time, especially in the so-called multimodal processing of olfactory and visual stimuli. From the practice of perfume consulting, it is known that a bottle that matches the fragrance in color and shape enhances the smell impression or the smell expectation positively. An attractive and associatively successful visual interpretation of the fragrance experience is particularly important.

Consumers usually come into contact with a fragrance through display windows, advertisements in print media, social media networks or television advertising, as well as displays, i.e. decorated promotion areas in the store, on the visual level. Here they see its bottle and its world. Only then do they smell it (Jellinek 1997). It is not by chance that many fragrances are also colored in matching colors that support the smell impression. More precisely: in colors that certain target groups, for example young women, love and find matching to the fragrance. In addition, the fragrance experience can be further enriched by the haptics of the form language of the bottle—but also by background music and other sensory stimuli.

Therefore, when presenting a new fragrance, edible ingredients are often offered for tasting—for example raspberries, if the top note is based on them. It is also no coincidence that new perfumes are often introduced to the press in cafés in a multisensory way. With the sight of different fruit tarts, the upcoming fragrance experience is made tasty and should figuratively melt in one's mouth in expectation.

If two or more coordinated sensory impressions, for example visual and olfactory, are processed together, a so-called "superadditivity of the stimuli" can arise. In almost all cases, however, the smell or the perfume, if it is not overlaid by taste impressions, wins in every combination, with this impression even being intensified. As brain research confirms, the joint processing of visual and olfactory stimuli even leads to increased brain activity (Pezoldt and Michaelis 2014), which—as recent studies show—can even be further increased, namely by unexpected, beautiful smelling in combination with other additional experience. This happens, for example, through a surprise in fragrance consulting, through individualization of perfumes, but also through their scarcity.

Since I will come back to this in more detail later, I will limit myself at this point to the description of the shortage. Unexpectedly beautiful smell in combination with scarcity often leads in practice to the following deliberately induced situation: The customer smells a new perfume. If he is very fond of the scent and asks for the price, the dealer claims that he first has to look in the store—or "in the back"—to see if the perfume is still available in the favorable size. This situation often triggers a hopeful expectation in most fragrance enthusiasts. Because often only the large and therefore also particularly expensive bottle stands in the shelf. If the small version is available, the customer feels a sense of relief: Thank God, it's there! Because the price of the perfume—even if it is not exactly low—can then be endured more painlessly. It can be assumed that during the entire situation—from the expectant tension to the relieved feeling of happiness—the brain activity is additionally increased in addition to the processing of visual and olfactory stimuli.

3.5 Perfume Choice: Findings from Marketing and Neuromarketing

- **On the trail of individual purchasing behavior**

The science of our purchasing decisions (Barden 2013) also offers many and exciting insights for fragrance advice.

Just so much: In recent years, a whole new field of application and research has emerged with fragrance marketing, which deals with the influence of olfactory stimuli on brands in the fields of communication, advertising and purchasing behavior. Scent is increasingly understood as a non-verbal form of communication. In this context, Meyer and Glombitza (2000) speak of an "invisible brand personality" that can make a brand unique (Rempel 2006). Most studies on fragrance market research deal with the influence of the olfactory system or the scent on individual purchasing behavior. In other words: To what extent do olfactory stimuli influence purchase processes? Among other things, the question is whether the preference for a certain luxury brand can be increased by a certain scent, for example the smell of leather. Another big topic is room fragrance. Is the visitor of a store more likely to buy if he is exposed to a certain scent in the room? This could already be confirmed by a study of the University of Paderborn in 1996. It examined the influence of fragrance at the point of sale (POS) in about 200 sports stores in Germany. The scents used in the stores are said to have increased the willingness to advise by almost 19%, the length of stay by just under 16%, the willingness to buy by almost 15% and the turnover by 3 to 6% (Pusch 2018).

Without going into the numerous studies that would go beyond the scope of this book, it can generally be said that consumption behavior is stimulated by scent, especially when the scent is matched to the respective areas of application. For this reason, numerous fragrance marketing companies offer special scents that are supposed to lead to a positive purchase decision.

Nevertheless, due to the numerous influencing factors, it has not been possible to come to a convincing answer to the question of who or what influences the choice of scent in the brain, in particular the individual purchase decision for a perfume. Nor are the answers to the questions of which scents work best on which people, with which products and in which situation, or which scents should be used for scenting or masking odors, often left to the discretion of fragrance marketing companies. Of course, one can expect that the intensified scent of fresh baked goods or coffee aromas will usually also trigger sales-promoting associations in the corresponding context. The scent of novelty, quality, and cleanliness, on the other hand, is more product- and situation-related. What you certainly want to achieve next to the effect on the purchase decision is the increase in the feeling of well-being and the feeling of well-being with the product itself. Basically, it is about arousing positive emotions that motivate consumers to buy. There are interesting findings from multisensory marketing, neuromarketing, but also from economics.

- **Multisensory marketing—synchronized choreography of the senses**

Multisensory marketing is based on the superadditivity of stimuli, whereby an attempt is made to address all five senses (visual [optics], auditory [acoustics], olfactory [olfactory], gustatory [gustatory], and tactile [haptic]) as coordinated as possible and in a holistic manner in order to optimize the experience. As already mentioned, it is assumed that the effect in the brain is higher when emotional and cognitive processing of incoming stimuli is carried out using coordinated stimulus modalities. Which individual sensory stimuli of different modalities are particularly suitable for superadditivity is also a field of research in art and media and by no means exhausted. In this book I will show in detail the connection between specific fragrance directions, colors, shapes, and experience. This connection is the basis for the Moodform Test©, which you can carry out as a self-test at a later stage. Marina Pusch shows in her work Multisensory Marketing in the Online Shop (Pusch 2018) based on Steiner (2017), to what extent color, form, space and movement supplement the found connections between scent, form, and color of the Moodform Test© (Mensing and Beck 1988; Mensing 2005).

Studies on the influence of individual sensory impressions on smell come to an interesting result: Although more than 80% of all sensory impressions are consciously perceived through the eyes, slightly more than 10% acoustically and only just under 4% consciously through the sense of smell (Pusch 2018), scents nevertheless play the decisive role for moods and feelings. In particular because we do not consciously perceive most scents and they are processed unfiltered in our emotional center.

That multi-sensory marketing is certainly the future of scent marketing is not disputed by most marketing professionals. Nevertheless, as can be seen from the example of Abercrombie & Fitch, it is difficult to implement and also no guarantee for lasting success. With the visual appearance, where lightly dressed men greet the customers at the entrance, there were complaints. Also, the loud music booming from the speakers caused important target groups to be lost or not even gained for sales. Also, the black floors and ceilings as well as the darkened stores, which are aimed at atmospheric room atmosphere and the sense of touch, were not everyone's taste. In addition, the Abercrombie & Fitch Signature fragrance "Fierce" came onto the market in a very concentrated form and mainly allergy sufferers avoided the stores. That was no wonder, because the company obviously sprayed the scent in the stores all day long and employees were even asked to spray it into the air conditioning. Perfumes or consumers are in a state of change with regard to their expectations of a scent, as we have seen above. So it is questionable whether—as currently planned—a scent box in the

3

private sector, which can be connected, for example, to visual media or the Internet, will ever be a commercial success. Like some printers, the scent box contains small cartridges with different fragrances that are emitted in different amounts and mixtures at the same time (Pusch 2018). Conceivable are essential oils, which together with other sensory impressions promote health and well-being in the context of accompanying fragrance therapies. For example, one could offer fragrance seminars, fragrance tests, or fragrance journeys with added value via social media, e.g. for Alzheimer's patients to smell better—and thus for more joie de vivre.

- **Neuromarketing—the connection between emotional systems and olfactory experience**

Perfume marketing, like marketing in general, is increasingly interested in findings from neuromarketing. Simply put, neuromarketing aims to study the effect of stimuli on marketing and sales from the perspective of the brain, or even simpler, to test the reaction of networks to certain stimuli. The focus will be on reactions controlled by emotions, especially by the limbic system. Accordingly, it is assumed that in humans the reasons for their actions are mainly influenced by emotional factors that cannot be easily uncovered by surveys. In order to test which brain regions and networks are activated by certain stimuli, imaging methods are used (such as functional magnetic resonance imaging), which will be discussed in detail later. But we must honestly say that the conclusion from brain reactions to actual experience is still in its infancy—just because individual brain regions and their networks, as we saw in ▶ Chap. 1, have different functions at the same time. So the amygdala is activated both by fear and by joy and knows all the facets in between.

Currently, neuromarketing is particularly focused on three large emotion systems that influence life and are used in consumer research and brand positioning. The German psychologist Hans-Georg Häusel (Häusel 2016) and marketing agencies such as "Konversionskraft" describe them as follows:

- **Balance system**: striving for stability, order, security, and belonging; avoiding uncertainty and fear.
- **Dominance system:** striving for power, assertiveness, status, and autonomy; avoiding suppression and foreign control.
- **Stimulant system:** striving for adventure, reward, and variety; avoiding monotony and boredom.

All three emotion systems result in an emotional space that has been visualized and known as the Limbic® Map for better overview.

In neuroperfumery and olfactory psychology, the three basic systems of this model developed by Hans-Georg Häusel are not unknown. Below is a first assignment to fragrance directions, which I will discuss in detail in their psychology at a later point:

- **Balance-System:** floral fragrance notes and floral notes such as the so-called Florientals and milk-like milk-mousse notes.
- **Dominance-System**: Chypre notes as well as leather and aromatic fragrance notes.
- **Stimulant-System**: Gourmand notes, that is, scents that the smell brain registers as sweet-edible and that act on the reward network. In the Limbic® Map they can be assigned to the Stimulant-System. Furthermore, fresh-green citrus fragrance notes fall into this system.

- **How and why do consumers choose a particular perfume?**

For the perfume shop, which is itself often perfumed, another, so far unanswered question arises: Who or what influences the choice of scent in the brain, i.e. the purchase decision for a perfume? The answer to this could have a decisive influence on the sales of individual perfume brands and on trade as a whole. In Germany alone, just under 50% of all women use perfume, most often in the concentration of an eau de parfum, followed by the lighter eau de toilette. I will introduce the perfume market in more detail later. Just so much should be said here: Only about every 20th perfume reaches the third year after its market launch (Schnitzler 2004). But even scents that survive this time and are sometimes still available are often hardly chosen or shown to customers.

In addition, from the perspective of marketing, there are general problems in the sales situation. It is assumed that 75% of all purchase decisions are made spontaneously. This is of course reinforced by appropriate advice. Often the customer has something else or another product category in mind and then, to stay with the perfume shop, discovers a scent. A perfume shop in Munich has perfected this almost to perfection. At the entrance, almost on the street, on a mobile display stand, the customer has the choice between different hair design products (e.g. decorated hair clips in different colors). The store is known for this. The fewest customers only leave the perfume shop with a hair product, at least they are given a perfume sample. But often they buy a perfume that is many times more expensive spontaneously, which fits the style and aura of the hair product. In marketing speak, this is called "upselling". To an affordable product like a hair clip, which is actually the main product for which the customer enters the store, an additional high-quality product such as a perfume is offered. Interesting about this: The price of the perfume often does not play a big role, but rather whether the perfume is experienced as matching at this moment. Cus-

tomers are happy to be advised on this in a perfume environment. The perfume shop also attracts customers who value advice. After all, consumers are flooded with an immense number of advertising messages that can quickly confuse them. Marketing professionals speak of 3000 messages that affect most of us every day, so our attention is correspondingly limited (Kirschberger 2015). That is why it is only too understandable that, due to overstimulation, around 30% of all purchases do not crystallize on a specific product until the point of sale (in our example the sales situation in a perfume shop). This means that the choice of perfume is also subject to impulse buying, and the influence of advice at the POS plays a significant role in the purchase decision-making process.

- **Factors influencing perfume choice—the contribution of economics**

There are very few studies and explanations for the perfume choice, which is probably the most central question for perfume practice. A study explaining the different influencing factors on the scent choice comes from Kerstin Pezoldt and her colleague Anne Michaelis from the Faculty of Economics and Media at the TU Ilmenau, Thuringia (Pezoldt and Michaelis 2014). Using the S-O-R model (Stimulus-Organism-Reaction model), they try to make the purchase decision for a perfume more transparent, i.e. to better illuminate the emergence of the perfume choice at the point of sale. Since many factors influence the perfume choice, as I will explain shortly, this is not an easy task. Therefore, I add some additional factors, and you, dear readers, will certainly be able to add something from your own experience.

With the S-O-R model, the perfume purchase, like any purchase of a product, can be understood as a process. Of course, one is ideally interested in predicting the purchase decision in the perfume

3

choice. But because of the complexity of the influencing factors, one has to keep the claims small—already because the consumer's brain is in a real fight with itself during the perfume choice, as studies of neuromarketing or brain research know. Accordingly, the analysis of the purchase decision using models such as the S-O-R can only make partial statements.

The S-O-R model analyses the purchase process over three large areas with influencing phases, here using the example of the perfume choice:

1. **Exogenous stimuli** (area S—stimulus)
   - **Marketing stimuli.** In this first phase of purchase, it is above all the perfume bottle with its material, its decoration, its embellishments, its colorfulness, its shape and its light refraction, and thus also with its haptic promise of quality and value, that stimulate us. Furthermore, the perfume brand, i.e. the brand name with slogan, and the general appearance, such as the outer packaging and the presentation as a whole, influence us. But the impression made by content such as the type of advertising (social media, print media, television, window and display advertising, ad motif with or without the support of famous personalities), in which emotions, moods, style, image, trends (e.g. gourmand notes for men), memories, expertise (e.g. that of a specific perfumer), origin (e.g. New York), occasion, special features (e.g. rare ingredients), availability (e.g. limited edition) and target groups (e.g. adventurers) for the use of the perfume are communicated, also plays a significant role. Of central importance is the communicated olfactory impression, e.g. by additional decorations in the window display with corresponding plants and objects, also in order to create a gender-characteristic expectation of the scent. Of course, the

price of a perfume can also influence the choice of perfume in advance.
   - **Environmental stimuli.** These include, for example, the influence of cultural or seasonal factors (vacation, holidays, etc.), but also time factors, up to days of the week or time of day. I will discuss these influencing factors in more detail, because, as we will see, it makes a difference when looking for a new perfume whether this takes place from Monday to Wednesday or from Thursday to Saturday. The time of day also has an influencing effect on those seeking out scents. Rich amber-oriental notes stimulate differently (more sensually) in the late afternoon than in the morning (more extravagant). Furthermore, climatic influences, in particular temperature and humidity, play a role. They all stimulate the potential choice of scent differently and influence the purchase of perfume even before smelling a perfume. The environment of a stationary perfume shop or the location (A-location or B-location) is also stimulated differently, as is the place of purchase (perfume specialty store, fashion boutique, pharmacy, drugstore, department store, grocery store or discount store). Of course, the social media presence also has an influencing effect when shopping online. In addition, it is decisive whether the respective store has a perfume department or only an area in which scents are presented (e.g. at the cash register or on a shelf). Does the store offer a perfume consultation or is one more dependent on self-service? With location and type of place of purchase, the building also acts accordingly, just as the store itself does with its style of furnishing. Even the view from outside into the store, which gives an initial impression of the personality of the sales staff, their

non-verbal behaviour or consultation style, influences the purchase of perfume. Certainly this has less effect on regular customers of a store, but especially new customers are influenced, especially when they are looking for a new perfume for themselves. **Conclusion**: With exogenous stimuli, marketing primarily wants to trigger emotions and associations in potential customers for the choice of perfume or create an environment in which a purchase is made.

2. **The buyer** (O—Organism area)

Since at the beginning of the consultation not much is known about the customer and his purchase decision, except if it is a regular customer about whom one knows more, this area is also referred to as the "black box".

– **Background factors.** In this phase of fragrance selection, fragrance socialization and thus memories of fragrances and odors that were often experienced as pleasant, neutral, or unpleasant in childhood influence fragrance preferences. Furthermore, the occasion plays a role for which one wants to use the fragrance (e.g. to experience oneself as more sensual and attractive), and whether it should have more of an effect on oneself or others. Fragrance selection is thus also influenced by mood, particularly by how someone wants to experience themselves or experience themselves anew. Furthermore, it is decisive whether one is looking for a perfume for oneself or as a gift, whether it is the re-purchase of a product that has run out or whether the customer wants to surprise themselves with a new perfume. Personal olfactory ability also plays a role in fragrance selection. This is in addition to the influence of one's own body odor, whether the fragrance adheres to the skin or smells too intense or too weak. Here,

personal usage habits come into play. As mentioned above, there are great differences in the way in which experiences with perfumes are reflected in the way they are used. Since many consumers have a real perfume bar at home, they may only be looking for a supplement or want to experience themselves in a completely different fragrance direction. Many consumers think that a fragrance must match their personality and style, which is a very subjective assessment. Other factors influencing fragrance selection are age, income, gender, the influence of the "best friend" or the social environment. Furthermore, perfumes are judged according to whether they can be worn in everyday life or only on certain occasions. There are even regional and national fragrance preferences. More on this in the penultimate chapter.

Health- and wellness related aspects are becoming increasingly important to customers. As mentioned above, they are looking for a fragrance with a more therapeutic effect than just smelling good. This also plays a role in fragrance selection in that the effect on self-experience through its modulation with corresponding self-therapeutic expectations is taken into account. All motives of fragrance users mentioned here as background factors will be discussed separately.

Conclusion: Fragrance users can have very different reasons for their fragrance selection. The purchase process is correspondingly influenced. It is certainly not wrong to say that the processes leading to the purchase of perfume include activation by the perfume itself, behind which are emotions, moods, motivation, and attitudes. And since it is precisely the close link between fragrance, mood, and emotions, which of course in-

3

cludes the subjective attractiveness of the fragrance impression itself.

– **Purchase decision.** In this phase of the perfume purchase, advice, brand awareness, atmosphere and the experience in the store play a role—and of course the perfume impression itself, with a speciality coming in with perfume: Those who would like to buy a perfume are limited by the ability of their own sense of smell in the number of possible perfumes that can be smelled at once with an untrained nose. Often, it is four to six intensely smelled perfumes, after which the nose, i.e. the brain, needs a break (about 10–20 min). But often, customers do not have enough time for that in the store.

It takes very different amounts of time until the customer comes to the perfume choice, if at all. Much depends on the customer themselves. The perfume choice is the quickest— usually five to ten minutes—when the customer has already identified a problem in advance. This can be a used up perfume or the search for a gift for a special occasion, where you already know the favourite perfume of the person to be gifted. The perfume purchase does not take much longer when the perfume has already been "pre-sold" through samples or perfume strips or through other advertising. This includes recommendations from friends. Although customers want to smell the one or the other to secure their purchase decision, the neuroperfumery and neuromarketing know: The perfume smelled first always has an advantage over the perfume smelled afterwards. Consequently, it is important for a quick purchase decision that the first perfume "sits". There are different techniques to determine this first perfume. One is a psychological col-

our-perfume test, which we will discuss later and which I offer as a self-test (Moodform-Test©). Another method is the strategic "perfume layering" that I have already mentioned and will discuss in detail later. With both methods, it is about creating an "aha" experience for the customer, for which I will give tips later on.

3. **Reaction/Response** (Area R—Response)

– **Final purchase decision.** This includes the behavior during and after the purchase. How the specific perfume and brand choice affects the decision between on- and offline purchase is in the hands of the retailer. This does not contradict the purchase decision just explained, where one has already decided to buy one or more perfumes. More and more customers are being advised in a perfume shop, also decide for a scent, but buy it online, often at a discount, from another dealer. So today you have to more and more between a "principled purchase decision" and a "final purchase decision", which takes place at a later time, distinguish. The final purchase decision also includes satisfaction with a product and thus a preliminary decision for the re-purchase. This brings us to another interesting phenomenon.

■ **Why we also like to wear the same perfume over a long period of time**

Another approach, which enriches the S-O-P model of economics, is presented in detail later on and originates from personality psychology. It starts with area O and offers, among other things, an exciting explanation of why many fragrance lovers also like to wear the same perfume over a long period of time. This approach tries to show the unconscious or semi-conscious dynamics of fragrance selection, which arise from the self or the self-experience and aim at approaching the ideal state. The self-discrepancy theory, which was developed by Edward Tory Higgins in the 1980s at Columbia University, is based on these dynamics (Higgins 1987). In analogy to his theory, the act of perfuming can be understood as a self-transformation process. The fragrance selection and the satisfaction with a perfume are strongly influenced by the experienced success. Satisfaction already arises when one can bring oneself from the "current self" (how one feels at the moment) with a perfume closer to the "ideal self" (how one would like to feel even more) at least to some extent. If a perfume can do this, it becomes almost the perfume of the ideal self. The perfume is then associated with a higher desired self and is therefore bought and used again and again.

One might actually assume that the price is decisive for the choice of perfume. Of course this plays a role, but next to other factors. Although neuropsychological studies have shown that the price activates the pain center in the brain and a product is only bought if its reward promise outweighs (Scheier et al. 2008). But it's not that simple with a scent.

You will read an unexpected thesis in ▶ Chap. 4. Because apparently we smell most of it twice in the brain: first unconsciously with the emotional center and there in particular with the Amygdala. She makes the decision, after the scent stimulus was played to her from the piriform cortex (PC) the fastest, whether she likes a scent emotionally or not. Then, in higher brain regions such as the orbitofrontal cortex (OFC), it is smelled consciously. Here, among other things, it is a more cognitive decision whether a perfume is worth its price. Can the amygdala be smelled past? At least it is physiologically possible. There are direct neuronal connections between the PC and the OFC and even vice versa, which are mainly activated during "olfactory learning". Theoretically, it would therefore be possible that we can smell without the emotional input of the amyg-

dala. This remains to be clarified for research and will probably only affect people like perfumers who associate olfactory ingredients with other sensory features when learning. Of course, it is also possible that only the amygdala and its network smell, without the scent impression being passed on to the OFC. Nevertheless: If it comes to perfume choice and price, both regions are activated.

Often, the amygdala and the OFC are in conflict with each other when it comes to choosing a scent. The amygdala feels comfortable with the perfume, but the OFC signals that you can't actually afford the perfume. Guess who wins when a perfume really fascinates the amygdala and the purchase is just barely possible financially. Right, the amygdala! Reason comes against feeling again and again—as so often in life—not a chance.

3.6 How to Smell: About Molecular Doormen and Ushers

Most treatises on the subject of smell and fragrance begin with the olfactory organ nose and the observation that the olfactory system is always activated and that it cannot be turned off as easily as the optical one. Unless you really hold your mouth and nose, but that doesn't work for long because you have to breathe. It is also often reported that living beings were able to smell before they could see and hear. The need for a nose arose during the transition from water to land life, because originally all vertebrates breathed through the gills. Then comes the hint that the size of the nose has no influence on the sense of smell. To set the tone for the topic, it is also common to talk about nose shapes and how they differ between people of different origin. They are said to have developed depending on the climate. Narrower noses

3

have arisen more frequently in cold-dry climates during evolution, while broader noses are typical of warm-moist climates. This is then the transition to the inside of the nose, because one of the tasks of the nose is also to make inhaled air warm and moist through contact with the mucous membrane.

The inside of the nose is also interesting for us fragrance psychologists. But it only gets really exciting for psychology when a fragrance stimulus reaches the smell brain, which is not counted as part of the nose. But that doesn't mean that the nose's ability to act as a big door to the world of smell should be downplayed. On the contrary, it is thanks to the nose that humans can distinguish more than 10,000 different smells. This begins with the fact that the nose has an olfactory mucous membrane that is approximately five square centimeters in size in both nasal cavities. Here are around ten million olfactory cells, which carry olfactory receptors on thin olfactory hairs. After suitable contact of the molecule of a fragrance substance with the olfactory receptor, the chemosensory stimulus from outside is converted into an internal electrical signal and forwarded to the smell brain. As you can imagine, the process is much more complex, and I therefore want to go into more detail here about smelling and how it works before we discuss the "Master of Perfume" ("„Maître des Parfums") or the struggle for the fragrance in our head.

■ **Prickling, biting, cooling …—what all contributes to the smell impression**

Two physiologically different systems are involved in olfactory perception: The actual olfactory system goes through the nose and the other, the nasal-trigeminal system, goes through the mouth and nose to a facial nerve (trigeminal nerve). This nerve produces, via the nose (nasal-trigeminal), sensations such as burning, prickling, biting, cooling, and sharp, tingling via the

mouth (oral-trigeminal). What we understand as smelling is the perception of volatile, airborne odorants. Non-volatile substances are perceived in the mouth cavity. However, many volatile substances trigger a smell sensation in both the olfactory system and the nasal-trigeminal system, such as mustard or onions, but also menthol.

The volatile odorants or fragrance molecules reach our nose primarily when inhaling or smelling. A second possibility is the indirect retro-nasal perception of the mouth-nose-throat connection. Here, as already mentioned, the odorants are released during chewing and then enter the nose-throat space when swallowing and exhaling. As we all know only too well, the mouth and nose-throat space are connected to each other, and so the odorants can be smelled by oneself via both routes.

■ **Odor threshold**

Let me continue with the smelling of fragrance molecules from the outside world, because everything that smells to us humans constantly emits small amounts of specific molecules into the surrounding air. These are particles of substances that consist of at least two atoms and that our nose and our brain can process and generate as an olfactory impression. Dogs and cats, which smell much better than we humans, would have pity on us if they knew how little we smell in comparison to them. Because much of what smells to them is below the odor threshold for us humans and can therefore no longer be perceived by us.

Molecules of individual substances have a different threshold value. The lowest known odor threshold is Thioterpineol discussed, a slightly citrus-smelling aromatic component of grapefruit juice. Of course, the individual threshold values vary depending on the individual smell of a person, but they also vary intra-individually. For example, a hungry person has a lower threshold value for scents that smell of food. However, smelling in the threshold range is not a pleasure because it is not possible to determine the scent, but only the sensation of an undefined smell.

For the sense of smell of humans, as already mentioned, about ten million odor cells (so-called olfactory sensory cells) are responsible, which are located in the upper part of the nasal cavity and renew themselves every one to two months. The odor cells have two functions: to recognize fragrance molecules in the air and to forward the fragrance information to the brain. The keyword for this performance is "chemo-electric transduction", which means that when smelling, a chemical stimulus is converted into an electrical signal. You can imagine the odor cells as telegraphs of the old days, which encode the messages of fragrance molecules in the air for the brain by converting them into individual electrical signals and sending them to the brain.

▪ Smell Police

In their work, capturing information from the scent molecules, the olfactory cells are supported by many thin cilia, which are surrounded by mucus and on which olfactory receptors sit. So that the molecules can pass through the mucus (also called mucus), nature has invented a transport system for them, the so-called odorant-binding proteins (OBP). These are small transport proteins that pull the odor molecules to the olfactory receptors. In the field of olfactory research, it is also assumed that the OBP act as a kind of smell police. They seem to be able to take over protective functions and to block the way to the receptors for certain odor molecules, for example toxic ones. If it were possible to control the OBP more specifically and effectively in the future, it would be possible to take olfactory action on the transmission of certain infectious diseases and volatile toxic substances at an early stage. A total of more than 200 different viruses are now known

to be able to cause upper respiratory infections and, like rhinoviruses (mainly responsible for colds and flu), to impair the sense of smell. In addition, the olfactory mucosa can be damaged by the viruses, which can lead to the shrinkage of the olfactory bulb.

▪ Slimy doormen

Mucus, as bad as it sounds at first, has a very important protective function and promotes smelling. Mucus helps to dissolve the fragrance molecules from the air inhaled. It moistens and warms the air we breathe and makes us smell nicer. An optimal viscosity of the mucus and thus best conditions for a nicer smell is at a relative humidity of around 75 to 80% and a mucus temperature of 35 degrees Celsius. Of course, a value of 60% humidity or less is recommended for the living room, but warm air contains more water vapor than cold air, and a higher temperature invites fragrance molecules to move more. This indirectly leads to more molecules getting into our nose and we perceive scents more intensely. Everyone knows the fullness of smell after a summer rain. But too wet and too dry air changes the smell again. Especially with too humid air, an olfactory overload can occur quickly. Then our sense of smell helps by interrupting the olfactory signal cascade that triggers the now hyperactive olfactory cells and their receptors after a while. This means that although the fragrance substances are still present, they are no longer or hardly perceived. In other words: an adaptation takes place. This is also known from your perfume when you apply too much, but at the same time you smell it less and less. But under which climatic conditions one smells best and stores his perfumes is different. As a rule of thumb: a perfume does not belong in the bathroom. As already mentioned in ▶ Chap. 2, temperature fluctuations and high humidity in the bathroom can quickly change the smell of an olfactory work of art. The bedroom

3

is usually the better place in most cases. Compared to other rooms, it is usually cooler, drier, and darker there with constant temperatures.

Back to the slime. It protects against dust, bacteria and viruses and, with antibodies contained in it, contributes to immunity. Mucus thus takes on a kind of doorman function in the nose.

■ **Ticket taker / Ushers**

The "smelling hairs" of the olfactory cells are called cilia. On them are, as already mentioned, olfactory receptors that are like a landing stage for the fragrance molecules. So far, about 1000 differently constructed receptors have been discovered in animals, for example in rodents, which are active and receptive to molecules. The human olfactory organ still has 350 different functional receptors. The US researchers Linda Buck and Richard Axel, who were awarded the Nobel Prize in Physiology and Medicine in 2004 for their research on smell and how it comes about, also examined the essential features of the olfactory receptors. In doing so, they and other scientists discovered different types of olfactory receptors. Olfactory receptors differ in that they only recognize very specific fragrance molecules and react less strongly, hardly, or not at all to those for which they are less responsible. This means that a type of olfactory receptor can react to many odors, but with varying intensity. For a molecule to bind to specific receptors, its chemical composition is important—and thus its surface structure. Because according to the most common theory of olfactory research, what finally triggers the sense of smell, only the right size and weight and thus the shape of a molecule allow it to be recognized by its specific receptor and that its information can be converted into a clear electrical impulse. In this sense, the matching fragrance molecule acts on its receptor like a key that fits into the lock. However, the reaction of

the olfactory receptors not only depends on the type of fragrance, but also on its concentration. At low concentrations, a receptor reacts less to a fragrance. It remains open in research how hundreds of different molecules find their olfactory receptor (one of 350). Perhaps the odorant-binding proteins mentioned above also act as ticket takers / ushers for the arriving molecules.

Most odorants have a molecular weight of < 350 g/mol (molar mass = unit system for the amount of substance, the unit is grams per mole). In order for smell to take place at all, the molecules must also be water- and lipid-soluble or lipophilic (fat-friendly) in order to penetrate the lipid-containing membrane of the olfactory cells or to reach the receptors through the aqueous mucus layer. In order to bind to the lipid membrane, the molecules must also be fat-soluble. Even if it is not quite clear what makes a molecule an "odorant molecule": It must be volatile or released during chewing and water-, lipid- and fat-soluble.

■ **First smell stations**

By activating specific types of receptors, a typical pattern arises in the first olfactory station in the olfactory bulb of the brain (olfactory bulb). In collaboration with other brain regions such as the piriform cortex, it creates the respective olfactory impression. It is calculated that humans, as already mentioned, can perceive around 10,000 different olfactory patterns, that is, they can distinguish odors. Since scents of lavender or rosemary, as we have already discussed, have 505 or 450 fragrance components alone, olfactory cells have to cooperate with their different fragrance receptors. Complex odors such as those of rose with more than 500 fragrance components create a "signal combination code" before the brain can decode it. Using the example of the rose, this means: The signal cascade that is created in the olfactory cells by the binding of the "rose molecules" to the receptors is sent to the brain in individual electrical impulses before it can be recognized there as the scent of a rose. This information transmission takes place via nerve fibers (axons) of the sensory cells, which are bundled together through the ethmoid bone (porous bone of the skull) into the skull to the olfactory bulb. However, the signal cascade is interrupted after a while in order to prevent an overload of stimuli. This means that although the odorants are still present, they are no longer or hardly perceived; adaptation takes place. Aldehydes, as they occur, for example, in "Chanel No. 5", are particularly affected by this, because one gets used to them particularly easily.

The olfactory bulb is considered an outgrowth of the brain and part of the olfactory cortex, from which central nervous processing of olfactory stimuli begins in other brain regions. The olfactory bulb itself works like a computing and relay center, in which a switching of the incoming information takes place. It collects the individual electrical impulses, which can originate from 100 or even 1000 different molecules, and puts them together. He does this with so-called glomeruli in the olfactory bulbs, about 2000 in number, which are also referred to as microregions in the olfactory cortex. From these microregions, the incoming stimuli are then forwarded to various regions in the cortex via pyramid-shaped mitral cells. Different Glomeruli work together, just like the olfactory cells, and generate a specific excitation pattern or the scent impression as a whole for the incoming electrical impulses. This is then compared with other sensory stimuli, such as visual stimuli, by further regions of the olfactory cortex, initially by the piriform cortex. However, the incoming information of the receptors is also significantly reduced by the mitral cells. So if only part of the excitation pattern and the lead molecule for the rose scent are evaluated, the brain already gets a first scent impression, in this case the one of a rose. As we have already discussed, our visual sense supports the expectation of smell via the piriform cortex, but some molecules give such a characteristic signal that the brain already gets a hint of the whole scent before it becomes a pattern. This can happen very quickly, because, as mentioned above, a first scent recognition can already take place in the area of less than 500 milliseconds, and a scent differentiation takes no longer than one to two seconds. To stay with the example of roses: The lead molecule or the lead substance for rose scent is geraniol. Whoever smells geraniol usually associates this scent immediately with roses. Although the scent of a rose can contain up to 500 individual substances, the smell of this substance is enough to recognize it for the first time. Even if the smelling person quickly realizes with and without visual support that something is still missing for a real rose, as we have stored its scent in our olfactory memory.

■ **One of the greatest wonders of nature**

The "chemoelectrical transduction" during smelling is a miracle of nature—perhaps one of its greatest. It is the speciality of many chemists and biologists. For example, Stephan Frings from the Department of Molecular Physiology at the Heidelberg Institute of Zoology and his colleague Clemens Prinz zu Waldeck have made it one of their research priorities (Waldeck and Frings 2005). The molecular basis of olfactory perception—that is, how we smell what we smell—has led in recent years to such a wealth of individual knowledge at national and international level (Reisert and Reingruber 2019) that I will only describe the process of transduction at the olfactory cells in very general terms here.

In "chemoelectrical transduction" at the olfactory cells, as with all cells in the body, the primary focus is on one function: they must communicate, and this is done via signals.

As I said, at the olfactory cells, that is, at their olfactory receptors, the information from chemical signals, based on their chemical composition or surface structure, size, weight and thus the shape of the molecules of a fragrance, is converted into electrical signals. However, the reaction of the olfactory receptors not only depends on the type of fragrance, but also on its concentration. With increasing concentration, the range of fragrances that trigger a reaction increases—the olfactory impression becomes more tangible and alive.

If a fragrance molecule now binds to its docking site (olfactory receptor), a so-called G-protein is activated or a so-called G-protein / adenylyl cyclase mechanism is triggered. G-proteins are a heterogeneous group of proteins within cells and the most important biochemical function carriers. They act, inter alia, as catalysts, transmit nerve impulses and thus enable movement and thus also smelling.

Which molecular reactions now occur in succession in "chemoelectric transduction" of olfactory perception? If biochemistry is not one of your strengths, it will now admittedly become somewhat complex in the next few lines. First of all, a G-protein is activated on the inside of a receptor cell. This G-protein then stimulates the enzyme Adenylyl cyclase. This enzyme is responsible for the forwarding of olfactory stimuli. This is achieved by the conversion of Adenosine triphosphate (ATP) into Adenosine monophosphate(cAMP). cAMP is a messenger molecule that can open the ion channels of the olfactory cell. This leads to depolarization: Positively charged sodium and calcium ions flow into the cell interior, negatively charged chloride ions flow out. This changes the electrical properties of the cell. The chemical stimulus becomes an electrical one. An action potential is generated, and the chemical signal that started the process is now sent as an electrical one to the olfactory bulb. It should be noted, however, that a single fragrance molecule is not enough to generate an electrical stimulus in the olfactory cells. There must be enough fragrance molecules flowing in the air so that they can dock at the olfactory cells or the olfactory receptors. If the olfactory cells catch enough fragrance molecules, the transduction generates such an electrical voltage that the fragrance stimulus is even amplified.

3.7   The Maître Des Parfums in the Brain or the Struggle for the Scent in the Head

With the transformation of the fragrance stimulus into an electrical signal, it becomes really exciting for every fragrance psychologist, because now it is about the actual sense of smell. Psychophysiologically speaking, it begins, as mentioned, with the olfactory bulb, the first station in the olfactory cortex, the so-called. primary olfactory areas or the primary olfactory brain. Here is the origin of central nervous processing.

But even with the importance of the olfactory bulb for smelling, the last word has not yet been said. Current research concludes that smelling also works without an olfactory bulb (Weiss et al. 2019), which astonished the experts and for which no proper explanation has been found so far. Whatever the final outcome will be—it will be particularly exciting for the further processing of the fragrance stimulus, psychologically speaking, for neuroscientists as well as for perfumers and fragrance consulting in the perfume shop. If the fragrance stimulus acts on our emotional center, an initial unconscious olfactory perception is created, which can then develop into conscious emotions, memories, and hedonic judgments or be perceived and artic-

3.7 · The Maître Des Parfums in the Brain or the Struggle ...

73

3

ulated as such. However, the majority of us remain unaware of the effect of fragrance stimuli.

It becomes really fascinating when smell stimuli reach higher brain areas such as the orbitofrontal cortex in the prefrontal cortex—a brain region whose importance for smelling was discovered by research not too long ago and which is also the seat of our self and essential for our personality. This region plays a particularly important role in the search and evaluation of a new scent. In other words: it is not the nose that decides, but the brain.

The orbitofrontal cortex with its special role in olfactory experience is located directly behind the eyes in the front brain, which is responsible for higher cognitive processes and belongs to the so-called frontal lobe in the cerebrum. The sense of smell itself, in which the scent impression is perceived, mainly belongs to the lower temporal lobe (temporal lobe). This is also where the olfactory tract ends. A large part of the olfactory brain overlaps with the limbic system, a phylogenetically very old part of the brain that has traditionally been considered the original seat of emotions. "Temporal lobe" is anatomically the region just in front of and directly above the ears. So the scent impression is developed on the way between the nose and the regions that are located above the ears.

Since the auditory center is also located in the temporal lobe, in which sounds are processed, and which is therefore responsible for the sensory quality of hearing, one would actually also expect a connection between smell and hearing—if only because smell is motion and thus also vibration. Maybe we could hear scents with a very finely tuned ear. In fact, there are languages and cultures—for example Russian and Chinese—in which one hears smells or in which one says that one hears a scent (Geiger 2019).

Anatomical proximity in the brain does not necessarily mean that these regions co-operate. Often they are relatively far apart, independent areas that take on tasks for the same or similar senses. In the processing of visual information supplied by the eye, for example, there are 30 different areas, some of which are spatially far apart.

You can now find the same thing more and more often with smell. It doesn't just happen in the smell brain, but—anatomically speaking—much further away. Even more, as I said: When making a decision— for example, for a new perfume—the smell brain may be involved, but a further away control authority then approves the decision. At least it tries. Welcome, therefore, to the orbitofrontal cortex (OFC) and the struggle in our heads about who gets to decide what and how much about smell and perfume! The main players are, as reported, the OFC and the amygdala—especially when choosing a scent at the perfume counter, at the so-called point of sale. Because very often the two are at odds. To resolve this conflict in practice, later in this book you will find tips on how to stimulate various brain regions and evoke enthusiasm when selecting a scent for a new perfume.

- **Why is there discord in the brain when it comes to smell?**

The orbitofrontal cortex, which is not counted as part of the smell brain, is today considered to be the maître des parfums in the brain due to its final decision-making power. You read that correctly: Our maître des parfums is not located in the smell brain or the emotional center. He basically controls smell from his home office in the cerebrum, that is, from outside. Of course, you can imagine that there are regions in the smell brain like the amygdala that have their own agenda and that the orbitofrontal cortex can hardly or only barely control. I'll come back to that later. But you can already guess that smell has its own rules and that regions in the cerebrum only receive olfactory information that the seemingly "lower" ones also release.

3

But what makes the orbitofrontal cortex so special when it comes to smell? Why is it considered the maître des parfums? It is the OFC that not only creates multisensory links, that is, it sensually couples the scent with other sensory modalities, but also determines the value assessment of a perfume and its bottle. If you have previously hoped to find the seat for a better smell, that is, the place of olfactory enjoyment, somewhere in the smell brain and thus in our olfactory center or in the temporal lobe, we must now say that it is only created through the interaction of various areas. Ultimately, olfactory enjoyment is controlled by the forebrain—at least it is tried.

3.8 Scent and Personality: How Brain and Personality Influence Olfactory Enjoyment

Increasingly, breathtaking findings from the psychology of scent and from neuroscience are showing how our brains or the personality structures inherent in us first influence our sense of smell and then our choice of fragrance and our experience of fragrance. The role played by scent socialization—what we learn, for example, smells good—and olfactory memory—the associations we make with certain fragrances—is great. But humans also have innate preferences for fragrances that are sometimes associated with the development of personality traits.

In magazines you often read that the choice of fragrance allows certain conclusions to be drawn about the wearer. Behind the classic questions such as "Which perfume are you wearing today? A warm fragrance, perhaps with woody notes, suitable for the season? Or something floral? Or were you in the mood for something citrusy and fresh today?" (Müller 2019) are mainly olfactory psychological relationships that are confirmed by the latest findings from neuroscience and personality psychology (Müller 2019). Since the turn of the millennium, it has become increasingly possible for scientists to use neuroimaging or, more precisely, functional magnetic resonance imaging (fMRI), to literally watch the brain smell and choose a fragrance. Some spectacular things came to light: individual brain regions seem to have certain fragrance preferences.

- **Programmed for Vitamin C?**

For example, the orbitofrontal cortex is particularly responsive to citrus notes (Romoli et al. 2012). Actually, one would have expected him to be a Maître des Parfums with a certain olfactory neutrality. In general, citrus notes rank fourth on the universal popularity scale of aromas—right after chocolate, vanilla and milk, which in turn fascinate other brain regions. This raises the suspicion that fragrance preferences also have a genetic anchoring. Perhaps nature has also programmed us for citrus aromas because the corresponding fruits contain the immune-supporting vitamin C that is so important for the immune system. Nature has gone one step further. The practice of perfumery has long been aware that citrus aromas attract fragrance users with extraverted experience desires and that they behave accordingly or want to behave—active, dynamic and open. In fact, the orbitofrontal cortex is also associated with the seat of the personality dimension extraversion. One might even argue that the personality traits associated with extraversion require a higher need for vitamin C and that our Maître des Parfums is programmed for it.

In addition, the invigorating effect of fresh, green, citrusy scents does not only have a positive effect on extroverts who want to keep a cool head. In aromatherapy many of these fragrance notes are considered mood-enhancing and even concentration-promoting. Traditionally, they were applied to the body as a light cologne or poured into a handkerchief, for example, to dab the forehead—the part of the head behind which the orbitofrontal cortex is located. In the perfume industry, this was mainly done using the most invigorating, but also the most ugly of all citrus fruits, the wrinkled bergamot. The smell of this fruit is probably known to many from the sparkling Dior citrus men's fragrance " Eau Sauvage " (Eau de Toilette) or the Malin+Goetz unisex fragrance "bergamot".

In the latter, bergamot is supported in the top note by other citrus fruits such as mandarin, lime and grapefruit as well as ginger and spicy mint in the heart note. This fragrance not only describes olfactory extroverted experience desires, but also evokes memories of light, sun, summer and the south, where such desires may have been more experienced. Our olfactory memory, which is located in the hippocampus, the so-called seahorse, and thus in the emotional center, works here for the Maître des Parfums and intensifies the olfactory impression or the fragrance experience.

Another personality trait, namely conscientiousness, has its seat in the prefrontal cortex, that is, in the immediate area of the orbitofrontal cortex. Characteristics of high conscientiousness are goal-oriented behavior, controlled-planned and structured approach. In fragrance psychology and accordingly in the practice of perfumery, a connection is known that must certainly be investigated more intensively empirically. In particular, male customers with high values on the personality scale with regard to conscientiousness—that is, those who describe themselves as particularly well-organized, reliable, forward-looking, planning and structured—seem to have an affinity for so-called fresh-aromatic, aquatic notes as well as for the fragrance direction "Fougère", which I will discuss later.

This is particularly evident in the classic fragrance "Cool Water" by the legendary perfumer Pierre Bourdon. With a top note that suggests a combination of seawater, mandarin, lavender, mint and other green notes, this fragrance particularly appeals to a personality trait and desire for experience in men that corresponds to the attributes mentioned above and is socially desirable in our culture. Fresh aquatic fragrances, in which unobtrusive, aromatic fougère notes—a green, for example, plant smell reminiscent of the barely smelling fern—also play a role, are not by chance for decades one of the most popular themes in men's perfumery. This is certainly also due to the fact that women influence the fragrance decision of their partners to a greater extent. Because they especially appreciate the personality trait of conscientiousness in men.

- **Edible vs. floral fragrances**

Currently, fragrance research in neuroscience, which I call neuroperfumery, is still focused primarily on the effect of individual fragrances and chords on the brain. These include, for example, those of chocolate, vanilla, cinnamon, peach, strawberry, orange, mango, and citrus (group 1) as well as jasmine, lily of the valley, lavender, and rose (group 2). In general, it has already been established with imaging methods that the brain reacts differently to the offer of food fragrances, that is, smells of edible (group 1) and floral fragrances (group 2). There are also studies on the effect of fragrance mixtures, that is, partial compositions of a perfume, on the brain. As one would expect, fragrance mixtures stimulate certain brain regions more than pure fragrances (Boyle et al. 2009).

However, the young neuroperfumery still faces many challenges that have an impact on the investigation of fragrance effects on the brain. I will only mention a few here. The length of the fragrance presentation seems to have an influence on the stimulation of the brain. Interestingly, imaging

3

methods (fMRI) suggest that fragrances that are offered to smell at shorter intervals stimulate the brain (e.g. the OFC and other regions) more than those that are smelled over a longer period of time (Han et al. 2019b). This is a good tip from neuroperfumery for fragrance advice. After only a few minutes, the brain is significantly less stimulated by a fragrance. This has also been shown in practice in the perfume shop, that the fragrance presented first has an advantage over the following ones. So the first decision for a perfume often already takes place within seconds after smelling. Not surprisingly, fragrance research also comes to the conclusion that the best effect on the brain can be seen within six seconds with a short repetition of the fragrance stimulation using fMRI methods (Georgiopoulos et al. 2018).

The influence on the extent and speed of neuronal processing also depends on whether the fragrance can be associated with something edible (Schoen 2018). This interests—as I will describe in more detail using the example of chocolate fragrances—certain brain regions and networks in particular. The processing in the brain is correspondingly even faster. Furthermore, hunger and satiety have an effect on the neuronal activity of a fragrance. They lead to reduced activity at a satiety level, but also to a change in localization in the brain (Small et al. 2001). However, there are also individual brain regions—or rather brain areas—that are relatively stable in terms of gustatory experiences. They are largely independent of the current experience or the current need for stimulation and do not show any change in signal strength.

However, there are still a whole range of other influencing factors that need to be controlled by brain research or neuroperfumery. The localization of brain activity also depends on right- or left-handedness, there are gender differences and further the

influence of individual fragrance experiences. What has also been shown in corresponding studies is that the excitement often only occurs in areas of one hemisphere, so that, for example, only parts of the right amygdala are activated. Accordingly, very precise descriptions are given in specialist articles as to where exactly the excitement takes place. Brain areas such as the amygdala or the OFC are distinguished much more precisely according to their areas (e.g. front, middle, rear) with hemisphere location (right/left) and involvement of other areas and networks with corresponding specification than I can show here for an overview.

Above all, research using fMRI methods allows very selected insights into possible stimulation of individual brain areas, as described by anatomy. With the prefrontal cortex, a distinction is made between the discussed orbitofrontal (located above the orbit of the eye), but also between a medial (in the middle) and lateral (slightly to the outside of the body) portion. For smell, the medial and lateral orbitofrontal region is of particular interest according to current state. The lateral prefrontal cortex is divided into dorsolateral (back laterally towards the back) and ventrolateral (front laterally towards the stomach) areas. This then leads to abbreviations in specialist articles, where only short forms such as VLPFC (ventrolateral prefrontal cortex) are communicated. With an r or l in front, it is then specified whether the corresponding area is in the right or left brain hemisphere.

Even more demanding are fMRI studies of the relatively small, almond-shaped and deep-seated amygdala. It is a paired area in the right and left hemisphere. Anatomy sees it as consisting of 13 individual nuclei, some of which are still divided into subunits. For better differentiation, three different areas are determined:

1. the centromedial nucleus group, including the central and medial nuclei,
2. the basolateral complex with the lateral, basal and basolateral nuclei and
3. the cortical nucleus group with the cortical nucleus.

The mere differentiation into right and left amygdala already shows a different specialization, i.e. responsibility, for example in the type of emotions and memories processed, with gender differences also emerging. For smell, according to current state, the medial nucleus located in the centromedial nucleus group (MeA) is of particular interest. It is a central node in the olfactory neuronal network and is also associated with pheromone perception.

Almost all of the areas involved in smell are therefore divided into various, often very specialized areas. In order to prevent losing sight of the details, I usually refer to the area as a whole in the following. However, those who would like to deal with an area in individual areas and functions in more detail will find information in the following sources that keeps me up to date with the latest research:

- PNAS (Proceedings of the National Academy of Sciences), ▶ www.pnas.org
- PMC (U.S. National Institutes of Health's and National Library of Medicine), ▶ https://www.ncbi.nlm.nih.gov/pmc/

So far, four regions or areas with networks in the brain that have an affinity for certain scents or, better said, fragrance directions have been known from neuroperfumery and fragrance psychology. I will discuss them in more detail. Simply put, these specific brain regions want to smell certain scents first in order to be stimulated by them. Of course, the current state of experience (e.g. hunger or craving for sweets) and thus the current need for stimulation play a role.

3.9 Olfactory Soothers of the Soul: Why Perfumes Do So Much Good

- **Can chocolate scents activate our addiction centre? Are milk or milk mousse notes soul comforters?**

An example of a specific olfactory stimulation is the network of the Insula, as I have already introduced it. It is its own brain region and works as a network mainly with the amygdala, the orbitofrontal Cortex, the thalamus and the Hypothalamus. It is now assumed that the insula is involved in the processing of emotions and arousal, including the awareness of one's own body states such as pain. In addition, one of its centers plays a major role in addiction. For example, imaging methods show that the insula and its network are particularly stimulated and activated by the scent of chocolate and sweet aromas, olfactorily and gustatorily delivered (Han et al. 2019a). It can therefore be said that the insula is by nature not averse to a certain "addiction" to sweet aromas, as we know them in perfumery, from Gourmand notes.

Even more: Studies with imaging methods not only show an increased activation of the insula network during the intake of drugs such as cocaine, but already during the desire for it (Risinger et al. 2005). It is quite obvious that we have a network in the brain, in which evolution has decided for us humans that a scent—even if mild—can have an effect on us like a drug. Even more: The olfactory anticipation of specific scents triggers an effect in a brain region that can even increase to a slight craving. In summary, it can be said: Specific scents have the potential to act specifically in our brain. They are even able to address and activate the addiction center specifically. This gives my initial statement "Perfumes—the most beautiful of all drugs" a deeper meaning than one could have initially assumed.

3

For the neuroperfumery the most exciting discovery at the moment is that there are brain regions or areas and networks with an affinity for certain scents—that can act as neuroscents especially targeting emotional responses -, to which they react with a specific arousal or effect. This olfactory affinity seems, as already mentioned, to be situational in many cases, that is, dependent on the current experience or the current need for stimulation. With the choice of a certain perfume, we therefore not only want to underline our personality, but also fulfill our experience, that is, stimulation desires. One can therefore look forward to further findings of the neuroperfumery, which show how, when and where something is liked to be smelled in the brain and what effect this has in each case.

The regions of the brain responsible for certain scent affinities, according to current research, are mainly located in the limbic system. This is our emotional center with a branching network that controls emotions. The latest research results on the amygdala, our deepest core in the emotional center, offer amazing insights into the world of smell. It has long been known that the Amygdala has a direct connection to our nose. Since it smells the fastest of all brain areas, the aforementioned hypothesis has arisen that we might smell everything twice. For this, the Israeli-American psychologist Daniel Kahneman, who was awarded the Nobel Prize in 2002, provides good arguments, as I will show later.

■ **The effect of milk or milk mousse fragrance notes**

The amygdala has its own olfactory interests and fragrance preferences and seems to be looking for more sweetness in perfumes than we are aware of. However, the conscious fragrance perception in the prefrontal cortex is rather rejected because it is perceived as not suitable for oneself. The feelings of the amygdala are mainly built up between mother and child during breastfeeding. Here a basic trust is created. Because through eye contact with the mother, the smell of mother's milk and the mother's breast, babies learn to relax. And now it gets really interesting: In recent studies it has been found that these relatively sweet skin milk smells can also have a calming effect on adults and often still provide almost childlike well-being.

In perfumery, these are the so-called milk or milk mousse fragrance notes. They smell like perfumes, but do not contain milk and are currently enjoying increasing popularity. Because more and more people who are heavily burdened professionally or privately consciously or unconsciously choose a fragrance that offers them a journey back in time to their carefree childhood; in a time with caring parents, and in which life was much simpler. So milk or milk mousse fragrance notes with their often light sweetness can be real soul comfort.

The amygdala responds to these smells. It is olfactorily calmed by a composition reminiscent of mother's milk, consisting of skin- and milk-warm scents with vanilla- and white musk-notes. For perfumers, enriching women's perfumes with vanilla and white musk has therefore become a quick way to increase the attractiveness of the fragrance. This is especially true for the after-smell on the skin. The perfumers are counting on a genetically programmed fragrance preference that is further reinforced by early childhood fragrance experience.

Now there is the entire fragrance direction of milk or milk foam in the most diverse variants—from less to quite sweet, which of course is relative for fragrance noses, but olfactorically ideal to address a larger group of users. An example of relatively less sweet is the perfume "Signorina Misteriosa" by Salvatore Ferragamo. Milk foam and black vanilla pod provide a warm, soothing fragrance impression. Even less sweet is the fragrance "Sweet Milk" by Jo Malone, which both men and women can wear as a balm for the loss of carefree days gone by. Clearer and therefore sweeter is the milk foam impression in the perfume "Vanilla Caramel" by Tutti Délices. Here, the warm smell impression is supported, among other things, by coconut.

So the perfumery always has to take into account the deep, almost forgotten emotions and experience desires of its users in order to create a fragrance experience. It also has something therapeutic, because it can work like a balm for the soul and thus directly and indirectly increase our well-being, our quality of life and even our health.

» *I know: Basic knowledge about factors that influence perfume choice and how to smell can sometimes be a bit dry. But be assured: Our exciting journey into the world of scents and perfumery, especially into the future of smelling, with insights into which scents make the brain feel really good, is about to begin!*

Summary

In this chapter, the different factors that influence perfume choice were introduced. The fact is: We smell more than we think, because we also smell with our tongue, skin and other organs. Smelling can touch emotional themes that date back to the time in the womb. For example, it is known that babies can still remember the smell of amniotic fluid, i.e. the primordial perfume surrounding them. Another aspect is the current state of research in the early detection of diseases, because our state of health also plays a role in the choice of fragrance. The core theme of the chapter was: Who or what decides the choice of fragrance? Various approaches were discussed, in particular the "S-O-R model", as well as findings from neuro- and multisensory marketing. With regard to a possible connection between fragrance preferences and personality traits, current findings from the still young neuroperfumery were presented. Finally, it was shown with two fragrance examples that certain fragrances—even if mild—can act like a drug.

References

Barden P (2013) Decoded – the science behind why we buy. Wiley, Chichester

Boyle JA et al (2009) The human brain distinguishes between single odorants and binary mixtures. Cereb Cortex 19:66–71

Busse D et al (2014) A synthetic sandalwood odorant induces wound-healing processes in human keratinocytes via the olfactory receptor OR2AT4. J Investig Dermatol 134:2823–2832

Geiger H (2019) Den Duft hören: Natur, Naturbegriff und Umweltverhaltenin China. Matthes & Seitz, Berlin

Georgiopoulos C et al (2018) Olfactory fMRI: implications of stimulation length and repetition time. Chem Senses 43(6):389–398

Han P et al (2019a) Sensitivity to sweetness correlates to elevated reward brain responses to sweet and high-fat food odors in young healthy volunteers. NeuroImage 11:116413

Han P, Zang Y et al (2019b) Short or long runs: an exploratory study of odor-induced fMRI design. Laryngoscope 130:1110

Hatt H (2006) Geruch. In: Schmidt F, Schaible HG (eds) Neuro- und Sinnesphysiologie, 5. Aufl. Springer Medizin Heidelberg, Heidelberg, pp 340–352

Häusel HG (2016) Brain View. Warum Kunden kaufen, 4. ed. Haufe, Freiburg

Higgins ET (1987) Self-discrepancy: a theory relating self and affect. Psychol Rev 94:319–340

Jellinek JS (1997) Per fumum. Semiotik und Psychodynamik des Parfums. Hüthig, Heidelberg

Kirschberger K (2015) Das musikalische Gehirn des Kunden: Wie Musik Werbung aus Sicht des Neuromarketings stärkt. Diplomica, Hamburg

Knoblich H, Scharf A, Schubert B (2003) Marketing mit Duft, 4. ed. Oldenburg, Berlin/Boston

Malik B et al (2019) Mammalian taste cells express functional olfactory receptors. Chem Senses 44(5):289–301

Mensing J (2005) Duft-Guide. Der schnelle Führer zu Ihren Ideal-Düften. In: Roller U, Spelman R (eds) Parfums – Edition 2005, 10. ed. Ebner, Ulm, pp 206–215

Mensing J, Beck C (1988) The psychology of fragrance selection. In: Van Toller S, Dodd GH (eds) Perfumery: the psychology and biology of fragrance. Chapman & Hall, London, pp 185–204

3

Mensing J, Viera D, Peek A (2017) Die Zeit am Rio Uneiuxi (Amazonas). In: Dux G (ed) Die Zeit in der Geschichte. Springer, Wiesbaden, pp 317–347

Meyer M, Glombitza P (2000) Innovative Marktforschung – Profilierung von Markenartikeln durch Duft. Planung Analyse 27(2):52–56

Müller A (2019) Der Maître de Parfum im Gehirn. München Süd, Nov 2019 – Jan 2020

Nölke SV, Gierke C (2011) Das 1x1 des multisensorischen Marketings. Multisensorisches Branding: Marketing mit allen Sinnen. Umfassend. Unwiderstehlich. Unvergesslich. Comevis GmbH, Köln

Pezoldt K, Michaelis A (2014) Parfümwahl am Point of Sale – Eine neuroökonomisch fundierte Analyse zur Ableitung relevanter Einflussdeterminanten auf Basis des S-O-R-Modells. Ilmenauer Schriften zur Betriebswirtschaftslehre, Ilmenau

Preuk M (2013) Doktor Hund. Wie Vierbeiner Krankheiten erschnüffeln. Focus Forschung, 22.09.2013

Pusch M (2018) Multisensorisches Marketing im Online-Shop: Akzeptanz einer Duftbox im Privatgebrauch. Hochschule Hof, Fachbereich Wirtschaft, Hof

Reisert J, Reingruber J (2019) Ca2+-activated Cl− current ensures robust and reliable signal amplification in vertebrate olfactory receptor neurons. PNAS 116(3):1053–1058

Rempel J (2006) Olfaktorische Reize in der Markenkommunikation: Theoretische Grundlagen und empirische Erkenntnisse zum Einsatz von Düften. Springer, Wiesbaden

Risinger RC et al (2005) Neural correlates of high and craving during cocaine self-administration using BOLD fMRI. NeuroImage 26(4):1097–1108

Rolls ET (2004) The functions of the orbitofrontal cortex. Brain Cogn 55(1):11–29

Romoli L et al (2012) fMRI study of smell: perceptual, cognitive and semantic components of cortical elaboration of 3 familiar aromas – lecture: German Research School for Simulation Sciences, Jülich

Scheier C, Verir S, Isenbart J (2008) Beitrag von TV-Werbung auf den Kaufentscheidungsprozess am PoS. Planung Analyse 2008(2):47–51

Schnitzler L (2004) Parfüm. Flüchtige Freuden. Wirtschaftswoche 2004(42):62–63

Schoen K (2018) Gegenüberstellung von Essensdüften und Blumendüften im Hinblick auf ihre Verarbeitung im mesolimbischen System – eine fMRT-Studie. Dissertationsschrift der Medizinischen Fakultät Carl Gustav Carus der Technischen Universität Dresden

Shepherd G (2013) Neurogastronomy: how the brain creates flavor and why it matters. Columbia University Press, New York

Small DM et al (2001) Changes in brain activity related to eating chocolate: from pleasure to aversion. Brain 124:1720–1733

Steiner P (2017) Sensory Branding: Grundlagen multisensualer Markenführung. Springer Gabler, Wiesbaden

Tamura K et al (2018) Olfactory modulation of colour working memory: how does citrus-like smell influence the memory of orange colour? PLoS ONE 13(9):e0203876

Trivedi DK et al (2019) Discovery of volatile biomarkers of Parkinson's disease from sebum. ACS Cent Sci 5:599

Waldeck C, Frings S (2005) Wie wir riechen, was wir riechen: Die molekularen Grundlagen der Geruchswahrnehmung. Biologie unserer Zeit. WILEY-VCH, Weinheim

Weiss T et al (2019) Human olfaction without apparent olfactory bulbs. Neuron 104(5):1023

Wu Y et al (2020) Stereo-olfaction in humans. In: Proceedings of the national academy of sciences, 22 June 2020

Welcome to the Neuroperfumery

Types of Wellbeing and How Fragrance Can Help

Contents

4

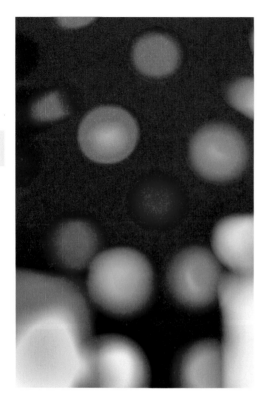

What does "feeling good" mean?

Seemingly simple questions are often not easy to answer. This also applies to personal well-being, which everyone defines differently for themselves and can also change according to need. Sometimes you feel good when you sit quietly by yourself, for example with a cup of tea and a book, on the couch, at other times you look for the happy company of friends. In general, one can say that one is filled with good feelings at the moment when one feels comfortable and does not think of fears, problems, or worries. But this raises a follow-up question. How does one know what makes one feel good at the moment? This is a particularly central question in perfume advice in a perfume shop. You want to feel comfortable in your skin, and the perfume you wear or choose should also contribute to this. Accordingly, you also choose your perfume according to

how you want to feel at the moment. Most of the time, however, one is not aware that with changing needs for well-being, olfactory needs also change and that the brain is then more open to specific olfactory stimuli at certain times, even actively seeking them out.

4.1 Latent Basic Expectations of Perfumes

Emotions, basic emotions and slowly building moods can be triggered by smelling. But it also works the other way around: Genetics, personality, olfactory socialization, learning experiences, physical and mental experiences, and the associated wishes influence olfactory perception and thus the choice of fragrance.

What latent basic expectations of perfumes arise from the mix of these factors? I would like to explain this using a study on the attraction of the shopping destination perfume shop. Most of the reasons for this may not always be fully aware of when entering a perfume shop. In a *BRIGITTE-*communication analysis, (Brigitte is a biweekly women's magazine in Germany), women between the ages of 14 and 70 were asked about large shopping destinations. The result: specific experience wishes are clearly in the foreground for perfume customers. This also applies—but to a much lesser extent—to other shopping destinations. When visiting a perfume shop, wishes for increased well-being, greater beauty and attractiveness are in the foreground due to the brands and products offered there. Of course, perfume customers also have other needs: competent advice, good brands, quality and/or a lot of shopping fun. But this is also offered by other shopping destinations—and sometimes even better.

The attainment of attractiveness and beauty as well as the increase of wellbeing are therefore the essential and at least conscious, latent basic expectations that one has of wearing a perfume. In other words: these basic expectations must also be the basic goal of a personal scent consultation. This sounds simple, but is anything but easy in practice. So, for example, every scent consultant in the shopping center perfume store has the goal that his customers feel comfortable during the consultation and also later with the new perfume. It is tacitly assumed that the expectations of wellbeing are relatively the same for all customers. The scent consultation should therefore be friendly and accommodating, the customer should ideally be able to sit comfortably on a couch, for example, and perhaps a drink should be served to increase the relaxed atmosphere. Psychology knows of various other means of increasing wellbeing—even standing. I will come back to this later.

But psychology also knows of another aspect: there is not just one type of wellbeing, but different types that people seek at different times. This means that the same customer can quickly have a different need for wellbeing depending on the form of the day or based on current experience and experience desires. Because everyone feels differently depending on the time of day and season, day of the week, and personal experiences. Sometimes you want more stimulation and entertainment, sometimes peace and quiet. Of course, there are also customers with particularly stable personalities who are only slightly influenced by the form of the day.

Seminars on sales psychology of perfumes teach how to recognize the experience desire of a customer and which scent is good for him in the current form of the day. The customer already gives the first hint when he enters the store through his non-verbal behavior. How to specifically match a perfume to it, I will explain later.

4.2 Olfactory Stimulation Needs of the Brain

As a psychologist who works in the perfume industry and often advises perfume seekers, I know four essential basic types of wellbeing wishes that are divided into four further subgroups. Behind each wellbeing wish is an affinity for specific fragrance directions/fragrance families, which the perfume industry uses for the classification of perfumes and which is also known to the neuroperfumery. According to current knowledge, the following areas of the brain or brain regions located in and outside the emotion center and equipped with their own "needs" are responsible for wellbeing. They work as a network or within the framework of three essential control loops, which we will discuss later, and play, as already mentioned, a great role in smelling and experiencing scent and thus in perfume advice. To put it simply, they are:

— **Hypothalamus** with a large network that extends from the Insula to the frontal cortex, among other things involved in the experience of enjoyment, reward associations, but also in addiction;
— **orbitofrontal cortex** located in the prefrontal cortex, among other things responsible for the personality traits of extraversion and conscientiousness in the network;
— **Hippocampus** (seahorse), among other things responsible for our stress-prone long-term memory in the network;
— **Amygdala,** among other things active as an emotional test center. It has direct access to the olfactory bulb (Bulbus olfactorius") via the piriform cortex, the main gate to the world of our sense of smell—main gate because we also "smell" with the skin (◘ Fig. 4.1).

Let me now go into the findings of brain research on the areas and networks concerned

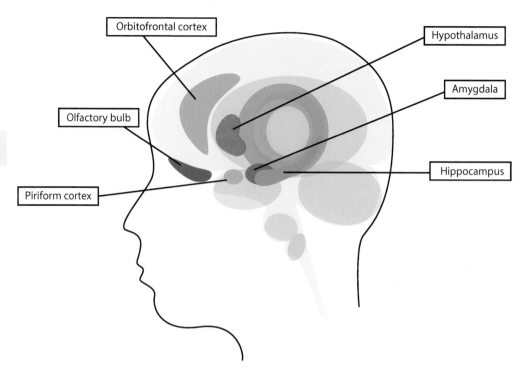

4

☑ **Fig. 4.1**   Brain regions

with the four directions of well-being and scent. I deliberately do not distinguish between the terms brain area Gehirnareal and brain region Gehirnregion because it is difficult to show a spatially delimited area of the brain, as one rather associates the term with the word "brain area", specifically for smell.

Imaging methods in brain research make it visible how and where, or rather, where a fragrance stimulates brain regions. In this way, we can quasi watch the brain smell and better understand which fragrance fits the state of excitement of which brain region or is even welcomed. Via the olfactory preferences of a certain brain region and its network, neuroperfumery gains more and more direct and indirect insights into the psychophysical relationships associated with the sense of smell—for example, which scents are ideal for well-being, but also what well-being means for a certain brain region and its network.

This way, brain research, i.e. neuroperfumery, can offer perfumers completely new insights and possibilities to create perfumes specifically for the stimulation needs of individual brain regions. For example, we can expect that the raspberry chocolate note of a fruity gourmand fragrance acts in our pleasure center, the hypothalamus as well as in the extended reward network, which includes the insula as well as the frontal cortex (forebrain). The gourmand scents associated with "edible" act like a reward and thus address the widely networked reward center. Imaging methods show that the frontal cortex is involved in the fragrance stimulation with the "edible" (Schoen 2018). It is, inter alia, responsible for the linking of reward associations and ultimately controls the behavior to get a reward.

The hypothalamus is located in the limbic system of our emotional center and is responsible for the perception and processing of odors, but also part of the reward center or, better said, the reward centers.

Important role in this also play the so-called happiness hormones Serotonin as well as especially Dopamine with their positive psychological effect. Serotonin is known for its contribution to normal brain function as a mood regulator. As researchers at Baylor College of Medicine found, olfactory stimulation seems to have a very positive effect here. It causes an increase in a serotonin transporter (Slc22a3—a protein) in astrocytes (a type of brain cell), allowing serotonin to be transported into the cells. The dopamine release is controlled in our brain, inter alia, via the hypothalamus and its network, which is also decisive for libido and communicates closely with the olfactory memory. Motivation, mood swings, and self-doubt, such as a low self-esteem, are often associated with a low dopamine level. Neuroperfumery now shows how we can keep our dopamine level high (at least support it) in a very natural and healthy way with scents that are tailored to our needs—in combination with other sensory stimuli, which I will discuss in more detail. Taking this a step further, in the future, "smelling better" could mean that certain perfumes as active fragrances or neuroscents, for example, specifically address the hypothalamus and its network in a positive way. These would be perfumes that bring the wearer closer to emotional wishes for experience through ingredients and fragrance development and provide for a demonstrably increased dopamine level. Of course, a "hormonal effect" of perfumes is legally prohibited in many countries (e.g. EU countries), but this does not exclude a slightly mood-enhancing effect.

As a scent psychologist one knows a whole range of wishes for experience that perfume lovers would like to experience as "mood modulation", so to speak as an emotional additional benefit, in addition to a wonderful smell through their perfume. These include, for example:

- *To feel more active and dynamic.*
- *To be relaxed and stress-free.*
- *To feel more sensual and cared for.*
- *To discover more creativity and power within oneself.*
- *To experience more success and recognition.*
- *To have more pleasure, fun, and joie de vivre.*

Dopamine also plays a big role in mood modulation for another reason: The neurotransmitter is, as mentioned, part of a network, the dopaminergic system, which extends into the entire limbic system and can thus further increase its positive effect. I will explain below how to achieve the fastest possible optimum increase in dopamine levels. For this purpose, a specific desire is literally fused with coordinated sensory impressions. This is the case, for example, when the desire for more pleasure, fun, and joie de vivre is supported by corresponding colors and music as well as tactile, gustatory and above all olfactory with selected fragrance notes. Almost everyone has a little "synesthete" in them who can use the fusion of the senses very well for themselves.

Back to the Hypothalamus. For him, feeling good through scent apparently takes place primarily by smelling pleasure. This is shown by its special affinity for imaginative, invigorating fruity-pleasure, edible-smelling gourmand notes. These are fragrance notes that usually smell like sweets and that the hypothalamus, like the Insula, associates with a delicious snack or dessert. The two most popular flavors in the world are chocolate and vanilla. Their effect on the brain has been discussed many times, with chocolate in particular said to make people happy. Whether the good-mood maker acts primarily through its components such as theobromine or tryptophan or also—like many other pleasures—through sugar and fat is not a central question for fragrance research. People have an extremely good sense of smell, in which many posi-

4

tive memories from childhood are stored. This also includes the effect of chocolate, especially if it was experienced as a reward. Even in adult chocolate lovers, the reward center in the brain is already activated at the sight and smell of the sweet delicacy, which in turn can lead to dopamine release.

According to the latest findings, the most intense dopamine release takes place when the neurotransmitter rises slowly. This is the case when the feeling of well-being increases step by step and turns out to be unexpectedly pleasant, as when one unexpectedly smells something nice. For the hypothalamus and its network, feeling good is especially unexpected pleasure with reward. A perfume with this effect has the chance to become a lucky scent and to be able to do much more than just smell good. In fragrance consulting, feeling good is an unexpected, invigorating-imaginative pleasure with reward—one of four types of well-being that is linked to a specific fragrance direction. Of course, these connections must be investigated in more detail above all by means of brain imaging studies with simultaneous interviews/assessments by the subjects.

Neuroimaging is another important term in this context. This procedure, which has been increasingly used and further developed since the turn of the millennium, has led to a major turning point in psychology. While behavior was increasingly in the focus of research after the First World War, researchers have now moved away from the models of the behaviorists and are focusing on consciousness. Basic discoveries about conscious perception have been made for over 200 years. This is because perceived colors, sounds, and smells are created in the head. Even if we are not all conscious of them or recognize them, the consciously experienced perceptions contribute to our consciousness. But even the stimuli that are not consciously perceived have an effect on us.

With the imaging methods, which do not interfere with brain processes, one can now look at the living brain at work and observe these processes. Many conclusions are still rather indirect, but there are numerous findings on which research can build. For example, differences can be seen in brain processing, depending on whether a stimulus, i.e. a sensory impression, becomes conscious or not. The American neurobiologist and cognitive scientist Bernard Baars was one of the first to draw attention to this: "If a stimulus is presented unconsciously, it activates areas in the cortex that are involved in the analysis of colors, sounds, faces, and the like. But if the identical stimulus is shown consciously, it also recruits regions far beyond the sensory cortex" (Baars 2003, p. 4).

This amazing observation of the sense experience also applies to smell. Green-citrusy notes are, as mentioned before, consciously experienced by a brain region that does not belong to the smell brain as fresh, cool, and invigorating, for example. The conscious smell impression thus arises outside the olfactory brain.

With the help of imaging methods, one can learn a lot indirectly and, conversely, conclude from this that our center for extraversion with its properties is particularly fond of the cool and invigorating freshness of green-citrusy notes. This fits in with the fact that extraverts also need free space to feel comfortable and appreciate targeted advice. As the seat of this personality trait, the orbitofrontal cortex at the forehead of the brain could be localized, as mentioned before. It is stimulated, although not alone, by green-citrusy, especially by bergamot-scented notes. In combination with extraversion, the personality dimension "conscientiousness" often plays a role. Here, the emphasis is more on planning, competence, striving for performance and thoughtful action. This personality trait is assigned to the prefrontal cortex, in which the orbitofrontal cortex is also located.

Conscientiousness also has a tendency towards a fragrance direction: "fresh-aquatic" with aromatic fougère notes. Of course, the relationships between personality, brain area, and fragrance preferences still need to be investigated in more detail. But there are already a number of references that point to these relationships. These relationships are currently still predominantly from the practice of fragrance psychology, but there are also first imaging studies on the relationship between personality traits and their location in specific brain areas (De Young et al. 2010).

Here is another example that shows the fragrance preferences of a brain area and its network when experiencing a specific feeling, namely increasing stress and sad mood: Clinical studies have long been able to use fMRI methods to show that the hippocampus shrinks as a result of chronic stress—as with other major psychological stressors (e.g. depression). Important brain regions with which the hippocampus works as a network when experiencing stress include the amygdala, the hypothalamus as well as the orbitofrontal and prefrontal cortex. As already partly shown above, the individual brain areas involved in the "stress network" have fragrance preferences. These increase with increasing stress. If this finally takes over completely, the fragrance experience collapses.

From the practice of fragrance psychology it is known that even mild stress affects the customer's choice of fragrance. Apparently the brain is trying to protect itself from overstimulation. The stress-prone hippocampus, seat of our long-term memory in the emotional center, seems to set its sights on the calming effect of floral, non-opulent flower notes like white flowers, clear roses, and gentle tea notes at first, when the stress level rises. The resulting associations according to the motto "That reminds me of something" relax this brain area. You probably know this from your own experience. But when stress turns into a sad mood and the amygdala is more involved, the preference for sweetness, as shown by the example of sucrose, increases

(Schneider 2015). But an increase in olfactory sensitivity was also observed, for example for rose scents like phenylethyl alcohol, a rose-like odorant. It can be perceived by people in a sad mood at low concentrations (Schneider 2015). The practice of fragrance psychology can confirm this. It also knows the preference for sweet, vanilla-flavored flower and gourmand notes, as I have already shown in connection with the Hypothalamus. We can look forward to future fMRI studies that, for example, investigate the odor acceptance and effect of sweet rose honey or vanilla Rose on the "stress network" in different moods such as sadness.

If stress and sad mood escalate to an irritable mood or even to depression, olfactory perception is reduced, but also taste perception. Studies have shown that compared to non-depressed patients, depressed patients have a reduced perception of sweet, sour, salty, and bitter taste substances (Canbeyli 2010; Rosenthal-Zifroni and Edelstein 1969). Obviously, this state has a negative effect on the prefrontal cortex, especially on the orbitofrontal cortex Kortex, which creates sensory impressions as the maître of perfumes and associates scents with our self-image.

Back to the first stress symptoms, when all you want to do is relax and let your soul dangle. Memory research shows that floral-flowery notes in particular invite fantasies and daydreams during stress. This relaxes the hippocampus and its network and stimulates it positively. Feeling good seems to be associated with relaxation for the hippocampus. This is also confirmed by aromatherapy. It knows different flower blossoms like Jasmine, roses or lavender, which—as they say in the perfume industry—olfactorily interpreted round, light and harmonious, create a comfortable, and relaxing feeling for many.

A network, the so-called "fear network" in the brain, regulates—as the name suggests—fear and anxiety. To relax, you need a feeling of trust. The main actor here is

4

the amygdala, which is strongly focused on facial expressions and has stored negative but also positive experiences and associations. Other important actors in the fear network are the so-called "Midline Thalamus Group" or the "Midline Thalamic Nuclei". They are a core area of the thalamus, which is located exactly in the middle between our hemispheres and is considered the gateway to consciousness. The thalamus generally has the task of forwarding information from the body and the sense organs to the cortex (cerebral cortex) as a relay station; specific thalamic nuclei play a role here. It is particularly the nuclei located around the so-called midline of the thalamus that are responsible for forwarding the fear memory and experience. They bundle information about the body's mood and emotional state and convey the stress level to higher brain regions, so that it becomes aware. This allows us to take the necessary measures to correct our mood—for example, to experience a relaxing feel-good scent.

Originally it was believed that—unlike hearing and vision—stimuli of the sense of smell are forwarded without detour via the relay station thalamus. However, recent studies conclude that the thalamus plays a significant role as an amplifier, corrector, and forwarder of olfactory information, even if the thalamus is bypassed for much of what is smelled. Even more: There is first anatomical evidence that the thalamus (specifically the mediodorsal nucleus) receives direct olfactory information from primary olfactory areas including the piriform cortex and maintains dense reciprocal connections with the orbitofrontal cortex (Courtiol and Wilson 2015)—even if the thalamus is bypassed for much of what is smelled.

It is reasonable to assume that smell only passes through the thalamus if at the same time discomfort, fear, or anxiety is experienced by the amygdala and its network. This also provides an additional control instance from nature for olfactory stimuli, which makes an olfactory check possible when smelling. In other words, it is evaluated whether a certain scent really suits you, is not associated with danger, but also what you want to smell more strongly in order to feel trust or a sense of well-being, for example.

For the skittish amygdala, the small, almond-shaped structure and the early warning system for dangers anchored deep in the emotion center, feeling good represents a feeling of trust experienced within the context of human affection. This is supported by scents that remind us on the one hand of warm-sweet, human skin scents, and on the other hand by the mother's own scent, her diet, sometimes sweaty skin and of course the vanilla-scented scent of mother's milk—a very personal, complex, and positive olfactory experience for every baby, if breastfeeding is experienced gently and protected. I will show later that this mixture of scents can not only have a psychologically calming, but even a psychophysically pain-relieving effect on adults. This primal olfactory experience is best reproduced in perfumery with two fragrance directions: with lighter amber-florals or so-called floriental notes that smell warm and empathetic, for example, sandalwood, and with the mentioned milk mousse notes that often, thanks to vanilla and white musk, radiate a similarly human warmth, attractive sweetness, and thus affection and emotional attraction.

◗ Figure 4.2 provides a simplified overview of the basic fragrance preferences of individual brain regions.

4.3 Well-Being Consultation: Fragrance Preferences of Individual Brain Regions and Their Networks

Young neuroperfumery is certainly only at the beginning of its findings. But, simply put, it already knows four—actually five—essential well-being needs for perfumery,

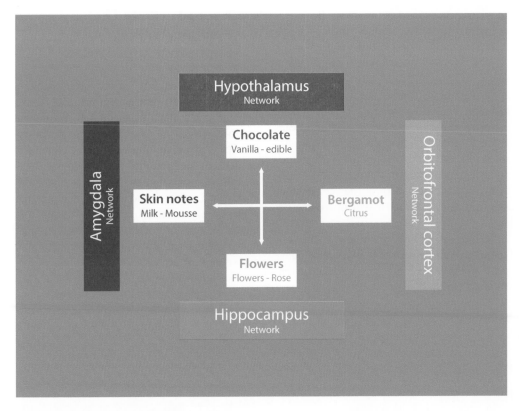

D Fig. 4.2 Basic fragrance preferences of individual brain regions

which correspond to certain basic fragrance directions and involved brain regions, as suggested by fragrance psychology:

1. **Feeling good as an invigorating-fantasy pleasure** and reward as part of a consultation with lots of shopping fun and surprise. The hypothalamus and its network, the dopaminergic "reward and addiction system", are involved. Base fragrance direction: Gourmand, "fruity-enjoyable".

2. **Feeling good as relaxation** in the form of harmony, anti-stress, peace, and recharge as part of a pampering consultation. The hippocampus and its "stress network" are mainly involved. Base fragrance direction: floral, "flowery-light".

3. **Feeling good as enthusiasm** in the form of extraverted freedom and unrestrictedness

as part of an efficient, target-oriented consultation. The orbitofrontal cortex is mainly involved. Base fragrance direction: green-citrus, "fresh-invigorating".

In addition, the prefrontal cortex with the personality dimension "conscientiousness" plays a role in this type of feeling good. Here, the need for feeling good is based more on experienced, systematic planning, good organization, thoughtful action, foresight, and responsibility. It can be referred to as **feeling good through competence** in contrast. Base fragrance direction: "fresh-aquatic" with aromatic Fougère notes.

4. **Feeling good as trust** and friendly affection as part of a consultation tailored to individual needs. The amygdala and its

4

"fear network" are involved. Base fragrance direction: slightly amber- florals ("floriental"), skin notes "warm-soft", with a certain sweetness playing a role.

The assignment of base-scent directions to different types of well-being certainly does not cover the wide range of perfume notes or scent directions in perfumery. There are other types of well-being with corresponding scent directions. Nevertheless, subgroups can be formed from these base scent directions alone, which I will introduce below.

Here is a simplified overview of the four types of well-being with associated base scent preferences and involved brain regions (◘ Fig. 4.3).

4.4 Neuro-Perfume Therapy: Findings for Fragrance-Supported Applications

The above statements about scent preferences of certain brain regions and their networks can be a huge step forward for neuro-perfumery and "neuro-perfume therapy", especially the discovery and development of neuroscents (active fragrances), on which, in particular, aromatherapy can build with further studies. Below I present some areas for neuroscent—or rather for aromatherapeutic oriented measures with neuroscents—supported therapy goals:

- **Neuroscent aromatherapeutic measures for the treatment of the "fear network"**: The above-mentioned findings suggest that sweet-warm-soft scents experienced, which

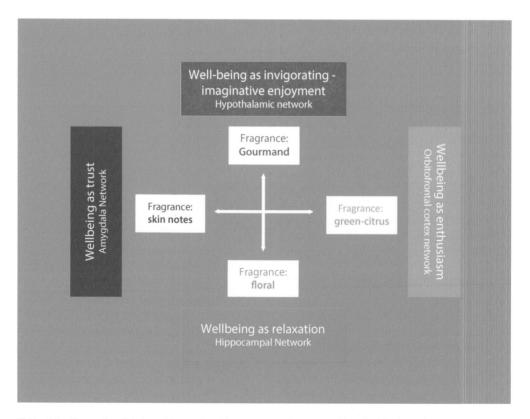

◘ **Fig. 4.3**   Types of well-being with associated base scent preferences and involved brain regions

are also similar to the smell of human skin, have a favorable effect on the amygdala and the thalamus in particular. They address a primal trust and can be used together with other therapy approaches in fear and anxiety states. Of course, there are still no specific indications from brain research as to which scent notes contribute particularly to the calming of the amygdala. In ▶ Chap. 10 I will introduce a fragrance-supported self-therapy with which I have had very good experiences in practice and which you can create and use for yourself according to your own needs.

— **Neuroscent aromatherapeutic measures for the treatment of the "stress network":** Imaging studies of olfactory perception provide first indications that preferences for and sensitivity to odors change with increasing stress. When first stress symptoms occur, floral scents like clear rose notes can offer some relaxation to the hippocampus and its network, also because they are able to trigger positive olfactory memories and associations. If stress increases and a sad mood sets in, the brain seeks out more sweet fragrance notes. They can be used as part of an aromatherapy-supported therapy in combination with other therapy approaches and act mood-modulating in this combination. If there is a real depression, fragrance can be an indication and a sign of healing progress. Since fragrance and taste impressions are experienced to a lesser extent in this disease, an aromatherapy-supported therapy can be used as part of exercises to regain sensory experience.

— **Neuroscent aromatherapeutic measures for mood enhancement through activation of the "reward network":** The brain knows fragrance notes—especially those that smell like food—that can also serve to lighten the mood, even make them almost addictive. Interestingly, the brain does not make a big difference in these notes for a positive stimulation between taste and smell. This was shown, for example, with chocolate scents. They could even serve as ideal diet companions because they have no effect on weight. The mood-enhancing effect of other food-smelling fragrances and thus many gourmand notes must of course be researched in more detail by research. Nevertheless, it is becoming clear that an aromatherapy-supported therapy with edible fragrances could play a positive role in the so-called mood modulation.

— **Neuroscent aromatherapeutic measures for the "self-assertion":** Low values of extraversion are associated with withdrawal, loneliness, or the desire to stay in a small group. This can extend to the inability and motivation for self-assertion, such as the defense of one's own boundaries and rights. It is known that green-citrus notes address extraverts in the orbitofrontal cortex with freshness, activity and dynamism. The question arises as to whether an aromatherapy-supported therapy with these notes would be beneficial, for example, as part of a self-assertion training. Probably yes, because these notes have an activating broadband effect on the brain. There are also successes in the context of psychological techniques such as the scent-supported power-posing. I will invite you to a self-experiment at a later point and present other possible uses of scents in self-therapy.

◘ Figure 4.4 gives a simplified overview of neuroscentaromatherapeutic measures or objectives to be discussed.

4.5 On the Dynamics of the Olfactory Experience: Wandering with Olfactory Preferences

The need for well-being and fragrance preference also depend on genetic disposition as well as on the current need for

4

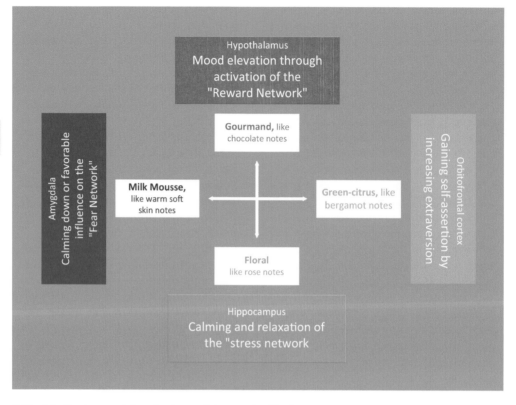

◻ Fig. 4.4   Scent-supported applications and therapeutic objectives

experience and stimulation or on corresponding wishes. They are therefore dependent on both situation and personality. Psychologists therefore distinguish between the current, changing state, and need of a person, the so-called "state", and the behavior across different situations, that is, the enduring personal property or disposition, the so-called "trait". This differentiation can be well transferred to the choice of fragrance and the associated feeling of well-being over time. So we can be fascinated by perfumes or fragrance directions that we would usually not choose, but which fit our current mood or the desired occasion and experience wish. Then we feel a certain situation-dependent need that we actually do not know in this form. Accordingly, the experience and the experience wishes can be more stable or situational.

You can also imagine the target direction of well-being and fragrance needs pictorially like a compass needle. It shows in which direction one wants to experience oneself—or the customer from the perspective of the perfume seller—more or less consciously; what one longs for; what attracts one magnetically; what one expects as a feeling of well-being; but also what one is now more likely to respond to as a perfume or fragrance direction.

Psychology knows four basic experience directions, better said: an experience space with four directions, which plays a role here. It consists of the dimensions relaxed/calm vs. stimulated/excited and introspective/inner vs. open/outward (◻ Fig. 4.5).

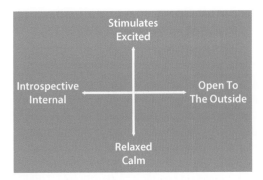

Fig. 4.5 Four basic directions of experience

Like the magnetic North Pole, one can be on constant wanderings in this experience space with one's well-being needs, that is, fragrance needs, but also quite stable due to one's personality. Unlike its geographical counterpart, the magnetic North Pole changes position with the magnetic South Pole faster or slower and sometimes not at all. The last reversal of the poles took place about 780,000 years ago. With humans, however, a change in mood takes place within seconds and minutes.

But personality can also change. Although around half of the personality traits are apparently inherited, numerous factors influence us throughout our lives. Nevertheless, life course researchers today assume that our essence changes only to a limited extent and gradually and slowly over time. This also applies to preferences for fragrances. The practice of perfumery knows, for example, that certain fragrance directions such as the fine-herbal chypre notes increase in popularity with female fragrance users as they age. This is certainly also due to various physical factors such as changes in nutrition and thus also in skin odor. Nevertheless, a change in self-experience with new experience desires leads to greater acceptance of fragrance directions for which one was less interested in younger years.

4.5.1 Smelling Better: Perfume as a Medium or the Transformation of the "Self"

Experience desires can be quite complex. In order to describe olfactory experience and to better understand consumers in their fragrance choices, perfume marketing often uses eight experience directions or—desires presented in ◘ Fig. 4.6 instead of four basic experience directions.

Each of these eight experience directions is associated with fragrance directions that I will discuss in ▶ Chap. 5. This means that these eight directions have a particularly strong influence on fragrance choice in terms of experience desires.

You can feel attracted to different experiences at the same time. This also applies to fragrance directions and their perfumes. Later on, I will explain how you can mix fragrances for yourself using perfume layering if you feel attracted to different experience directions. You combine fragrances from different fragrance directions and create a very complex mood or feeling. This is what makes creating with fragrances so unique. You can use them to accentuate your own olfactory needs and desires that contribute to a desired feeling of well-being

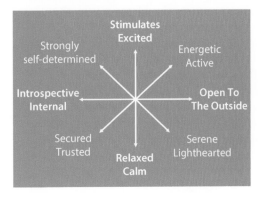

Fig. 4.6 Eight basic experience directions

or desired well-being and bring you subjectively closer to your desired experience.

Psychologically speaking, a process of transformation takes place. Fragrances tailored to your own needs become an offer of transformation to your own self. The "current self" (how I experience myself at the moment) should be brought closer to the "ideal self" or the ideal experience (how I would like to experience myself and be more). Regardless of whether this goal is achieved, this movement, this process towards the desired ideal state is experienced by our self as pleasant and attractive. It is the beginning of "smelling better". Against this background, the promise in the title of this book can be better understood. Beautiful SCENT as "smelling better" always leads to an increase in one's own self, because in the ideal self, the self of a higher self is always already contained—who one actually is, where one belongs, what one is entitled to, but also how one wants to be seen by others.

A perfume consultation should therefore fulfill the following latent basic expectation: When entering a perfume shop and looking for a perfume, one expects more or less consciously that the current, personal needs for well-being will be taken into account. Even more: One expects that one feels more attractive through the fragrance and the associated feeling of well-being, and experiences the ideal self. In the perfect fragrance experience, this is particularly supported by the fragrance experience. The perfume plays the role of a medium. It offers the "current self" a transformation offer that is to lead it towards the ideal experience and thus into the near future. If the perfume succeeds and one experiences an increase in experience through the fragrance, "smelling better" occurs. You feel more attractive through the perfume, at least for yourself.

4.6 Smell Open: How Evolution Promotes Certain Types of Smelling

In personality research, the number 5 has been popular for almost 30 years. So human behavior is often described with the "Big Five Personality Test", which analyzes a corresponding number of personality traits. These are extraversion, agreeableness, conscientiousness, neuroticism, and openness to new experiences.

Neuropsychology has been increasingly interested in recent years in mapping the "Big Five" to specific brain regions using imaging methods, i.e. functional magnetic resonance imaging (fMRI). This can already be shown quite well for some personality traits. For neuroperfumery, these discoveries are particularly exciting. As already mentioned, the orbitofrontal cortex (OFC) and its network seem to have specific preferences for smells, as we know from neuroperfumery. Imaging methods have shown that citrus notes (aromas) stimulate this brain region in particular. The connection between extraversion and the preference for these fragrance notes has long been noticed in the practice of perfumery. The history of perfumery also has evidence for this.

Fresh-green-citrus notes like the smell of bergamot are and were particularly appreciated by extraverted personalities, for example the composer Richard Wagner (1813–1883). The conductor Herbert von Karajan (1908–1989) is said to have been almost addicted to it. So it is said that he, the lover of fast cars, was always surrounded by a fine citrus cloud during his races on racetracks. The fragrance direction was apparently discovered by his third wife, the former Dior mannequin Eliette Mouret, during his time as musical advisor to the Orchestre de Paris in the years 1969 to 1971. In 1966, Dior had brought the men's fragrance "Eau de Sauvage" from the fragrance family "Citrus-aromatic" onto the market. It had been created by Edmond Roudnitska, one of the great perfumers of the time and, according to contemporary witnesses, a decisive, extraverted personality. As early as 1955, he had created the Dior women's fragrance "Eau Fraîche" with an extraverted citrus note.

How the preferences for certain areas or regions of the brain come about and, above all, how individual differences can be explained, is still unknown. So as a fragrance psychologist one can possibly attribute the preference of extraverts for fresh-green-citrus notes to the influence of early childhood experiences and the associated fragrance socialization. But then they would have to be similar for all extraverts. Ultimately, one comes to the conclusion that preferences for fragrances must also be inherited, so to speak, genetically determined in us.

The same problem arises for neuropsychology when it comes to the question of why personality traits vary in their intensity. Despite all the environmental influences that certainly play a major role in the development, this must be due to genetic roots.

Are you still reading eagerly? Because now it's really exciting!

A recent study by an international research team from the UK, USA, and Italy (Riccelli et al. 2017) has looked at interesting differences in cortical brain anatomy—the outer layer of the brain. They focused on measuring three things: the surface area of the cortex, the thickness, and the amount of folding in the cortex. They compared these measurements with the levels of the Big Five Personality Traits. This is based on the theory of "Cortical Stretching". This theory suggests that the brain evolved faster than the skull and needed

more space. This space was needed for higher brain functions such as self-awareness and certain personality traits. The extra space could only be created by the brain becoming more folded and thinner in areas where more processing power was required.

The scientists actually found such relationships. They found that the brain had evolved the most space for the personality trait of "openness" in terms of openness to new experiences in the prefrontal cortex, which as I showed earlier, is also important for smell. This area was found to be larger in surface area and more folded with a thinner cortex. Therefore, the evolution had a particular interest in the development of openness, which is associated with the following characteristics: enjoying collecting new experiences as well as being curious, imaginative, and inventive. People with these traits also seek a certain amount of excitement and variety. They also enjoy exploring other cultures without prejudice. In contrast, people who have this trait less developed tend to prefer conventions. They are more one-sided in their interests and rely on what is tried and tested.

In contrast, evolution seems to have less interest in the further development of extraversion. The researchers conclude that the corresponding area is smaller in relation to openness and has a thicker cortex. This could be because extraversion has a number of highly branched networks that extend into the visual cortex at the back of the head. However, previous studies have also found that cortical thickness in the orbitofrontal cortex is correlated with extraversion (Rauch et al. 2005).

Can we now draw conclusions about smell from the observations that evolution favors properties with increasing size of the surface of the area, less thickness and more folding in the cortex? In the form that evolution supports certain types of smell more

4

in humans than others? Based on the theory of "Cortical Stretching" one must at least say that evolution sets certain features in humans that are particularly worthy of promotion. And these include special types of smell, which I would like to describe as smell openness.

It was in the interest of evolution to further develop the smell brain in humans—and this especially outside the original areas. Only in this way could the Maître des Parfums, which is located in the orbitofrontal cortex and thus in the prefrontal cortex coordinating all other senses, develop out of the actual "olfactory brain". With this, evolution certainly also wanted to create an openness to smell in order to be able to merge smells with other senses for an increased and intense experience. Furthermore, it must also have been important for personality development and thus for self-preservation that open smell is supported by favored personality traits. In this case, evolution seems to have been less interested in supporting individual fragrance directions with smell openness than in making them available to all fragrance directions and types of smell.

Smell openness promotes the imagination and curiosity of humans and contributes to approaching foreign cultures without prejudice. Especially the smell of one's own group, the familiar barn smell, was and is not only a means of differentiation in the animal kingdom. Nature apparently wanted to counteract the associated focus on one's own living space and one's own group and open up for other impressions. Because it has always been in the interest of nature that different genetic material is mixed, just to promote the immune system of the offspring. But openness was also important to her, so that new experiences and ideas could be adopted from others. For this, people had to be open to other foods,

but also to other skin smells—they might even find them attractive.

I think you agree with me: Nobody should miss this chapter. The research results on "Cortical Stretching" are simply spectacular. Just like the findings of brain research, i.e. neuroperfumery, for personal well-being. Insights that fascinate every perfume lover!

Summary

In this chapter, we first looked at the expectations of consumers who enter a perfume store for a fragrance consultation. We saw that it is the acquisition of attractiveness and beauty as well as the increase in well-being that play a role. Building on this, findings from neuroperfumery on the relationship between fragrance directions, well-being, brain regions and their networks for neuro-olfactory or rather for aromatherapeutic measures with neuroscents (active fragrances) for the treatment of anxiety and stress as well as for reward and increase in self-assertion were presented.

We saw that when smelling, especially when "smelling good" with perfumes, a more or less conscious transformation takes place. At the end of the chapter, a spectacular theory for smelling was discussed: "Cortical Stretching". It states that the brain evolved faster than the skull and therefore needed space. The only way it could do this was by folding the brain over itself to create more space, and by becoming thinner in areas with corresponding performance. More space was needed for higher brain performance such as self-image and certain personality traits. One of these favoured personality traits is "openness" in the sense of openness to new experiences. If we transfer this observation to smelling, we can conclude that evolution is interested in our "smell openness".

References

Baars B (2003) How brain reveals mind. Neural studies support the fundamental role of conscious experience. J Conscious Stud 10:9–10

Bremner JD et al (2000) Hippocampal volume reduction in major depression. Am J Psychiatr 157(1):115–118

Canbeyli R (2010) Sensorimotor modulation of mood and depression: an integrative review. Behav Brain Res 207(2):249–264

Courtiol E, Wilson DA (2015) The olfactory thalamus: unanswered questions about the role of the mediodorsal thalamic nucleus in olfaction. Front Neural Circuits 9:49

De Young CG et al (2010) Testing predictions from personality neuroscience. Brain structure and the big five. Psychol Sci 21(6):820–828

Rauch SL et al (2005) Orbitofrontal thickness, retention of fear extinction, and extraversion. Cogn Neurosci Neuropsychol 16(17):1909–1912

Riccelli R et al (2017) Surface-based morphometry reveals the neuroanatomical basis of the five-factor model of personality. Soc Cogn Affect Neurosci 2017:671–684

Schneider S (2015) Der Einfluss von trauriger Stimmungsinduktion auf die Riech- und Schmeckschwellen gesunder Frauen. Dissertation der Medizinischen Fakultät der Friedrich-Schiller-Universität Jena

Schoen K (2018) Gegenüberstellung von Essensdüften und Blumendüften im Hinblick auf ihre Verarbeitung im mesolimbischen System – eine fMRT-Studie. Dissertationsschrift der Medizinischen Fakultät Carl Gustav Carus der Technischen Universität Dresden

Steiner JE, Rosenthal-Zifroni A, Edelstein EL (1969) Taste perception in depressive illness. Israel Ann Psychiatr Relat Discipl 7(2):223–232

Insider Knowledge Perfumery

How to Smell Your Way Through the Perfume Jungle, What we Know About the Effects of certain Ingredients, Fragrances and the Psychology of Perfume Users and Their Experience, How Perfumes are Remembered by Consumers, and Much More

Contents

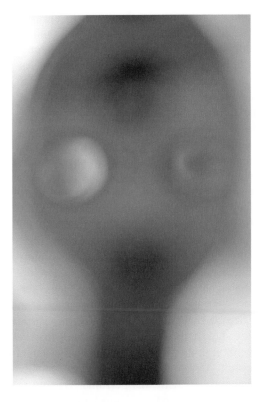

One might think that creating a perfume is basically smelling good ingredients and merging them. That is certainly also true, but perfume is also an industry that is very strategically oriented. With around 2,000 new fragrances every year, it takes a lot of insider knowledge to conquer the market with a perfume or to attract the attention of fragrance users. Often consumers are not aware that, like in the gangster movie Ocean's Eleven, specialists with different talents are behind a new fragrance, who have been hired for a "perfume coup", i.e. for the development and successful market launch of a perfume.

5.1 Odorless … or 10,000 Smells Smell

Certainly not everyone has the same access to smell and does not react equally to certain smells and situations. You can also suffer from anosmia and therefore not perceive any or only very weak smells. In Germany, about 3 to 5% of the population are affected by this. These figures apply if there are no increased infectious diseases. There are indications, for example, that a loss or reduced sense of smell and a loss of taste, for example with Covid-19, were observed more frequently. Therefore, anosmia is a valuable indication for the early detection of a health condition.

The absence of the sense of smell is very stressful because the enjoyment of eating and drinking is restricted. Joy and sensuality also arise through smell. However, the sense of smell can gradually return. Because olfactory cells have the ability of cell division, which is an exception among nerve cells. They renew themselves every 60 days or so.

Many men are convinced that they can smell worse than women, have a restricted sense of smell, or simply often smell wrong. The former is called hyposmia, the latter dysomia, which rather means a distortion of the sense of smell. There are reasons for this. Women can usually smell better, but men often have an olfactory expertise, for example when tasting wine. Actually, one only associates this with the sense of taste, but recent studies have shown that—as already mentioned—there are even olfactory receptors on the tongue. So we also smell with the tongue, even though it is not yet completely clear how far this is possible. If this should be the case, this also applies to men who describe their beer.

The sense of smell can be trained like hardly any other sense. Optimists speak of 5,000 to 10,000 smells that can be distinguished. That also explains why one quickly becomes "speechless" when smelling. The approximately 200 adjectives that e.g. the German language has to describe emotions, basic emotions, and moods are not enough for the olfactory experience.

5

■ **Taxonomy of the Invisible**

Describing scents for oneself and others has certainly been a need since the beginnings of perfumery over 9,000 years ago. But it quickly became a problem because our ancestors lacked the words for it. To this day, it is a challenge to communicate smells in detail. This is shown in the attempts of perfumers to classify scents. For over 200 years, the fragrance industry has been working on a Taxonomy of the Invisible. For example, there is the suggestion to distinguish between 44 fragrance classes, for example in fruity, aromatic, almond-like, minty, citrus-like, sweet, vanilla-like, soap-like, metallic, animal-like or floral.

In order to achieve a quick and easy orientation in the perfume industry, creative noses like to categorize their fragrance creations, i.e. perfumes, according to fragrance families or fragrance directions. Floral, oriental, chypre or fresh-green-citrus notes are common subdivisions. Usually 8 to 16 fragrance directions are distinguished for women's notes and 6 to 12 for men's notes. These will be introduced in more detail and with their olfactory psychological meaning in this chapter.

The assignment of perfumes to fragrance families is based on ingredients and fragrance character. But that doesn't make it any easier, because alone in the fragrance family "Floral" there are over 10,000 perfumes on the market, some of which would have to belong to two or more fragrance families because of their originality. Therefore, the individual fragrance families are further subdivided, e.g. into "floral-powdery", which makes it even more complicated. This is especially true for some fragrance families, which are subdivided into seven and more subgroups.

When assigning perfumes to groups, another difficulty becomes apparent: Scent—and above all perfume—are also always in motion and therefore not easy to classify. The scent develops from the initial impression of about 10 to 15 minutes, the so-called top note, which shows itself after half to one hour in the heart note and finally in the base note, which can sometimes still be smelled for days. Even with perfumes that are created more linearly, with a powerful top note running through to the base, there is movement. By the way, perfumes always become faster. So there is a trend in perfumery that is made for impatient noses. You want the top note to develop faster, have a clear character, then stay that way for as long as possible and, in a transferred sense, become the "earworm".

Perfumes have always been created for the zeitgeist. In the perfume world, there have also been quieter times when more time was given for the top note to develop. The legendary perfume "Parure" by Guerlain is an example. It came out in 1975, in a decade that was less hectic in comparison to today. It took about 20 minutes for the combination of plum, rose and citrus fruits to fully develop in the top note. Patience was required with this perfume. When first applied, it smelled dark and blurry, until finally a gentle wave was released that became a sensual and elegant masterpiece that was then difficult to classify into one fragrance category. In such cases, the fragrance category is determined by the ingredients. For "Parure", most noses agreed on the fragrance category "chypre".

5.2 Perfumery is Not Additive and There Is No Room for Argument

The assignment of perfumes to groups is also difficult because fragrance concentration plays a role. Depending on the concentration, the same fragrance can be experienced differently, although there is an amazing phenomenon: Higher concentration, that is, more fragrance oil mixed with alcohol, does not necessarily mean better, heavier or more intense. Perfumery is not additive. So the higher concentrated perfume can smell lighter than the same fragrance with less concentration. Also, the more concentrated perfume can collapse, or it does not breathe anymore, or certain ingredients suddenly come to the fore. They can change the smell so that it gives the impression that the fragrance now belongs to another fragrance family.

Perfumes are willful. But that's what makes them so interesting. You should not argue about them. The legendary "Cool Water" by Davidoff with only 16 ingredients smelled unfinished to many when it was introduced. It was thought that this "summer fragrance" for men would not survive the first winter on the market. But in the perfume world, too, the saying goes: The dead live longer. No wonder, because there was never agreement in the perfume world. Often, creations are judged like the work of a dentist. "Who made that?" it is then said. Even rather simple topics, such as what concentration is considered an eau de parfum or perfume, can cause arguments. Nevertheless, in the following, I try to create some clarity on this subject—which I have already dealt with in detail in ▶ Chap. 2:

- Fresh-green-citrus notes are often found as so-called **Eau de Toilette** (EDT) in lighter concentration with 6 to 8% fragrance oil (in alcohol with some water), but there are also Eau de Parfum (EDP) versions with around 10 to 14%.
- For **Eau de Cologne,** the concentration is only 3 to 5% fragrance oil (in alcohol with some water).
- The **Body Mist** is particularly light with a concentration of around 3% fragrance oil (in alcohol with some water).

The fresh-green-citrus fragrance, especially in lighter concentration, develops quickly but does not last long if the base is not reinforced, for example, by woods.

In contrast, there are the richer and most popular fragrance concentrations in Germany:

- **Eau de Parfum** (EDP) in a concentration with 10 to 14% fragrance oil (in alcohol with some water) and
- **Parfum** in a concentration with often 15 to 40% fragrance oil (in alcohol with some water).

There are actually no legal regulations for the concentration indications of perfumes such as EDP or EDT, because—as already said—a lot of fragrance oil does not necessarily mean a lot of smell. So don't be surprised if a fragrance with a concentration of 8% fragrance oil is already offered as an Eau de Parfum.

5.3 Perfume Jungle: How to Smell Your Way In

- **Classification systems, fragrance families, or fragrance directions**

There are now apps for mobile devices that provide an overview of fragrance families and are therefore good for systematically getting to know the respective perfumes in a perfume shop. One example is the app "Symrise—Genealogy of Fine Fragrances", which provides an overview of the most popular perfumes—both from today and the past four decades. Fragrance lovers can quickly and easily find extensive information by grouping into feminine and masculine perfumes divided into fragrance families with subgroups, for example under ▶ https://www.symrise.com/newsroom/article/the-symrise-genealogy-of-fine-fragrances-is-going-digital/.

Very detailed information on fragrance groups and divisions can be found, for example, in the "Fragrance Wheel" of the well-known Australian perfume collector Michael Edwards, as well as in detailed fragrance families under ▶ www.fragrantica.com and ▶ www.fragrantica.de. The English and German-language websites of Fragrantica have become an online encyclopedia of perfumes,

5

a perfume magazine and a community of perfume lovers. For years, Fragrantica has offered so-called user-oriented classification systems ("user-driven classification systems"), in which consumers can also rate perfumes based on their olfactory experience. The Fragrantica website, founded in 2007, captures over 80,000 perfumes (2023). The English-language system of Fragrantica distinguishes seven "Olfactory Groups" (Aromatic, Chypre, Citrus, Floral, Leather, Amber, and Woody), i.e. fragrance families, which are divided into up to seven subgroups. In addition, the year of market launch is given for the perfume examples. This information is also given by other classification systems, such as that of Symrise.

At Fragrantica you can classify a little more in detail and see which perfumes and when they came to which market; to which categories (designers, colors) the creations belong; which brands or personalities are behind them; whether the scents are created for women, men, or both; for which season and which occasion they are recommended by other fragrance lovers. The user of the Fragrantica website can call up the bottle and packaging as well as a fragrance description as a visualized "fragrance pyramid" with pictures of ingredients of the top, middle, and base note by clicking on individual perfume names several times. The entire system is open to consumers. They can rate the perfumes and post their own comments. Whether it is always only about company-independent persons, however, is left open.

User-oriented classification systems offer manufacturers and consumers a direct and quick access to the world of scents. However, the danger is that with a very detailed classification with many subgroups you will quickly lose overview.

In my perfume & perfume insider workshops I discuss perfumes and their meaning in different perfume markets, as I will show in ▶ Chap. 14. Here is a first overview of the fragrance directions/groups fragrances:

Fragrances
Female notes:
1. Chypre—leathery
2. Fresh-green-citrus/Aqua- & Ozone notes
3. Gourmand-fruity
4. Floral-powdery
5. Floral-aldehyde
6. Floriental (Amber floral)
7. Amber-oriental
8. Woody-aromatic

Male notes:
1. Fougère
2. Fresh green citrus/Aqua- & Ozone notes
3. Gourmand-fruity
4. Leathery
5. Amber-oriental
6. Woody-spicy

In order to better describe individual perfumes within the fragrance directions/groups in terms of their experience and their effect, a distinction is made between the tendencies "lighter" and "richer" or "intense".

Certainly, the eight fragrance directions for women's notes and the six for men's are quite broadly divided. Only the fragrance direction "Chypre" knows perfumes that, in addition to the clear Chypre notes, one of the following subgroups could be assigned to for a more detailed determination:

Subgroups of "Chypre"
– Chypre-fruity
– Chypre-amber
– Chypre-floral
– Chypre-woody
– Chypre-leathery
– Chypre-aromatic
– Chypre-Fougère
– Chypre-Gourmand

However, such a detailed classification quickly loses overview—especially if you want to study the importance of a fragrance family in a fragrance market. The fragrance family "Chypre" is the smallest segment in many markets with relatively few new releases each year, and the subgroups would then be sparsely populated for an overview of a fragrance market.

For floral notes with many new releases each year, however, a detailed classification makes sense.

For this fragrance direction, one could further subdivide as follows, creating new fragrance families in the process:

Subgroups of "Floral"
- Floral-powdery
- Floral-aldehydic
- Floral-fresh
- Floral-aquatic
- Floral-Gourmand
- Floral-warm
- Floral-amber or woody/musky

Another subdivision is also useful for the "Amber" (also called "Oriental" or "amber-oriental") fragrance family, which also inspires new fragrance families:

Subgroups of "Amber"
- Amber-woody
- Amber-spicy
- Amber-Gourmand, Amber Floral, Amber Fougere, Amber Vanilla, and even Amber Citrus. Why do we use the term amber or amber-oriental in this book and not oriental? "Oriental" has become an outdated fragrance term on its own. More and more fragrance houses are replacing it with "amber" or adapting the term "ambery" to modrnize the language.

For a detailed analysis of perfumes or a fragrance market, one would also have to further differentiate "Gourmand" from "fruity" in order to, for example, distinguish between the fragrance families "Fruity-floral" and "Fruity-Gourmand".

Similarly, one would also have to further differentiate citrus and divide it into fragrance families such as "Citrus-aromatic" and "Citrus-Gourmand".

- **The three-level model of perfume diagnostics**

In the following, I show a fragrance classification system scent space with 16 fragrance families or directions before we discuss the most important ones in detail below:

A fragrance classification system with 16 fragrance families
1. Gourmand-fruity
2. Aromatic-fruity
3. Aromatic-fresh
4. Aquatic-fresh
5. Citrus-fresh
6. Aquatic-flowery
7. Floral-fresh
8. Floral-aldehyde-clear
9. Floral-powdery-balsamic
10. Floral-aldehyde-rich
11. Floral-oriental/floral-amber or fl-oriental (floral sensual)
12. Milk/milk mousse -musk
13. Amber-Oriental-warm
14. Woody-earthy
15. Chypre-leathery
16. Chypre-fruity

In this fragrance classification system with 16 fragrance directions, ladies' and men's perfumes can be sorted together. It is multidimensional with three levels, and I would therefore like to introduce it to you as a **Three-Level Model** for perfume diagnostics. To say it right away: Perfumes can only be on one level or have characteristics of two or three levels in this model.

The **base level** (first level) is based on the four basic fragrance directions:
- Gourmand-fruity,
- floral-powdery-balsamic,
- amber-oriental-warmand
- citrus-fresh.

The base level is called that because many current perfumes in our cultural space can be mapped on this level, and also because our olfactory brain knows these four basic fragrance directions and they are associated with specific neuronal stimulation needs of individual brain regions and their networks.

There are perfumes that have characteristics of two fragrance directions on this level, e.g. Gourmand-fruity with citrus-fresh. In general, the fragrance direction "Gourmand-fruity" has gained importance in many fragrance markets (such as the German or the US) in recent years and is now one of the pillars of the perfume industry.

Above that **(second level)** is the fragrance space with the four directions:
- aromatic-fresh,
- floral-fresh,
- floral-oriental/floral-amber or floriental (floral sensual) and
- Chypre-leathery.

On this level, above all, the large fragrance family "flowery" can be further differentiated (fresh vs. warm). Here, too, new fragrance directions are introduced with "aromatic-fresh" and "chypre-leathery", which are particularly typical for individual fragrance markets. However, the first and second level in combination also show fragrance trends, for example with the partly increasingly complex chypre notes, where "chypre-leathery" is combined with "gourmand-fruity".

On the **third level,** perfumes can then be further specified. The eight fragrance directions of the third level are:
- aromatic-fruity,
- aquatic-fresh,
- aquatic-flowery,
- flowery-aldehydic-clear,
- flowery-aldehydic-rich,
- Milk/milk mousse-musk,
- woody-earthy and
- chypre-fruity.

This level is particularly beneficial for flower notes that can only be described on this level or that have characteristics of the first and/or second level due to their complexity. The third level is ideal for the description of fragrance, especially for many niche perfumes that, for example, combine "woody-earthy" with other fragrance characteristics.

This fragrance classification system with a total of 16 fragrance directions make it possible to describe a very complex perfume in its fragrance development or impression in quite a precise way. Often, only two fragrance directions on one level are enough to "map" a new perfume creation, for example, or to position or compare it with existing creations on the market according to its fragrance character. This means that this classification system can be used to particularly well describe perfumes that fall into two or more fragrance families at the same time as so-called "crossovers" and thus capture their fragrance development or fragrance dynamics. It turns out that the third and second level do not necessarily have to reflect the top note of a perfume. Modern perfumery also loves surprises, and thus complex crossovers can be captured in a creation that is, for example, assigned to the basic fragrance direction "Gourmand-fruity", but in which fragrance facets from the "Chypre-leathery" as well as from the "aromatic-fruity" area also play a role after a certain time.

◘ Fig. 5.1 shows the visualization of a self-created crossover perfume, as one of the participants in my perfume insider workshops experienced his scent. The example makes it clear why one describes a complex perfume in a systematic fragrance description, i.e. "perfume diagnostics", bet-

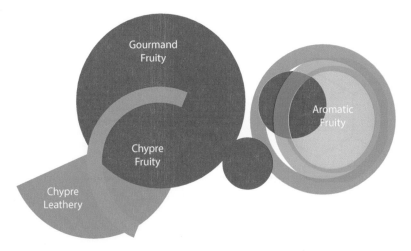

◘ Fig. 5.1 Crossover perfume

ter on different levels in order to recognize its uniqueness.

A good example of a complex crossover is "Virgin Island Water" by Creed. The perfume, which smells like an iced coconut swimming in a rushing stream, has other facets, including a light white rum and a milk note. The perfume has basic characteristics of the fragrance family "Gourmand-fruity" as well as "Aquatic-fresh" with clear citrus chords, to which the mentioned rum and milk note join to the surprise.

◘ Fig. 5.2 shows the visualization of how "Virgin Island Water" can be positioned on the first level. But it is also clear that the scent can only be described in its richness on the third level (which is not shown here) with "aquatic-fresh" and "milk/milk mousse musk".

To be even more precise, one can use the numbering of the directions, that is, the olfactory impression, to additionally describe the temporal course of a perfume. This will also do justice to the division into top, middle, and base notes, as it is traditionally visualized as a perfume pyramid.

■ **Perfume as a sculpture**

Many perfume lovers will agree that the dynamics, novelty, and uniqueness of a fragrance can only be approximately represented by a fragrance pyramid. The claim of fine perfumery has always been to create new olfactory impressions for noses and thus to provide olfactory surprises. If you ask perfume lovers to design a fragrance as a sculpture (e.g. using modeling clay sets with different colors; see also ▶ Sect. 12.4, "Experience Perfumery"), often very unusual different-colored shapes arise. Here are three examples: In the perfume "Angel Share" by Kilian, a gourmand fragrance, it is a cognac note that presents itself as a spherical headnote with cinnamon and melts into the base note with praline, milky vanilla, and sandalwood. With the alcohol-free creamy gourmand fragrance "Sandalsun" by Hermetica, you can describe the perfume in the form of a foamed "latte macchiato" consisting of two to three different layers; the layers of hazelnut and vanilla rotate comfortably over a "wooden" layer. The heart of the gourmand perfume "Lune Féline" by Atelier des Ors, in whose bottle small pieces of 24-karat gold leaf float, smells green and associates a leaf that grows from an imaginary "cinnamon-vanilla-cardamom plant".

Perfumers are also brave and quite innovative with many creations, which is dif-

5

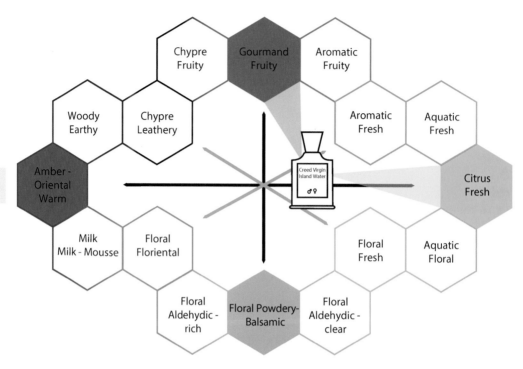

□ **Fig. 5.2** "Virgin Island Water" by Creed in the fragrance classification system with 16 fragrance directions

ficult to express with a fragrance pyramid, but a 3D fragrance sculpture can visualize—especially when familiar fragrance notes are used. Here are three examples: Rose scents are rarely created for men. With "L'Homme À la Rose" by Maison Francis Kurkdjian, Damascus roses and grapefruit create such a bright impression in the top note that you want to modulate it as a green-glowing comet's tail. With the perfume "Body Paint" by Vilhelm, you experience a real hot-cold contrast through cleverly coordinated, but familiar, vegetative notes that offer a lot of inspiration for a perfume sculpture to visualize the energy of "Body Paint". "Colonia Futura" by Acqua di Parma offers a similar large contrast. The lemon, bergamot and grapefruit used, for example, are expected to be a refreshing tonic. But pink pepper also causes real waves of heat to arise, which are wonderfully unexpected in our brains. This creates a double sculpture that winds around itself in different colors. As a per-

fume lover, one would therefore wish that new scents were also presented as a sculpture graphic to make users even more curious about the fragrance experience and visualize the perfumer's idea.

- **Olfactory psychological mapping in marketing**

As you can see, it quickly becomes complex with the 16 fragrance directions described above, and it is easy to lose sight of the fragrance if you do not have the knowledge and nose of a perfumer—especially if you then still distinguish between "lighter" and richer ("intense").

Therefore, marketing uses short descriptions of the individual fragrance directions to position perfumes in a fragrance classification system.

Furthermore, the descriptions of the fragrance directions can be further illustrated by the basic experience directions visualized in Fig. 4.6. This creates a fragrance psychological space with 16 fragrance directions (◘ Fig. 5.3).

◘ Fig. 5.4 shows an example of how individual niche perfumes (e.g. "Tannhäuser" by Drops Barcelona, "Enigma" by Roja, "Tuberose in Blue" by Altaia, "Rose trombone" by L'Orchestre Parfum, "Leather forever" by DE Gabor, "Sublime Balkiss" by The Different Company, "Mimosa Tanneron" by Perris Monte Carlo, "1A-33" by Schwarzlose Berlin, "Voyage Onirique du Papillon" by Salvador Dali, "Sweet Rose" by Rosendo Mateau or "Cardinal" by James Heelay) can be positioned or mapped in this olfactory psychological space. A two-dimensional space is shown, which is based on the three fragrance levels, which would visualize (here only indicated) the differences between the individual perfumes even more clearly.

In marketing, one sometimes wishes for a quick overview of a fragrance classification with only a few basic directions—especially when comparing perfumes from one or a few brands with each other. Here is an example of a fragrance classification with only four fragrance directions, which can also be assigned to women's and men's perfumes. This fragrance classification also allows the representation of crossover fragrances Crossover-Düften, which combine characteristics of two fragrance directions. Furthermore, one can position perfumes rather lightly or richer or typical or less typical for a fragrance direction by positioning typical representatives of a fragrance direction more to the outside (to the end of the scale). An example would be "New York Nights" by Bond No. 9, which has more characteristics of a gourmand note and would therefore be positioned

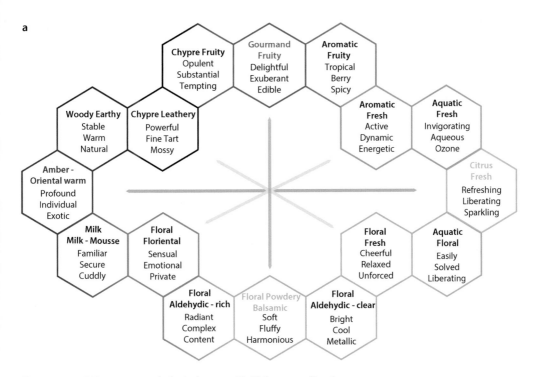

◘ **Fig. 5.3** **a, b** Fragrance psychological space with 16 fragrance directions

5

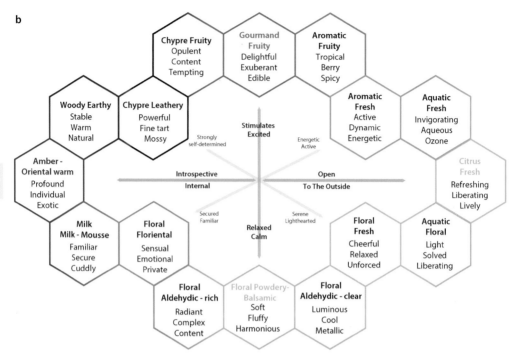

. Fig. 5.3   (continued)

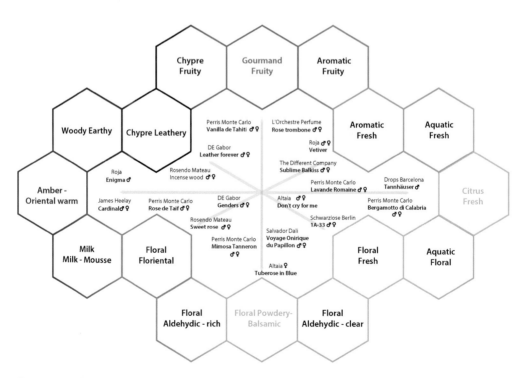

❑ **Fig. 5.4**   Olfactory psychological space with 16 fragrance directions and niche perfume assignments

further outside than "Nolita" from the same perfume house which has also typical notes of the floral-powdery-balsamic fragrance family (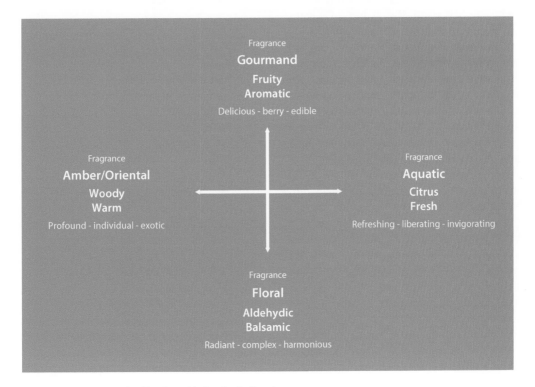 Fig. 5.5).

However, as you can see for yourself, four basic directions are a minimum in order to be able to position perfumes meaningfully. In particular, large fragrance families such as "flowery" are no longer adequately represented. Let us agree on a classification of eight fragrance directions for women and six for men's fragrances for a more detailed discussion of fragrance families.

At my perfume workshops I present the individual fragrance directions with certain women's and men's perfumes. I deliberately show more established, but also trend perfumes. I also concentrate more on individual brands when it comes to examples in order to show their strategy or that of their portfolio within the framework of a fragrance mapping. But before I go into more detail below and name the different fragrance directions with concrete perfume examples, I would like to give you some tips on how to sniff them.

- **Tips for buying perfume—how to sniff the best and get to know fragrance families and their perfumes**

For each fragrance family, you should discover three to four scents for yourself that you can smell in every specialty store. Some perfume stores now offer up to 1000 perfumes, with a trend towards more. About 30% of them are new.

Even if you have a preference for a particular fragrance family, it can happen that you do not like a perfume belonging to it. This is because it has one or more ingredients that are not positively associated with you personally. Perfumes should therefore always be tested first on the scent strip and only then on the skin. It is also recom-

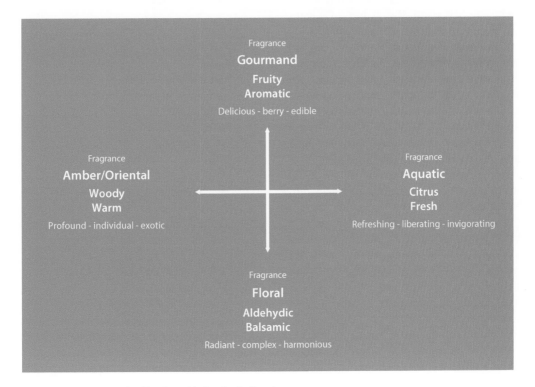

Fig. 5.5 Fragrance classification with four basic directions

5

mended to take a walk with the scent—if time permits. Nothing beats fresh air to get to know scents.

Since you can only smell a limited number of scents one after the other, the rule of thumb is: test a maximum of four to six scents and then ideally take a break of 15 to 20 minutes. Of course, a trained nose can smell more scents in a row than an untrained one. There are also days when you can smell worse due to stress and fatigue. In addition, the length of the break also depends on the olfactory environment. In some perfume stores, the air is literally full of fragrance. Basically, it helps to smell coffee beans in between. Then you can usually distinguish the next scent better after a few minutes. In addition, you should first sniff the perfume before smelling it. This way you can sort out the scents that are not an option in advance.

However, one should not forget: Smelling is also strenuous for the brain, which is why the first scent always has an advantage. Experience shows that you can quickly get over-smelling yourself with four to eight scents—depending on how intensely you smell them—despite the break. At the latest from the sixth smelled scent, the smell brain/olfactory brain starts to be really overwhelmed. After all, it has to run a small film more or less consciously for each scent. The result: you get impatient. As a rule, a feeling of hunger also sets in, and the enjoyment of the scent finally turns into the opposite.

To enjoy the top, heart, and base of a perfume properly, you should not test more than one or two scents per day and per arm on the skin—a total of four perfumes on both arms (in the armpit and on the wrist)—and preferably at the weekend when the head is freer.

If you want to avoid wrong purchases, you should take your time when choosing a perfume. Often a scent develops in a different direction after 20 minutes than the top note

promised. And the base note only shows its true character after two to three hours.

■ **Perfumes behind fragrance families**
Here is an overview of the fragrance families with perfume examples, whereby I discuss the eight fragrance families for women's notes and the six for men's notes in their respective fragrance characteristics in section 5.4 in detail.

Women's fragrances
1. **CHYPRE-LEATHERY**
   Typical examples:
   - Tendency lighter: "Blackout" by Derek Lam, "212 Splash" by Carolina Herrera
   - Tendency more intense: "Bleeker Street" by Bond No. 9, "Royal English Leather" by Creed
2. **FRESH-GREEN-CITRUS/AQUA- & OZONE NOTES**
   Typical examples:
   - Tendency lighter: "Millésime Imperial" by Creed, "Sag Harbor" by Bond No. 9
   - Tendency more intense: "Mint & Tonic" by Atkinsons, "High Line" by Bond No. 9
3. **GOURMAND-FRUITY**
   Typical examples:
   - Tendency lighter: "Something Wild" by Derek Lam, "Virgin Island Water" by Creed (already discussed as more of a crossover scent)
   - Tendency more intense: "New York Nights" by Bond No. 9, "Obscure Oud" by Phuong Dang
4. **FLORAL-POWDERY-BALSAMIC**
   Typical examples:
   - Tendency lighter: "Warm Cotton" by Clean, "Très Chère" by Mizensir
   - Tendency more intense: "Nolita" by Bond No. 9, "Rive Gauche" by Yves Saint Laurent

5. **FLORAL-ALDEHYDE-RADIANT**
Typical examples:
   - Tendency lighter: "Paris" by Yves Saint Laurent, "Vicolo Fiori" by Etro
   - Tendency more intense: "Trésor" by Lancôme, "Stockholm" by Vilhelm
6. **FLORIENTAL (Floral sensual)**
Typical examples:
   - Tendency lighter: "2am Kiss" by Derek Lam, "Narcisse" by Chloé
   - Tendency more intense: "Heliotrope" by Etro, "Oscar de la Renta" by Oscar de la Renta
7. **AMBER-ORIENTAL**
Typical examples:
   - Tendency lighter: "Casmir" by Chopard, "Ambre D'Alexandrie" by Boucheron
   - Tendency more intense: "Untamed Oud" by Phuong Dang, "Oud Save the Queen" by Atkinsons
8. **WOODY-SPICY**
Typical examples:
   - Tendency lighter: "Juniper Sling" by Penhaligon's, "Ideal Oud" by Mizensir
   - Tendency more intense: "Santal de Mysore" by Serge Lutens

Men's fragrances
1. **FOUGÈRE**
Typical examples:
   - Tendency lighter: "Weekend" by Burberry, "Silver Shadow Altitude" by Davidoff
   - Tendency more intense: "Icon" by Alfred Dunhill, "Viking" by Creed
2. **FRESH-GREEN-CITRUS/AQUA- & OZONE NOTES**
Typical examples:
   - Tendency lighter: "212 Men Splash" by Carolina Herrera, "Icon Racing" by Alfred Dunhill

   - Tendency more intense: "Allure by Chanel, "Himalaya" by Creed
3. **GOURMAND-FRUITY**
Typical examples:
   - Tendency lighter: "Remix" by Emporio Armani, "Jump" by Joop!
   - Tendency more intense: "Pirates' Grand Reserve" by Atkinsons, "Fleur du Male" by Jean Paul Gaultier
4. **LEDRIG**
Typical examples:
   - Tendency lighter: "Pure for Men "by Jil Sander, "Aventus" by Creed
   - Tendency more intense: "Knize Ten" by Knize, "Avant-Garde" by Yohji Yamamoto
5. **AMBER-ORIENTAL**
Typical examples:
   - Tendency lighter: "For Him" by Narciso Rodriguez Musc, "Individuel" by Montblanc
   - Tendency more intense: "Le Male" by Jean Paul Gaultier, "Colonia Ambra" by Acqua di Parma
6. **WOODY-SPICY**
Typical examples:
   - Tendency lighter: "Chrome" by Azzaro, "Bois du Portugal" by Creed
   - Tendency more intense: "Davidoff Classic" by Davidoff, "Yohji Homme" by Yohji Yamamoto

5.4 Psychology of Perfume—The Fragrance Families From the Perspective of a Fragrance Psychologist

In the following, I will show what makes fragrance families or fragrance directions special from a psychological, sociological, and marketing perspective. In order to understand the fascination of the individual fragrance directions in their essence, I will describe them ideally. An analysis using ideal types is a common means of theory formation in the social sciences. Ideal types do not necessarily reflect reality, but they are closer to a phenomenon. In this context, I name perfume classics, but also fragrances from the above overview. They express the typical psychology behind the fragrance experience for the respective fragrance direction. I start with the "feminine" fragrance directions.

5.4.1 Feminine fragrance directions

- **Fragrance direction: CHYPRE-LEATHER**

Traditionally, **Chypre notes** are fine-herbal-fresh-aromatic interpreted fragrance notes with woody, partly quite amber-oriental (warm-sensual) echoes in the base. Modern interpretations have many increasingly leathery (often reminiscent of soft unisex leather gloves as well as leather seats of a brand new sports car, but also something of juniper and thus of gin), as well as floral (especially rose or jasmine chords), but more and more often also gourmand chords, e.g. with apricot or forest strawberry, which give the fragrance direction next to fine-herbal also something fruity. They have been spreading independent femininity, verve and self-confidence since 1917, when Francois Coty created a whole fragrance family with his Chypre perfume inspired by men's perfumery, and are considered the fragrance direction of emancipation.

Leather perfumes have been enjoying more popularity for years and one should actually lead them as their own fragrance direction. Olfactory psychology, however, can compare them with classic Chypre notes because of their character. They share with them a certain "I-strength" and an aura of female independence. Meanwhile, leather perfumes are especially offered by niche perfume brands (mostly small fragrance brands with a limited number of sales outlets), which are based on classics such as "Royal English Leather" by Creed (1781), "Tabac Blond" by Caron (1919) or "Cuir de Russie" by Chanel (1924) and offer further developments, such as Rose & Cuir by Frederic Malle (2019).

Today's leather notes come in olfactory facets, which remind of velour over suede to exotic leather and are enriched by fresh-spicy as well as aromatic, woody, smoky, floral, and fruity, but also tobacco notes. Often leather scents still have a hint of aristocracy, thus betraying their origin in the glove perfume of the 16th century, which could only be afforded by the rich as a fragrant fashion item.

From the glove perfume and the perfume of other leather products, their own nuances or directions of fragrance have arisen. For example, the "Spanish leather", which smells leathery-floral or leathery-fruity (for example, "Spanish Leather" by Truefitt & Hill, 1814), as well as the "Russian leather", which rather reminds one of the somewhat leathery-smoky smell of military uniforms in combination with leathery-spicy-woody to untamed nature (such as in Knize Ten by Knize, available since 1925).

In 2020, there were 50 women's fragrances from the "leathery" fragrance family on the market, another 500 unisex fragrances and just under 300 men's fragrances. The "chypre" fragrance family named about 160 fragrances for women, 180 for men, and 270 as unisex fragrances at the same time.

It only makes limited sense to distinguish chypre fragrances from leather fragrances as feminine and masculine, because many of them can smell more feminine or more masculine when smelled on the skin. Both fragrance families became attractive for women in particular with the emancipation movement in the first half of the 20th century.

Women no longer wanted to smell only after the classic feminine floral notes, and one began to orient oneself also on the men's perfume, which then influenced the women's perfume from the early 20th century.

Typical classic chypre fragrance examples are:

- Tendency lighter: "Alliage" by "Estée Lauder"
- Tendency more intense: "Miss Dior" by Dior, "Mitsouko" by Guerlain

Psychologically speaking, the often already fine-herbal Chypre notes for women are successful scents. More or less consciously, their wearer seeks success and is willing to take things concretely and with vigor. Thanks to their self-confident, performance-oriented and competent aura, Chypre notes are popular in job and career interviews. Chypre is particularly trendy with women when they are asked in an organization about economic problems and should radiate cohesion and stability. Men have traditionally never been big fans of classic Chypre notes on women because they then smelled too masculine to them. But the more a woman wants or has to feel and the more she has to lead and implement her own claims, the more she finds Chypre scents attractive, especially in combination with leather chords.

In the past 15 to 20 years, Chypre notes have been appearing on the market again in Europe and America. What is interesting is that the strength and expressiveness of modern Chypre perfumes for women are somewhat slowed down or more hidden. Today's notes have become more sophisticated, subtler, and sexier. So now they also please men. This was achieved through crossovers, a combination of Chypre with Gourmand fruit notes or with velvety leather chords as well as a combination with sensual, sweet-spicy and amber-oriental notes—for example with "Miss Dior Chérie" (EDT) by Christian Dior (2005) as well as with the Eau de Parfum (2011) of the same brand as well as with "Gucci Guilty Absolute pour Femme" by Gucci (2018).

In terms of fragrance psychology or rather: fragrance sociology, the modern Chypre notes reflect the complex success claims of a modern woman in terms of professional and family life as well as her desire for more sensuality and attractiveness.

- **Fragrance direction: FRESH-GREEN-CITRUS/AQUA- & OZONE NOTES**

This refers to two different, but closely related fragrance directions. These lively, invigorating, and refreshing notes have been an integral part of modern perfumery since 1670, when the inedible citrus fruit **bergamot** was introduced from Calabria. The scent was obtained by pressing the peel of the green fruit. This work used to be done by large families, as the peels had to be pressed by hand. Today, this fragrance family has branches in many different directions. Various fruits such as grapefruit,orange,lemon, or tangerine are combined with leaves, grasses, and aqua or ozone notes.

Aqua or ozone notes remind us of sea water and sea breeze and often bring an aromatic component to the scent through plants such as sage,rosemary, or lavender combined with other spicy notes, creating a new fragrance family: **"aromatic"**. This is especially the case with men's, but also with women's perfumes, where the aromatic is combined with fresh, green, citrus, or aqua & ozone notes, such as "Scilly Neroli" by Atkinsons (2016) or "Wall Street" by Bond No. 9 (2004).

As pure aromatic scents, in 2020 there were over 180 perfumes for women available on the market. Over 450 more were listed as unisex, such as "Blue Mediterraneo Foglie di Basilico" by Acqua di Parma (1999) or "Eau Parfumée au Thé Bleu" by Bvlgari (2015).

In the category of fresh, green, citrus scents, in 2020 there were around 250 perfumes for women on the market, and just under 500 unisex scents. If we include the unisex scents of the aqua & ozone notes including the aromatic notes, we get a veritable large fragrance family for women and men.

Typical examples of "fresh, green, citrus" or aqua & ozone scents are:
- Tendency lighter: "Aqua Allegoria Herba Fresca" by Guerlain, "O-Zone" by Sergio Tacchini, "Eau Fraîche" by Bvlgari
- Tendency more richer: "Concentré d'Orange Verte" and "Hermès Eau de Citron Noir" by Hermès

Psychologically speaking, Aqua- & Ozone notes stand for taking a deep breath, having less stress and getting your head free. The fresh-green-citrus notes are appreciated as invigorating. You want to feel fit, sporty, and dynamic. At the same time, you want to refuel with new energy, as you quickly get exhausted. The light perfumes of this double fragrance direction correspond especially to the need for freedom and independence as well as the desire to feel refreshed, alive, active, and open. This more extraverted Unisex fragrance direction mainly attracts somewhat impatient people, which is especially true for men. Carriers of this fragrance direction would rather act than wait. They don't want to unnecessarily lose time and momentum. People who often experience this mood and for whom it is part of their personality take their life more forcefully into their own hands, they would rather be drivers than passengers, love freedom and independence.

This desire for experience is particularly evident in the **aromatic citrus fragrance notes**, which are increasingly establishing themselves as a large, independent fragrance family from the crossover of both fragrance directions and have been particularly trendy since Corona as a pick-me-up in home office. In 2020, 280 perfumes for women, 670 for men and just under 1000 unisex fragrances could already be assigned to this fragrance family, e.g. "Cactus Garden" by Louis Vuitton (2019), "Mugler Cologne Fly Away" by Mugler (2018), "4711 Acqua Colonia Green Tea & Bergamot" (2020) and "Orangerie Venise" by Giorgio Armani (2019) or "Orange De Bahia" by Boucheron (2019) as well as "CK one Summer 2020" by Calvin Klein (2020), which additionally has a salty Aqua note.

- **Fragrance: GOURMAND-FRUITY**

These are also two different, but closely related fragrance directions. Pure gourmand perfumes are newcomers to the perfume industry. They are now offered in a wide range of variants—from amber-oriental-gourmand to gourmand-fruity-light. In many gourmand perfumes, familiar fragrance aromas of classic desserts predominate. Although there have been vanilla fragrances since the 1970s, the "chocolate-underlaid and pineapple-scented" Angel by "Thierry Mugler" from 1992 is considered the first example of gourmand perfumes. The second genre of perfume is characterized as "fruity" and is interpreted primarily in the variants "fruity-flowery", "fruity-culinary" and "fruity-exotic". It is particularly popular in Germany and the USA. However, in classic French perfume, "fruity" is assigned to other fragrance families. But the success of this fragrance, which began in the late 1970s with "Valentino" and has continued since the 1980s with "Liz Claiborne", "Calyx" and "Fruit Cake" by Demeter, justifies classifying "fruity" on its own, but also in connection with "Gourmand" in its own fragrance family.

In total, 1400 perfumes fell into the trend fragrance spectrum "Gourmand-fruity" for women in 2020, with this fragrance direction also overlapping somewhat with the floral or floral-fruity. Some fragrances, as we will see, are already available for men, and over 300 can be classified as unisex.

Typical examples of "Gourmand-fruity" fragrances are:
- Tendency lighter: "Envy Me" by Gucci, "Ralph" by Ralph Lauren, DKNY by Donna Karan
- Tendency more richer: "L'Eau Cheap and Chic" by Moschino, "L'Eau Jolie" by Lolita Lempicka

Psychologically speaking, especially the new fruity fragrances such as Osmanthe Liu Yuan by Le Jardin Retrouve for women and men (2023) stand above all for fun, pleasure, spontaneity, and the more or less conscious desire to break out of routine and everyday life as well as to live out fantasies. You want to enjoy them to put yourself in a good mood, spread joie de vivre, and not have to commit to anything. In this mood you are open to new experiences, sometimes even a little reckless, because the world without joy would seem dull and sad to you. For this you spray yourself above all with the rich, fruity gourmand notes of this fragrance direction. So you want to raise your mood and reduce stress. People who often experience this mood and for whom it is part of their personality are always up for a surprise. They should live where there is a lot going on—as long as there is no boredom. Often the name of the perfume already reveals where the fragrance journey is going, e.g. with "Amor Amor Love Festival" by Cacharel (2020), a fragrance with blackcurrant, Coca Cola, and vanilla. With "Red Cherry" by Castelbajac (2020) red berries, cherry, vanilla, and praline promise a lot of fragrance pleasure. "Clean Classic Summer Day" by Clean (2020) promises to enjoy olfactory coconut water. "La Petite Robe Noire Intense So Frenchy" by Guerlain (2020) focuses on the olfactory seduction of candy floss, raspberries, and rose.

■ Fragrance direction: FLORAL

The floral theme is the heart of perfumery. As mentioned above, the range extends from "floral-fresh-fruity" to "floral-aldehydic" to "floral-powdery-balsamic" and finally to "floral-warm". This makes "Floral" a large fragrance family. In total, there were about 5200 floral perfumes for women on the market in 2020. Just under 40 floral fragrances were addressed to men, and over 1500 can be classified as unisex fragrances.

In order to do justice to the countless compositions, they should at least be divided into two fragrance families for women's notes. This also makes it easier to understand the different floral preferences from a psychological point of view.

■■ Fragrance direction: FLORAL-POWDERY-BALSAMIC

These are timeless, modern, rather warm and soft floral notes. Despite their rich and somewhat narcotic expression, they do not appear too opulent. Classics of this fragrance are the perfume flower children of the late 1960s and early 1970s, because something flower power can be smelled in all of them. The new perfumes of this fragrance are extremely likeable.

Typical examples of fragrances are:
- Tendency lighter: "Climat" by Lancôme, "Skin" by Clean
- Tendency more intense: "Chamade" by Guerlain, "Warm Cotton" by Clean

One might think that flowers are peaceful and stand for friendly affection. Psychologically, however, this fragrance direction is about a hidden or very quiet rebellion. You want to feel good, be yourself, be with yourself, set yourself apart, live your own style individually and not swim in the wake of others. For people for whom this mood is often experienced and for whom it is part of their personality, there is actually the need to do exactly the opposite of what is expected. They oppose constraints and regulations. In the depths of their hearts, they reject a rational, performance-based view of the world. Your own perfume thus becomes a self-protection and buffer between your own world and that of others—like a protective cuddly jacket, in which you snuggle into the soft lining. It is the powdery, typically white musk in this family of fragrances and their perfumes that evokes these feelings above all in the after-smell. In "Ancestry in Paris" by Amway (2018), for example, the fragrance consists of a symphony of tuberose, which merges with amber and white musk. Powderiness that makes the skin smell human-sympathetic for the wearer herself is also richly smelling in "White Musk Flora" by The Body Shop (2019). "Delina" by Parfums de Marly (2017) gives the skin smell with musk, cashmeran and frankincense warmth, well-being and more depth—ideal to hide in yourself. A classic of the floral-powdery originates from the year 1954. The 1950s were the years of the classic women's role and thus the silent revolution, before it became louder again in later decades. The time of the silent revolution includes the launch of "White Rose Natural" by Shiseido (1954). Rose, ylang-ylang, and musk form the shield here to the outside world.

■■ Fragrance direction: FLORAL-ALDEHYDE-RADIANT

It was an old dream of perfumery to make flowers shine even more than in the sunshine. With the discovery of aldehydes this became possible. The history of perfume names the perfumer Pierre Armigeant as the first to use aldehydes in a perfume. "Rêve D'Or"L.T. Piver in the edition of 1905 should have been the beginning of the aldehydes' triumphal march in fragrances. One of the first great perfumes with the radiance of aldehydes was created by Ernest Beaux for Chanel and presented by him as the fifth variant. Whether the high dose of aldehydes in "Chanel No 5" is due to a mixing error, as has been rumoured again and again, is difficult to prove. The fact is: In 2020 there were approximately 350 aldehyde perfumes for women and about 70 unisex fragrances on the market.

Perfumers love the diversity of aldehydes. They come in countless fragrance variants or in the most different shimmering colours, like those in a well-stocked paint box. With them, as in the scent of Chanel No 5, you can almost create an icy fragrance impression that is reminiscent of snow and melting snow, to which, for example, sensuality in contrast is then set with jasmine. But you can also let the fragrance building block melonal (a fragrance ingredient of the perfume oil manufacturer Givaudan) create the association of green melon and cucumber. If fragrance building blocks such as adoxal (Givaudan) or farenal (Symrise) are used in the creation, suddenly floral-marine notes are in the air. With citronellal, a plant-derived fragrance ingredient, it smells of lemongrass with a little rose mixed in, and with triplal, a fragrance ingredient of the perfume oil manufacturer IFF, the association of green grass is created. And then there are aldehydes that smell, for example, of mandarin, cinnamon, anise, wax, or even metal. So it's no wonder that perfumers love the palette of idiosyncratic aldehydes so much.

Typical classic fragrance examples for "flowery-aldehydic" women's notes are:

- Tendency lighter: "Paris" by Yves Saint Laurent, "Super" by Estée Lauder
- Tendency more intense: "Trésor" by Lancôme, "Madame Rochas" by Rochas

Psychologically speaking, this fragrance is about self-esteem and recognition. You want to feel more elegant, exclusive and stylish. It is the desire to surround yourself with small treasures, the details of which others easily overlook. You want to celebrate yourself and the moment.

People who often experience this mood and for whom it is part of their personality are particularly attracted to things that embody timeless beauty and high quality. Everything loud and short-lived is rejected. Accordingly, both in private and in professional life, particular value is placed on success, respect, and what is achieved through one's own efforts.

The house of Chanel has recognized this need for lovers of Chanel No 5. In 2018, for the first time since its launch in 1921, the perfume was filled in a new bottle as "Chanel No 5 Parfum Red Edition". The red bottle came as a limited edition from luxurious Baccarat crystal. Other perfume houses also like to bring out limited perfume editions for the floral-aldehyde fragrance, e.g. "Iris Poudree Limited Edition 2018" by Frederic Malle. As early as 2015, Lancôme came out with its version: "Climat L'Edition Mythique". A collector's item is also the collection "Armani Prive Laque" by Giorgio Armani (2019)—also a perfume of the fragrance "Floral-aldehyde" for women and men.

■ **Fragrance: FLORIENTAL**

A primal theme of perfumery is the seduction of the senses—one's own senses as well as those of others. Seduction knows lightness, tenderness, imagination, and has its own sensual secret. For this, perfumes were created that—as the name says—are located in the border area between "floral" and "amber-oriental".

It is not a small fragrance direction. In 2020, 3400 women's perfumes of the fragrance direction "floral-oriental" could be assigned, in addition there were about 60 men's fragrances and just under 1500 notes that can be called unisex. All of these fragrances are characterized by a rather sweet, sensually warm and powdery base that conveys emotional impressions of smell and acts accordingly on the senses.

Since the seduction of the senses probably only began in the late afternoon in the past, the name of one of the first great and still available perfumes from this fragrance direction was no coincidence: "L'Heure Bleue" by Guerlain from 1912.

Typical examples of "Floriental" for women are:
- Tendency lighter: "Ombre Rose" by Jean-Charles Brosseau, "Narcisse" by Chloé
- Tendency more intense: "Loulou" by Cacharel, "Oscar" by Oscar de la Renta

Psychologically speaking, the Florientals, i.e. the combination of "floral" and "amber-oriental" or the floral-warm perfumes as well as the milk-based or milk-mousse notes mentioned above, which with their often light sweetness are also true soul comfort (I will discuss this in the context of self-therapy with scent), are about sensual indulgence. The perfume should caress soul and senses and radiate a beneficial, generous warmth that exudes inner peace and balance. It is the desire to feel cared for at the same time and to live out unrestrained dreams and romantic fantasies.

People who experience this mood often and for whom it is part of their personality usually have a keen sense for people and situations. This is usually manifested in their love for animals, environmental awareness, and social thinking, as well as in their rejection of a purely sober and rational standpoint.

The olfactory experience that the florientals trigger can be further enhanced and is then paired with a touch of nostalgia and gentle melancholy, with a hint of world-weariness in the air. The secret of many ladies' perfumes of the floriental fragrance family lies mainly in the merging scent impression of three ingredients: vanilla, ambra, and tonka bean—which almost every perfumer has at their disposal as a basic, so to speak, as a ready-made perfume ingredient, in order to use it in their perfumes as a great sensualizer. At least the flattering bases smell of this triumvirate or have it as a model, even if today many other and more ingredients are used for the scent impression. "Casmir" by Chopard

5

(1992) is a good example. The base note consists of a lot of vanilla with tonka bean and ambra, which was further reinforced by the perfumer Michel Almairac with musk, patchouli, and sandalwood for additional warmth and more depth.

I will discuss the scents of vanilla and amber in more detail later on. Here is some background information on the tonka bean. What makes tonka such an attractive scent enhancer is a powdery-sweet, balsamic, warm scent impression. Tonka, the tree with the black fruit of the tonka bean, grows mainly in South America and came to France in 1793, where it was cultivated. The tonka bean is not entirely innocent. It contains the dangerous coumarin, which is harmful to health if taken in large quantities and is now only used synthetically. The processing of the tonka bean is also not entirely without: In the classical processing, the beans lie in rum for 24 hours. This also makes the bean irresistible, especially as an aroma it not only entices gourmets to enthusiastic cries. The combination of vanilla taste and something flowery-spicy rum aroma also results in a super smell composition. No wonder that tonka exudes a pleasantly sweet sensuality with a hint of eroticism as an absolute in perfumes. In the 1980s, the aforementioned "tonkaresierte" "Loulou" by Cacharel came onto the market. The packaging still shows a red flower today, which would like to lure the viewer into a deep tropical forest before smelling it. Something forbidden and dangerous is in the air, which one can hardly resist, just like the forbidden fruit. That is the secret of the Florientals. What begins dreamily sensual develops, like the octagonal bottle of "Loulou", which reminds of Aladdin's lamp, complex, profound and—as already said—not without danger. In the fairy tale, the princess ordered Aladdin to look at her on a whim, but this was strictly forbidden and he only barely escaped.

■ **Scent direction: AMBER-ORIENTAL**

As said above, "oriental" has become an outdated fragrance term on its own. More and more fragrance houses are replacing it with "amber" or adapting the term "ambery" to modernize the language. Still, let's call this scent direction "amber-oriental" since the Orient, the birthplace of modern perfumery, has repeatedly influenced the world of fashion and scent in different waves. In 1910 it was the famous Parisian designer Paul Poiret (1879–1944) who brought the Oriental style back into fashion. He was inspired by the Ballets Russes, an innovative dance group from St. Petersburg, which performed the play Kleopatra in Paris in 1909. The sensually fantastic costumes of the dancers fascinated. Every theater evening was a sold-out event, and the Parisian ladies began to dress "à l'orientale", with turbans, feathers and darkly made-up complexion. Stars of the dance group were Ida Rubinstein and the "god of dance", Vaslav Nijinsky, who enchanted women as well as men long before Josephine Baker took the stage in Paris in 1925.

As is typical for fashion designers, for Paul Poiret fashion was the same as perfume trend. So his perfume "Chez Poiret" was the final crowning of his collection. The perfume itself was influenced by Guerlain's oldest and still available fragrance "Jicky" from 1889, one of the first women's perfumes (originally created for male noses) with synthetic ingredients.

In 2020 there were just under 650 amber-oriental perfumes for women, about 330 for men and over 1800 unisex fragrances.

Typical examples of "amber-oriental" for female notes are:

- Tendency lighter: "Poison Girl" by Christian Dior, "Allure Sensuelle Parfum" by Chanel
- Tendency more intense: "Cinnabar" by Estée Lauder, "Opium" by Yves Saint Laurent

Psychologically speaking, this fragrance is about extravagance, individualism, and inwardness. You are looking for a rich, exotic, and deep world that challenges you to demonstrate a distinctive personal style and to express your artistic potential. People who often experience this mood and for whom it is part of their personality, reflect a lot, deal with things in depth and therefore also often set themselves apart. They need privacy and above all freedom. They actually only want to share their own world with a few good friends.

Typical for amber-oriental perfumes is their deep warmth. Accordingly, fragrances with a warm scent like sandalwood are often combined with other woods, whereby the scent, like the classic "Shalimar" by Guerlain (1925), appears as a great contrast. In "Shalimar" you can smell something citrusy and fresh at first, and the surprise is all the greater when, in the course of the fragrance, it becomes increasingly deep in the direction of warmth and mystery with a scent of incense. But that is precisely the nature of amber-oriental perfumes. They want to surprise and fascinate and therefore play on many chords. From the contrast of "fresh" and "warm" something passionate, unique and therefore strong arises, a "femme fatale". She has always existed and always offered women a second or one of the multiple "I"s of transformation, for example into the demonic seductress, as we know her from mythology, for example Delia, Pandora, Helena, or the sirens. Diminished and more subtle, but with fascinating eye contact and modern female self-confidence, this type of female staging was particularly "en vogue" in the silent film era of the 1920s, a time to which "Shalimar" also belongs and which was inspired by women like Gloria Swanson.

Swanson was an American actress of the silent film era, very self-confident and successful. At the beginning of her film career she earned 13.50 US dollars per week and then negotiated her salary to 22,500 US dollars per week. She also had her very own ideas about love and marriage and was married six times in total.

Swanson was one of the most glamorous women and style icons of the 1920s. Female viewers in particular were fascinated by Swanson's elaborate dresses. With Rudolph Valentino, the epitome of the southern lover, she played in the 1924 silent film drama "Beyond the Rocks". The German title was "You shall not covet your neighbor's wife", which summed up the plot. Swanson is recognized in the drama by the narcissus scent of her perfume. The name "narcissus" comes from the Greek ("narcao") and means "narcotic", which refers to love, even self-love. There are hundreds of types of narcissi. They also play an important olfactory seduction role in various amber-oriental perfumes, such as "Must de Cartier" by Cartier (1981), "Samsara Eau de Parfum" by Guerlain (1989), "Classique Eau de Parfum" by Jean Paul Gaultier (1996) and "Boudoir" by Vivienne Westwood (1998) as well as in "Coco Noir" by Chanel (2012). As the saying goes: You have to love yourself before you can love someone else. Amber-oriental perfumes help with that.

- **Scent: WOODY-SPICY**

"Soft and pleasant scents of exotic woods harmonize with warm, slightly bitter spices like pepper, cloves, nutmeg, or cinnamon". This statement sounds very tempting to many perfume-lovers. They love natural smells and perfumes with few ingredients. The trend originally comes from aromatherapy. But one could also say that "woody-spicy" is a permanent trend in perfumery since its beginning about 9000 years ago.

In modern perfumery, the fragrance "woody-spicy" was originally reserved for men's perfumery—until women discovered it during World War II with the launch of the naturally fresh-woody smelling fragrance "Replique" by Raphael in 1944. It was a commitment to clear reality and

5

thus the opposite of dreamy floral notes. In 1976, the harmonious and uncomplicated fragrance "Jovan Woman" by Jovan came onto the market with deep woody-spicy warmth—fair in price and somewhat sexy and elegant in effect. Because wood was "in" during the 1970s, not only in furniture. Meanwhile, this fragrance has cult status. Women today wear men's fragrances quite naturally, which smell particularly fascinating and extraordinary on female skin.

There were 60 fragrances for women from the "woody-spicy" fragrance family in 2020, about 400 for men and over 560 for unisex.

Typical examples of "woody-spicy" fragrances for women with unisex character are:
- Tendency lighter: "Juniper Sling" by Penhaligon's, "Nutmeg & Ginger" by Jo Malone
- Tendency more intense: "Santal de Mysore" by Serge Lutens, "Sandalwood Absolute Oil" by Clive Christian

Psychologically speaking, this fragrance direction is about cleansing the senses, starting over and focusing. Its fascination lies in a subtle strengthening, behind which more or less consciously the desire stands to find its own power place. You were stressed and feel a little burned out. Those who experience this more often rely on inner strength and stability, want to grow in their own rhythm without many promises. Or to put it another way: "There isn't anything wrong with being uncomplicated."

5.4.2 "Masculine" fragrance families

What characterizes the fragrance families for men's fragrances?

There were over 60,000 perfumes on the market in 2020. They are increasingly overlapping in terms of the division into women's and men's notes. This is also reflected in the fragrance families. Accordingly, the psychological profiles of the women's and men's fragrance markets are becoming more and more similar. However, some fragrance families are still typical for the men's fragrance market, such as the fragrance direction "Fougère". But even in this fragrance direction, more and more perfumes are coming onto the market that very specifically address female fragrance users.

- **Fragrance direction: FOUGÈRE**
These are aromatic, spicy, slightly mossy, and herbaceous scents that traditionally derive from the olfactory impression of ferns, which actually hardly smell. Most of these notes radiate determination and clarity and convey a well-groomed, masculine impression. They are often associated with success, motivation, and stability.

Typical fragrance examples are:
- Tendency lighter: "Weekend" by Burberry, "Silver Shadow Altitude" by Davidoff
- Tendency more intense: "Nightflight" by Joop!, "Polo" by Ralph Lauren

In fragrance development, marketing and sales, it is common to work with target groups that are defined by various socio-demographic characteristics, but also by their self-experience and their experience desires. These are, as already mentioned, ideal types that do not necessarily reflect reality, but come closer to a phenomenon. The phenomenon of fougère notes is that they have been dominating men's perfumery in Germany for years—even if they are occasionally overtaken by the Amber-Oriental fragrance family, for example.

Fougère fragrances are of course available in perfumeries, but often also at other points of sale such as men's fashion stores. One example is the ZARA fragrance "Midsummer Collection Deep Fougère" for men, which has been available since 2017.

What scents men use is strongly influenced by women. Therefore, the fragrance direction "Fougère" must express a masculinity that—in our case—is considered attractive by women (and not only). So you can also assign certain type descriptions of their carriers to the individual men's fragrance directions. So you can characterize many lovers of Fougère notes as **capable pragmatists**. They like to see themselves and their ideal self-image as responsible, practical, organized, systematic, efficient, and solution-oriented. Psychology would describe this personality trait as high conscientiousness.

In an ideal user of Fougère notes, at least the willingness to succeed and perform is dormant. He likes to approach tasks concretely, correctly, and systematically, without hiding. And he directly tells his environment what he expects from him to achieve a common goal. Men in this group are usually very family-oriented, maintain friendships and also club life. They work without asking for a long time and work reliably, persistently, and disciplined. Career choice and hobbies are mainly in the areas of construction, technology, planning, and administration. His environment would describe such a man as follows: "He is capable, practical, and organized. He likes to make plans that he implements efficiently and systematically, he likes to work quickly, but tends to overload himself. If he has a goal in mind, it is not easy to dissuade him. "

Sensory, this type of man is particularly attracted to quality, functionality, good solution, or handling. It is important to him that something feels solid and compact, durable and resistant, and enriches his own experience in a beneficial way. He especially likes things with double benefit that offer a good price-performance ratio. Overall, this is the ideal type of many men worldwide.

■ **Scent direction: FRESH-GREEN-CITRUS/ AQUA- & OZONE NOTES**

These are stimulating, invigorating notes, mainly from the green-vegetable and citrus fruit areas, whose lively, cheerful, refreshing aura I have already described in the ladies' perfumes. As fresh-green-citrus men's or unisex fragrances, however, they radiate even more dynamism, fitness, and activity.

Typical examples of scents are:
- Tendency lighter: "212 Men Splash" by Carolina Herrera, "Connect for Us" by Esprit
- Tendency more substantial: "Allure Homme Sport Cologne" by Chanel, "Himalaya" by Creed.

The ideal type that goes with it in men's perfumery is the **sporting dynamo,** who overlaps in some personality traits with the capable pragmatist. In his self and ideal self, however, the sporting dynamo is even more goal-oriented, active and risk-taking as well as more interested in outdoor and performance. This is particularly evident in the aura of the aromatic citrus notes mentioned above, which have already established themselves as a separate fragrance family in men's perfumery with around 670 fragrances in 2020.

● **Scent direction: GOURMAND-FRUITY**

These are lively, spontaneous, enjoyable fragrance notes from the "Fruity-Gourmand" spectrum. As an independent fragrance direction, these notes are relatively new on the men's fragrance market and are usually intended. Their translucent fruitiness, for example, is reminiscent of tangerines, but with an edible component such as hints of a dessert. But there have been intense gourmand notes for men for a few years, often in combination with Amber-Oriental wood notes. This includes, for example, the fragrance "Pirates' Grand Reserve" by Atkinsons, which surprises with a rum chord. It is reminiscent of the ring-shaped, French yeast pastry "Baba au Rhum". Strictly speaking, the creation of enjoyable fragrance notes is an ancient theme of perfumery. So the attractiveness of the first perfume creations was already increased with sweet resin and honey notes. Not a few top perfumers of our time have taken this over and like to use something sweet honey tobacco in their creations to increase the attraction for both male and female noses—in other words brains.

Typical fragrance examples are:
- Tendency lighter: "Remix" by Emporio Armani, "Jump" by Joop!
- Tendency more intense: "Pirates' Grand Reserve" by Atkinsons, "Fleur du Male" by Jean Paul Gaultier

The ideal type of man who goes with it in men's perfumery is the **spontaneous multitasker.** In his self and ideal self, he is modern, spontaneous and experimental, and oriented towards trends and new products.

Scent direction: LEATHERY

These are expressive compositions that are predominantly dominated by leather notes with a spicy-ambry base.

Typical examples of scents are:
- Tendency lighter: "Pure for Men" by Jil Sander, "Polo Black" by Ralph Lauren
- Tendency more intense: "Knize Ten" by Knize, "Avant Garde" by Yohji Yamamoto

The ideal type that goes with men's perfume is the **latent freedom fighter.** In his ideal self, he is self-determined, ego-strong and self-confident with a desire for more freedom and independence. You could imagine him on a motorcycle on Route 66 with a black leather jacket.

Scent direction: AMBER-ORIENTAL

These are extravagant, erotic, deep, individualistic notes that are rounded off with warm and sensual noble woods and unfold particularly when worn on the skin.

Typical examples of scents are:
- Tendency lighter: "Musc for Him" by Narciso Rodriguez, "Individuel" by Montblanc

- Tendency more intense: "Le Male" by Jean Paul Gaultier, "Obsession for Men" by Calvin Klein

The ideal type that goes with men's perfume is the **creative individualist.** In his self-image, he sees himself as innovative, unconventional, style-conscious, and artistically oriented.

Scent direction: WOODY-SPICY

This is about cultivated, distinguished, private, classically oriented notes mainly from the woody fragrance spectrum.

Typical fragrance examples are:
- Tendency lighter: "Chrome" by Azzaro, "Bois du Portugal" by Creed
- Tendency more intense: "Davidoff Classic" by Davidoff, "Yohji Homme" by Yohji Yamamoto

Wood is a classic theme of men's perfumery. It is associated with men in a different way than with women who wear this fragrance. The perfumery knows a whole range of woody notes such as sandalwood, rosewood, agarwood or guaiac wood. They give a perfume depth, but also warmth. As a rule, woody notes, if they are not interpreted too spicy and not too fresh, give a men's fragrance a certain calm, harmony, and balance—almost one would like to say: inner sovereignty. An example is the fragrance classic "Habit Rouge" by Guerlain as an eau de toilette for men. In addition to a fresh chord, the fragrance has rosewood in the top note. In the heart note, sandalwood with a little spiciness comes into play, which further harmoniously combines with resins and moss impressions in the base. The fragrance, which is overall warm, light, and a little spicy and has sweetness, is considered elegant and unobtrusive.

The ideal type that one could associate with it is the **sensitive protector -** an elegant, unobtrusive gentleman who is helpful, fair and considerate, but also socially and environmentally conscious.

5.4.3 Not So Easy: Determining Target Groups in Marketing

However, with ideal types you quickly reach your limits in marketing. Insiders of the perfume industry are therefore intensively concerned with the following topics:

- How can marketing determine target groups?
- How can one gain knowledge about fragrance users that helps in the development of a new, fascinating perfume?

In order to be able to plan, develop, and optimize, marketing in the perfume industry—as in any industry—wants to understand users better first. As a rule, there are more questions than answers. Typical questions of perfume marketing not only relate to the reasons for the choice of fragrance—that is, who the users of certain perfumes are and what characteristics they have—but also to how they can be addressed specifically. These insights are particularly interesting for perfume re-development in order to position a fragrance in such a way that a brand can win new target groups. Often, however, one also wants to revive a perfume that is already on the market and thus fascinate existing target groups and win new ones. In both cases, it is not enough to only determine socio-demographic characteristics such as age and other typical characteristics of users. Of course, one knows, for example, that age correlates with certain fragrance preferences. But from this no fragrance concept can be derived that could be implemented in advertising. Marketing can not collect enough knowledge about users for the development of a new fragrance concept. With this knowledge, the perfumers will also be briefed in the creation of a new perfume to fascinate the noses, that is, brains, of potential target groups.

In the past, marketing has traditionally focused on lifestyle preferences when gaining knowledge about fragrance users and then categorized or "clustered" consumers, that is, assigned groups. This method has so far mostly only brought mixed success. So quickly target groups emerged such as, for example, the Sporty, who are said to prefer particularly fresh-green-citrus or Aqua- & Ozone notes as perfumes—an understandable approach for the theory formation and the work with ideal types. The only problem is that most fragrance users—even the so-called Sporty—usually have eight to twelve fragrances at home in their fragrance bar, which belong to different fragrance directions. Of course, you can always ask for the current favorite perfume and thus derive the membership in a target group. Often, however, this assignment turns out to be unstable. Because there are both fragrance users who wear a certain fragrance and thus a fragrance direction over a long period of time exclusively, and more and more fragrance lovers who change their perfume and thus the fragrance direction like music depending on their mood. This makes the determination of target groups in marketing more complex than initially thought and was initially underestimated by many marketing managers who came from other areas.

Clustering by gender has not always proven to be helpful in understanding fragrance choice or in determining target groups. While this does influence—just as age and income—the fragrance choice and is suitable for initial approaches to target group determination, the relationships remain also blurred—solely because women have always worn men's fragrances as well. In addition, it is no longer possible to clearly distinguish what is supposed to smell masculine or feminine. An increasing number of fragrance users now find such assignments outdated.

Clustering by income group does not necessarily bring one any closer to the desired information. Certainly, stationary perfume shops have more financially strong customers with an average age of "45+". However, the statistical evaluation of these criteria is difficult because the customers of traditional family-run perfume shops form a rather large, homogeneous target group. This means that they are about 10% of the population, usually the slightly better-off people of a city in the age group "40+". An analysis of target groups based on gender, age and income is therefore likely to be more worthwhile for larger chains such as Douglas, Sephora or dm, where customers are on average more heterogeneous.

Even in the luxury perfume segment, it is becoming increasingly difficult to cluster by lifestyle and user preferences—not only because many luxury brands make premium perfumes affordable with perfume special sizes of only 10 milliliters, but also because the understanding of luxury has changed for some customers in the perfume shop. Certainly there are customer groups who choose niche perfumes in the upper price segment. But this does not say anything essential about the specific fragrance choice. Maybe there could be a certain connection derived from the fact that in recent years many niche perfumes with the fragrance "woody" have come onto the market. So one could assume that the preference for such premium perfumes from the fragrance "woody" is correlated with a target group that is no longer quite young and has a higher income. However, these results would hardly be suitable to provide marketing with a deeper understanding of target groups.

Lifestyle and user preferences also change. Not only millennials increasingly assess a new perfume according to what it can do for them, how it works for them. With this they expect more or less consciously an active perfume, to which one develops a loyalty, because it offers an experience that one can share with others. This has the effect that fragrance users rather buy this type of perfume than established and often expensive luxury brands, which often play with the themes of elite, wealth, and separation. In particular, the wealth does not correspond to the identity of today's millennials. Maybe it's something you aspire to, but as a fragrance it's not what you need to be, for example, happy, spontaneous, or carefree and to share this feeling of life with others.

The luxury aspect of more and more perfume brands is increasingly being relativized by expanding their distribution. So the same perfumes can be found not only in the noble environment of a stationary perfumery, but also as special sizes in the local drugstore, even in the grocery store. Also, brands cannot do much about legally prescribed non-binding selling prices of their perfumes if authorized dealers with corresponding depot contracts offer their products 20 to 30% or even cheaper in online stores.

5.5 Perfume Makers—The Teams in the Perfume Industry

When consumers sometimes decide in seconds whether they like or dislike a new perfume based on the top note, they are usually not aware that behind every fragrance development is a small armada of experts who usually work on the fragrance and its launch for 8 to 16 months. With the larger brand manufacturers, it is usually a team of the following members:

- **Perfumer**: Usually two to three perfumers are briefed for each fragrance development. There are an estimated 2000 trained perfumers worldwide, of which e.g. about 50 work in Germany. As a rule, they are employees of fragrance suppliers or perfume houses/-manufacturers.
- **Bottle designer:** Mostly independent companies design the bottle with cap in 2-D and 3-D with technical dimensions and provide a "mockup", a life-size prototype for presentation purposes or product samples.

- **Glass and cap manufacturer:** They develop the technical data for the tooling or tool. The focus is on glass, i.e. the bottle production, which is suitable for fully automatic or at least semi-automatic filling of the fragrance into the bottle.
- **Packaging designer:** They develop the outer and inner packaging of a perfume with a printing company. The correct reproduction of color gradients is checked using test prints, as well as the appearance of the packaging under different lighting conditions. This is followed by print approval.
- **R&D**(Research & Development): This is about fragrance stability and development of formulations for additional products of a perfume such as shower gel or body lotion. Formulations for deodorants are usually taken over by independent companies.
- **Evaluation**: This is often a two- to six-person team of experts for specific fragrance markets that advises mainly on the international launch of a perfume. They are responsible for the evaluation and assessment of a fragrance, suggests together with the perfumer how a new fragrance is to be created—in other words, smell—in order to address consumers of different markets and regions, and stimulates fragrance optimizations based on results from market research tests.
- **Market research**: This is about concept and fragrance tests, product and advertising acceptance or optimization, purchase intention, brand image, target group determination as well as celebrity tracking—that is, whether and how a prominent person can support or make a perfume attractive to their fans and other fragrance users.
- **Marketing**: This department is responsible for the concept and business plan as well as for the control of development and production costs (so-called "Cost of Goods"), for positioning, launch, and rollout strategy, briefing of suppliers including fragrance houses that provide the fragrance oils, as well as for the development of marketing materials such as displays, giftsets or samples, coordination and agreement with management and the licensor, for example a celebrity and their team.
- **PR orPR agency:** Development of press material, implementation of press events.
- **Media agency**: Together with marketing, it develops on- and offline communication and advertising strategies such as print ads and plans their placement.
- **Promotion Department/Agency:** Development and coordination of special events and promotion activities with the trade.
- **Sales/Distribution**: These departments are responsible for the presentation of a new fragrance to the trade, for discussions about conditions and bonus agreements. For well-known perfume brands and in larger countries, this is a team within a team. It consists of sales managers and area managers, supported by local travel assistants.
- **Trainer**: They train employees in the trade about the characteristics of a perfume as well as its ingredients and explain the corresponding additional products. This is often also taken over by travel assistants.

In addition, teams from production, store design and decoration, material purchasing, logistics, warehousing and finance, customer service, from the legal department for registration and protection of trademark rights, as well as independent consultants for the various areas mentioned above are needed.

In addition, as is the case with most companies, of course, with such a large number of employees, a complete infrastructure including management, human

5

resources, receptionists, and not least cleaning staff is required.

It is no wonder, then, that the development costs for perfumes of larger brands reach millions—and that only a fraction of the new launches can be kept on the market in the next few years. In order to minimize the risk, the brand manufacturers have developed a number of methods and strategies.

5.6 Fragrance Evaluation—A Creativity Killer? How to Discover Perfumes With High Potential

How would you proceed if you wanted to launch a perfume on the market and your perfumers presented three very interesting fragrances to choose from?

To answer this question, you could of course follow your personal taste and gut feeling or ask friends and colleagues. However, it is better to use a scent evaluation—especially if you want to keep the risk low, a lot of money is at stake and the final perfume should be very well received in different markets and countries. You would then carry out a regular perfume market research survey / consumer market research for your three perfume candidates in order to increase the decision-making certainty for all those involved in the project. For this you need another perfume: a benchmark.

A benchmark is a comparison and reference value. When benchmarking, companies compare their performance with that of their best competitors. A typical comparison is the acceptance of one's own new products with already available products that sell well.

For this reason, benchmarks are particularly interesting scents in perfume market research. When smelled in a blind test, they achieve a very high acceptance with consumers—why, one actually does not always know exactly. Possibly because they smell familiar and known in a blind test, are already popular, evoke learned associations and/or correspond to the current taste. If you want to bring a new perfume to the market, it is often compared with a benchmark from a certain fragrance direction. Large perfume manufacturers have secret lists of the current benchmarks against which their perfumers have to compete with new creations.

Perfume consumer market research is a science in itself and requires a lot of finesse. Because you always have to be careful that the comparison of scent acceptances does not restrict the creativity and individuality of a new perfume. On the other hand, for large perfume brands, international perfume introductions are investments of several million euros. In order to obtain the most meaningful results when testing a scent, various methods are constantly being developed.

A popular method of scent evaluation is the Top-2-Box analysis. It does not aim at the average opinion of all test persons on a scent. Rather, it wants to find out how many people rate the scent in a blind test as "I absolutely love it" or "I like it very much" or "I love it very much" or "I like it very much" For example, on a seven-point scale as the two highest possible positive statements. For this purpose, the perfume is smelled on scent strips and on the skin for a certain period of time. As a rule, four, sometimes even more than six evaluations take place: the first impression when spraying and after 5 to 10 minutes on scent strips, then the same on the skin and then the evaluation of the heart note after 20 minutes on the skin. This is followed by the evaluation of the base note on the skin after about 60 minutes. The next day, the memory of the perfume is evaluated during a telephone conversation. This time-consuming process is usually only carried out if three or four perfume candidates were neck and neck in a previous screening.

When top-2-box scores of about 12% of the test persons are given for a fragrance, which were pre-selected according to certain criteria as perfume users (target group), and one ideally still beats the benchmark in a blind test, one is already very satisfied with the tested perfume. So it doesn't matter if other consumers reject the same fragrance at the same time. You are particularly satisfied with a test candidate when the fragrance already shows up very well on the scent strip, but then gets better and better for the test persons in development. To discover this, you need a fragrance rating scale that measures something finer. Therefore, a nine-point scale is often used in fragrance evaluations in the USA.

There are many other techniques of fragrance evaluation, with each responsible person in market research having his own proven methods. In addition to fragrance rating on a scale, I also like to use so-called projective test methods—just because in international fragrance tests, Americans find it easier to say "I love it" than Germans say "Ich liebe es" or "I like it very much".

Projective test methods are usually more speechless, or less dependent on the language skills of the test persons. Fragrances are for example currently assigned to momentary favorite colors and corresponding mood pictures, which in turn say something about the test participants' wishes for experience.

Personally, I also appreciate the digital conjoint analysis. Here, test persons can work creatively online with their new favorite fragrance, which they have smelled blind and then selected. They are offered to choose a bottle for the perfume and determine its color. The same applies to the packaging including design and name. Often it is less about the artistic performance achieved, but about the degree of passion with which a test person approaches his new fragrance.

The market research on fragrance choice will be enriched even more as soon as emotion-recognizing computers and techniques can better recognize and evaluate emotions off- and online. For this there are already promising approaches such as the recognition of emotions and moods via the type of mouse movement, the eye movement, the facial expression or the speech melody as well as the determination of speech speeds and pitch and their deviations. For some time now, attempts have been made to infer the state of mind from physiological test methods, for example, which measure brain waves.

The most promising are currently the imaging methods such as magnetic resonance imaging, which is leading to new insights in neuroperfumery. All this will open up new possibilities for fragrance market research. She is, like the whole perfume industry, on the eve of a revolution.

But is the fragrance evaluation a creativity killer? And does fragrance market research work at all? For premium perfumery, these are really legitimate questions. Despite all market research methods, many perfumes fail with the users. The flop rate has not decreased significantly, especially with new introductions. For most premium perfume launches, a variety of fragrances are usually screened, but in-depth fragrance evaluations are often only carried out to a limited extent due to time, cost, and other reasons. Market research is also suspected of being a creativity killer or at least limiting creativity in relation to innovative perfumery. This often leads to curious situations. So perfumers often do not pass on the fragrance they believe to be the most creative to the evaluation, but the one from which they expect the best test result. Therefore, many perfumers often have true fragrance treasures in their drawers as formulations that they only release or mix for someone when they know that their fragrance is being used for a project as it is.

In the past, testing was only done on a small scale—if at all. Consumer tests were even considered to be creativity killers. I

5

still remember my early days at the company Coty/Lancaster. Officially, fragrances from Davidoff,Chopard,Jil Sander and Joop! were never allowed to be tested using market research methods. Nevertheless, it was secretly fully supported to evaluate and optimize the acceptance of the respective fragrance proposals in different markets. So I had a "backstage" huge influence. Of course, it was also my top priority not to restrict perfume creativity.

So I had to mediate between the participants, but also between the fragrance preferences of the markets, with a lot of sensitivity. This was not always easy because, for example, fresh does not mean fresh. So I found four to five different types of freshness in the markets where a fragrance was to be launched. Only depicting the understanding of freshness in Germany in a perfume would have meant losing potential Spanish fragrance users.

The whole thing was further complicated by the fact that different types of freshness in different markets are also associated with different associations of colors, moods, and desired experiences. That was and is of particular interest to advertising agencies for perfume advertising.

In addition to the evaluation methods explained above, a more depth-psychologically oriented market research and product development method has developed over the years—that is, a psychological method for perfume consulting in the perfume industry, which above all offers insights into the experience desires of customers such as consumer groups and the associated aesthetic preferences such as fragrance, color, and shape. This method is known as the "Color Rosette Test" and the "Moodform Test". I will come back to this method, which also reveals a lot about individual fragrance choice as a psychological fragrance test, later on and will ask you to do a psychological self-test so that you can check the individual relationships for yourself.

5.7 Perfume Flankers: Entertainment for Impatient Noses

Every year, about 2000 perfumes come onto the market in the German-speaking world. Of these, about 3 to 5% manage to establish themselves over the years. This is not necessarily due to market research, but to various other reasons—such as the brand itself or the launch strategy. Often, too long (distance between two fragrance launches), but mostly too closely timed new fragrance introductions negatively affect the interest of consumers. This is especially true for closely timed so-called flankers. For a new fragrance, the bottle design of a fragrance already on the market is used. This only differs from its predecessor in part in color, decoration, packaging, name, and fragrance—sometimes only by other fragrance nuances and a new name.

Many fragrance brands don't even plan for all new launches to last long on the market. Flankers are typical of this. They are designed as part of a brand strategy to offer a limited, temporary olfactory enjoyment to certain target groups with a limited development budget and accordingly small market research budget. In this way, flankers can successfully revive perfume brands. These include summer editions of big perfume classics. The best example of this is the fragrance classic "Angel" by the perfume house Mugler with its limited summer fragrance "Angel Eau Croisière" inspired by summer cruises and fruity cocktail aromas for the 2019 season. Often, the nose-entertaining editions surprise with their will to live and even survive further seasons.

Flankers also offer the opportunity to adapt classics to the zeitgeist, changed fragrance preferences and new uses, such as a lighter perfume. In doing so, one hopes and plans that they will establish themselves as a complementary, permanent part of the fragrance range. Mugler also offers a good example of this. In summer 2019, a new "Angel Eau de Toilette" was also presented, replacing the 2011 version. This is a younger, fresher, and fruitier interpretation of the legendary Angel perfume from 1992, which made a name for itself worldwide, especially in the concentration as an Eau de Parfum.

The image of a flanker is associated with the risk that the new perfume will be deprived of personality and uniqueness. Younger consumers usually don't mind this—and others less and less. Because they also expect their favorite fragrance brands to come up with something new on a regular basis. For this reason, brand managers will continue to set flankers—and certainly even intensify them—for purely economic reasons. The production costs for flankers are usually considerably lower than for a new fragrance. You don't have to develop a new bottle, the five-digit euro costs for design and glass production tools are eliminated, the existing bottle is usually "only" decorated with a new name and graphic for a flanker, and the new perfume is differently colored. So the financial loss of a flop can be more easily accepted if it should occur.

Flankers that do not damage the classic in terms of sales, but rather draw attention to it again, are ideal for keeping a brand in the consumer's mind and attracting new customers. This is especially true for younger target groups. For example, you want to bring them closer to a classic that they only know from the big sister or the mother.

Regular, relatively timely perfume launches have therefore become a tried and tested strategy to keep the brand up to date. This is especially true if you have impatient consumers as your target group—but also because you want to re-inspire the fan club of a perfume classic with scent and brand.

The sales team or field service of a brand also love flankers. Often one fears disadvantages for one's own brand if new launches are too far apart. You don't want to lose shelf space or shelf presence with the dealer, i.e. in the stationary perfume shop, and ideally stay decorated on an equal footing for customers. Also, the sales team knows: If it has nothing new to offer the dealer, another brand will do it. Because the dealer himself lives from fragrance novelties, which on average make up 30% of his sales. And flankers are also part of this.

But you have to be careful. Too many launches in quick succession can water down a perfume brand—a downward spiral that takes on a momentum of its own, as big brands like Hugo Boss know. It is difficult to get out of it again because new things are quickly overlooked by the consumer. On the other hand, quick successive launches can also be part of an ongoing success story of scents. " Bond No. 9" has shown this: the same bottle is decorated again and again to match the current scent theme. A bottle design collector's edition is created, from which true fans of the brand do not want to miss any of the many perfumes in their collection.

The launch of a new perfume therefore requires a sense of touch, but also talent and ability to be self-critical. A new perfume must also really offer something new. This is quickly overlooked in the hectic pace of entertainment marketing. 30 years ago it could take three to four years for a big brand to launch a new perfume. Today we often have the impression that the perfume industry has turned into an entertainment industry. Teams working on a new perfume have to become faster and faster. Only a few years ago, 18 months from the start to the delivery of a new perfume were considered ambitious. Today, a development is only allowed to take 12 or even less months. The reason for this is also the short-lived attraction of celebrities, with whom brands conclude licensing agreements to market perfumes under their names.

5.8 Perfume Recipe: Successful Ingredients of Fine Perfumery

Perfumery, better understood as fine perfumery, understood as the creation of perfumes, is the art of combining different fragrant ingredients and combining them into a harmonious whole—sometimes even deliberately to their opposite. The pure enjoyment of fragrance, usually without any additional benefits or functional applications, is in the foreground. However, this does not mean that other areas of perfumery, such as functional perfumery, for example in the scenting of shampoos and cleaning agents, are not creative. On the contrary: Sometimes it is even more difficult to create an attractive scent for something that does not smell good in its original state.

Mostly, in fine perfumery, for Eau de Parfums (EDP) is created. They make up just over 50% of all fine fragrances on the market. Each perfumer has his own handwriting when creating. Typically, you can recognize it by which ingredients he likes to use. They are often used in whole bases, so compositions of different ingredients, as it were, as a perfumery finished product in perfumes. The experience of the respective perfumer, his previous success and what comes across to target groups, but also what other perfumers think, as well as the current trend play a role. Each perfumer hopes that research will provide him with new fragrance building blocks that he can use exclusively for as long as possible. The latter often caused completely new fragrance impressions and great fame for the perfumers throughout the industry. But chance also plays a role in the use of ingredients. Suddenly a fragrance that has arisen through accidental mixing smells surprisingly good.

But perfumers also like to use certain ingredients for another reason. You want to make it as difficult as possible for others to copy your fragrance. Perfume fakes, that is, counterfeits, now also pose a major problem for the perfume industry through Internet trading. The trade in perfume counterfeits is constantly increasing. In Germany alone, it is said to be worth several hundred million. Often the ingredients used are also questionable from a health point of view. So a perfume that looks like one by Christian Dior can contain a very high proportion of methanol. This not only leads to skin irritation, but can also damage eyes and nervous system.

In order to ensure that it is an original, perfumers or fragrance manufacturers also use invisible markers in their perfumes, which are hardly perceptible. With special methods they can then be made visible as a fragrant copyright. More and more often, fragrance brands also equip their perfume packaging with invisible codes that are as invisible as possible in order to protect them from the so-called grey market and thus from distribution through unauthorised sales channels. The code can be used to identify which official dealer supplies or distributes the goods via such a sales channel, for example Amazon.

Newcomers to perfumery always ask which ingredients they should get to know first. Almost every perfumer has their own preferences. So it can quickly add up to 1,000 recommendations. However, findings from fragrance research show which ingredients, fragrance notes or complex structures, so-called fragrance chords, one should become familiar with or smell for oneself. These include fragrance building blocks that are partly responsible for the success of international perfumes (Vasiliauskaite 2019). For these findings, fragrance notes that can be found in thousands of perfumes were examined in a veritable Sisyphean task. A total of 1047 different notes were examined in 10,599 perfumes by fragrance users in terms of their attractiveness. However, no account was taken of factors such as age, gender, origin, etc., which influence the choice of fragrance.

The focus of the study was on fragrance notes of popular women's perfumes such as "Light Blue" (D&G), "J'adore" (Dior), "Euphoria" (Calvin Klein), "Chanel No 5" (Chanel) and "Chloé" (Chloé). What was found at first is no surprise. The ten most used and popular ingredients are no strangers in the perfume industry. In order of decreasing importance, they are: musk, jasmine, bergamot, sandalwood, amber, rose, vanilla, cedar, patchouli, mandarin/orange.

It is certainly an art to distinguish oneself from other scents on the market with these ingredients alone. That is why it is interesting to find out which ingredients can make a particularly big difference in the fragrance character. This is indeed possible with many, but five are used internationally: anise, iris, orchid, bamboo and clove.

It is also interesting how individual ingredients are experienced in combination with others as fragrance chords. Particular favorites include:

- Vanilla,Oak moss (or the aftersmell of oak moss)
- Ylang-Ylang, aldehyde, jasmine
- Amber,Musk, jasmine
- Lily of the valley,Musk, jasmine

But also the combination of jasmine/mint and musk/vetiver/vanilla have a very positive effect on perfume evaluations. In other words, it is worth smelling the listed ingredients alone and in combination to get to know the basic DNA of many perfumes in a first step.

Of course, these results from consumer research would have to be evaluated by chemical fragrance analysis with a gas chromatograph. Because whether a fragrance really contains the indicated ingredients can only be finally clarified by means of a gas chromatogram. It provides information about the ingredients of the used essential oils both in terms of quality and quantity.

Nevertheless, subjective consumer impressions are very informative, and fragrance professionals also read the comments on perfumes with interest, for example on ► fragrantica.com or in blogs.

---

20 ingredients that the Central European fragrance market loves

At my perfume and perfume insider workshops we smell 20 ingredients that form the DNA of many women's and men's perfumes, especially for the German fragrance market:

1. Chocolate
2. Vanilla
3. Milk note (caramelized with/without white Musk)
4. Bergamot (furthermore Grapefruit, Mandarin, Lemon, Orange, Lime)
5. Rose
6. Jasmine
7. Ylang-Ylang
8. Lily of the Valley
9. Aldehyde
10. Ambroxan
11. Oud
12. Lavendel
13. Vetiver
14. Zimtrindenöl
15. Cumarin (Tonkabohne)
16. Patschuli
17. Eichenmoos (the simulated smell of oakmoss
18. Myrrh
19. Heliotrope
20. Cashmeran

5.9 Scent Secrets of Ingredients— Smelling Exercise for a Better Smell

In my workshops we not only smell the 20 ingredients listed in the overview, but we also explore their secrets from differ-

5

ent perspectives. In fact, every ingredient comes with a small secret, and it's fun to smell it from different angles—in other words, smell angles. For this reason, I would like to encourage you with an example, Oud, to do a smelling exercise—which of course you can also do with the other ingredients mentioned above—to get to know their different effects on yourself and others. Oud as an ingredient can be obtained in different qualities from the stationary trade as well as from the Internet.

■ **Oud**

The rare fragrance ingredient derived from an agarwood tree infested with fungi or from its wood was first mentioned in India in the Hindu Vedas, one of the oldest surviving texts in the world. The wood of the tree was referred to as "wood of the gods". The fragrance ingredient derived from it, Oud, is the secret of many exquisite Amber-Oriental perfumes and has been a trend fragrance in our culture since the 1970s. Depending on the quality, the price per litre for the natural oil derived from the infested trees ranges from 50,000 to 80,000 US dollars (2 ml of pure essential oil for getting to know it is already offered on the market for around 60 US dollars). There are special "Agarwood" conferences attended by Buddhist monks, perfumers, biologists, and collectors, which attract lovers of the wood from more than 30 countries. Today, Indonesia, Malaysia, Vietnam, Cambodia, Thailand, Laos and Papua New Guinea are the main production countries of Oud, and Singapore is the main trading centre. Originally, it was probably the ancient Egyptians who were the main customers and who obtained it directly from Arabia, possibly already from India. It is believed that the fragrance ingredient or essential oil was already used more than 3000 years ago, especially in death rituals for a continued existence after death, which the ancient Egyp-

tians believed in and prepared for in this life (Nazis et al. 2018).

Now for the smell test: Try to "smell" Oud from the following six perspectives. I'll start with the description of the fragrance character. Then we'll smell synesthetic associations. This is followed by a description of how the ingredient might work on gender, then what it might be best mixed with, and finally when it might be best worn, e.g. in the evening. Have fun with this way of smelling better!

— *Fragrance character:*
Woody-sweet, with a slightly burnt-caramelised note that can have a light honey tobacco undertone depending on the quality. Attention: Those who smell too quickly, who do not give the fragrance time to develop, can feel a sharp impression at first.

— *Synesthetic association:*
A piece of dark wood that is square and slightly weathered, soaked in the sun, with dark green, violet, and gold Arabic engravings set with small points.

— *How does it affect me (as a woman)?*
Rich, complex and warm. Already when applying, you can feel how the day, so to speak, turns into an exotic night. An adventure is in the air that you have and control yourself. The longer you smell, the more sensual fantasies arise that you want to live out spontaneously and unconventionally.

— *How does it affect my counterpart (man)?*
At first masculine or unisex, it associates strength and independence. But when men smell Oud on female skin, the associations turn around: a mysteriously erotic woman is described who decides what she wants. Oud almost turns into an elixir under body heat. For many it smells latently like "love unwashed", which flutters fantasies.

— *What is the best combination?*
With Rose Absolute, it becomes feminine-elegant and gets even more class.

— *When is the best time to wear it (as a woman)?*

In summer, even better in winter, in the late afternoon or evening. To seduce, because there is a subtle-attractive power in Oud. If a woman wants to show in her application for creative jobs that there is some artistic potential in her and she also has the stuff for "art director", she can also use an Oud perfume in the morning. For example, applied in the hollow of the knee, a sophisticated aura develops over the day through body heat.

— *Typical perfumes with the ingredient Oud:*

"Midnight Oud" by Juliette has a Gun, "Black Rose Oud" by Trish McEvoy, "Armani Privé Oud Royal" by Giorgio Armani, "Oud Save The Queen" by Atkinsons.

■ **Smelling quickly makes you speechless**

If you found it not quite easy to describe the scent in the exercise above, that is completely normal. The problem, why smells are so subjective with perfumes, remain personal impressions of each one and sometimes simply make you speechless, has a special reason: We had and have in our culture and our language area often no or too little interest to develop our own language world for smells. Descriptions of scents can hardly be derived from the term for a single concrete object, such as "lemon". Perfumers rather use a wide range of expressions such as animalic, green or amber-oriental. They stand for a smell that can come from several sources. To be honest, our language of scents, measured by 10,000 smells that could be distinguished, is rather trivial.

In fact, there are cultures that are further along in their language of scents than we are. Like the Maniq, a people of hunters and gatherers in southern Thailand, who describe smells with abstract expressions, which however are not suitable for other sensory impressions (Wnuk and Majid 2014). There are many reasons why there is little interest in developing a language of scents in our cultural area. Certainly, one reason is that already in antiquity male opinion leaders underestimated the importance of smell for our experience and our quality of life. The sense of smell was seen as a lower, less valuable sense. So research was rather left behind. In addition, the fact that perfumes always had something dubious for the great philosophers, who were all male, added to it. The scholars oriented towards Greek philosophy and culture for a long time took over the views of Plato, Aristotle and Co. without modification.

5.10 Perfumers: The Perfume Industry Owes Almost Everything to Women

For a long time the opinion of women, the main target group of perfumes, remained almost unregarded. And that, although the perfume industry and also the greatest successes in its history are due to them. Because it is widely unknown that the perfume industry has been and is still being run by women. So ten female noses have created or co-created 700 global perfume hits in recent years:

— **Ann Gottlieb** has created or co-created 38 very successful perfumes, including "CK one".

— **Calice Becker** created over 100 top perfumes, e.g. "DKNY Delicious Ripe Raspberry".

— **Nathalie Feisthauer** can look back on over 45 of her created perfume hits, including "Versace Blonde".

— **Nathalie Lorson** has created over 166 fragrances, e.g. "Cool Water Deep".

— **Mathilde Laurent** created over 40 perfumes, such as "Aqua Allegoria Herba Fresca".

— **Marie Salamagne:** Almost 50 perfumes bear her signature, e.g. "Amo Ferragamo".

— **Juliette Karagueuzoglou:** Over 35 perfumes, e.g. for Coach or Jimmy Choo, have already been created by her.

— **Honorine Blanc** has created 70 perfumes, e.g. "Black Opium".

— **Christine Nagel** gave over 120 top perfumes her fragrant life, e.g. "Sì" by Giorgio Armani.

— **Sophia Grojsman** has created over 35 world-famous perfumes, e.g. Trésor by Lancôme.

Even a look back at the history of perfume—as I will show with the first recorded perfumer and the Egyptian rulers' obsession with scent—reveals the enormous influence women have had. Without their skilled influence, France would never have become the olfactory superpower it is today. Instead, Italy would have become the sole perfume power in the Western world as early as the 16th century. The shift from Italy to France as the epicenter of the perfume industry was due to a wise marriage from the French perspective. The reason for the change was that the future King Henry II married 14-year-old Catherine de' Medici in 1533. Until then, enjoying scent in France was often in the form of small scented bags, called "coussines", or in shaped clay bottles, known as "Ölselets".

Few could have guessed what would happen next and how incredibly lucrative it would become for the perfume industry: Catherine de' Medici introduced functional perfumery to France, which began with the scenting of gloves. This made her one of the first in Central and Northern Europe to economically invest in this art in Europe— even before the then newly born Elizabeth I., who led England to become a world power, introduced it there some 30 years later. For this, Catherine brought scented gloves from her homeland of Tuscany. The perfume used for this purpose covered the unpleasant smell of tanned leather much better than anything else on the market at that time. To keep this secret, Catherine brought her own Florentine perfumer Renato Bianco or Rene de Florentin, as he was also called, and her glove maker with her to France. Her personal perfumer also had the privilege of being able to open his own shop in Paris—unlike many of his colleagues who still offered their scents from a handcart.

Katherina's glove maker was able to brilliantly conceal the persistent smell of tanning with a new perfume. The business of Rene de Florentin developed rapidly in Paris, even though he and Katharina were not the only ones driving the art of glove perfumery forward. In addition to the Italian, there was also a Spanish way to perfume gloves. Individual cities in Italy, such as Venice, also had their own techniques for leather scenting. Nevertheless, Paris became the center of functional perfumery for gloves. Their ideal perfumery is probably due to Marquis Muzio Frangipani or Don Cesare Frangipani, both descendants of an old Roman family, who developed a perfume based on bitter almonds during their stay in Paris, with which gloves could be perfumed even better.

The story goes on. The perfumed gloves must have smelled so good that pastry chefs were stimulated and created cream desserts with this wonderful almond smell. Thus probably the popular Frangipani cream (frz. "Crème Frangipane") was created, which is still used today as a filling for almond-based baked goods. Normally, the perfume industry is inspired by trends from the flavor industry, e.g. by ice cream with iced almonds or iced salty caramel, which then comes onto the market as a perfume, e.g. "Salt Caramel" by Shay & Blue as a pure olfactory pleasure. In this case, however, it was the other way around.

As a result, in the 16th century it also became possible to perfume other everyday objects with a strong own smell, which were more difficult to take on a scent. In 1656, the Society of Glove and Perfume Masters ("Maître Gantiers et Parfumeurs") was founded in France, and with it the first association for functional perfumery.

» *And? Do you already feel like a perfume insider and aspiring fragrance psychologist after this chapter? Yes! You are on your way, and the following chapter will take you one step further.*

Summary

We first looked at the psychology behind fragrance families or fragrance directions in this chapter. In this context, different fragrance classifications were presented for the positioning of perfumes or for the "fragrance mapping" with 4, 8, and 16 fragrance directions. To give examples of how to map fragrances in perfume marketing, the fragrance classifications were visualized with brand and product examples.

Furthermore, occupational groups were introduced which work on the development and launch of a perfume on the industrial side. Someone who is not an insider of the industry may think that a fragrance success is only due to a perfumer. This is certainly true in some cases. But most of the time, perfumers work in a team, and for example, valuable suggestions from evaluation or marketing are often used to create a new fragrance. In this context, we have discussed the strategy of a specific type of perfume—the flanker. This type of perfume is gaining market importance, and hardly any perfume brand believes that it can do without it.

At the end of the chapter we went into the role of women who have made the perfume what it is today. So ten female noses have created over 700 global perfume hits in recent years.

References

Nazis PS et al (2018) The scent of stress: evidence from the unique fragrance of agarwood. Front Plant Sci 2019 10:840

Vasiliauskaite VE (2019) Social success of perfumes. PLoS ONE 14(7):e0218664

Wnuk E, Majid A (2014) Revisiting the limits of language: the odor lexicon of Maniq. Cognition 131:125–138

Insider Knowledge Perfume Industry and Trade

What Opportunities the Perfume Industry Offers and What You Should Know Before Launching a Perfume

Contents

6

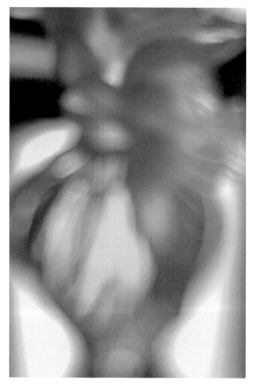

Maybe you've already thought about creating and marketing your own perfume?

Even if you don't have that in mind, this chapter, like the next one, is a treasure trove of insights from practice. Social media platforms today offer innovative ways to market a perfume as a perfume influencer. To promote his scent excitingly, one can learn a lot from the most successful influencers. This chapter will first reveal who the makers in the perfume industry are, what conditions the trade expects for listing a perfume, and how a perfumer can find his niche with his scent.

6.1   The Makers of the Industry

Surprisingly, the job title "perfumer" is not legally protected. Therefore, hundreds of thousands call themselves that—rightfully or not. However, there are probably only around 2000 professionally trained perfumers worldwide who have studied, for example, at the ISIPCA (Institut supérieur international du parfum, de la cosmétique et de l'aromatique alimentaire), a vocational training institution on the campus of the University of Versailles and member of the Université Paris-Seine Higher Education Association, (► https://www.isipca-school. com) or one of the schools or training courses of well-known perfume manufacturers such as Symrise or the Swiss company Givaudan—both of which are among the world's largest producers of flavors and fragrances—have visited. In particular, the Ecole Givaudan in Paris is considered one of the most demanding perfume schools. Every year, only a very few of the hundreds of applicants are selected for the four-year training (► www.givaudan.com/fragrances/ perfumery-school).

■   **The training to become a perfumer**

The beginning of every training to become a perfumer is often compared to learning how to write. In the first months, the aspiring perfumers have to learn about 500 raw materials, like an alphabet, usually 150 natural and 350 synthetic substances. They then know the chemical composition of the raw materials, and the aspiring noses can then describe with their own words and pictures how and what they smell. In the same step, they also learn to mix raw materials, first two, then three, four and more, until they can create an entire recipe, i.e. perfume, that can contain anywhere from 30 to 60, but also like "L'Air du Temps" 120 ingredients. In parallel, they learn which fragrance preferences certain markets have and how classic fragrances, but also new perfumes are parfumistically constructed. Part of the training is also chromatography (a process for the chemical separation of substance mixtures) for the analysis and production of scents with their quality control.

Details about training to become a perfumer are also given by the German Association of Raw Material Manufacturers (DVHR, ► http://duftstoffverband.de). The DVRH is a member of the International Fragrance Association (IFRA, ► http://www.ifraorg.org/), which is a true treasure trove for perfume lovers. The IFRA also offers training positions. In addition, there are other associations that provide information on training, such as Cosmetics Europe—The Personal Care Association (► https://www.cosmeticseurope.eu/) and the International Association of the Soap, Detergent and Maintenance Product Industries (AISE, ► https://www.aise.eu/).

In addition to Symrise and Givaudan, a whole range of other companies (perfume or fragrance manufacturers) offer high-quality training to become a perfumer. Here are just a few: Firmenich, IFF, Mane, Bell Flavors & Fragrances, Robertet Group and Takasago.

■ **Perfume manufacturers**

Professionally trained perfumers mainly work in Paris, Grasse, New York or Geneva for one of the following fragrance suppliers, which are also referred to as a fragrance house or better as fragrance manufacturer or perfume manufacturer, but also as flavor manufacturer. In short, they are the manufacturers of fragrances and food additives.

The "Big 4" from this group together have a almost oligopolistic world market share of no less than 63% (2022). These are the companies **Givaudan (18%), IFF(22%), Firmenich (11%) and Symrise (12%)** (► https://www.statista.com). Together with another five fragrance manufacturers, even close to 75% of the world's perfume production is in their hands. Almost all top perfumers in the world are employed by them. These are internationally operating companies that dominate the world of fragrances mainly from their offices in Paris, Geneva and New York.

This group of perfume manufacturers significantly influences current fragrance trends. In their large research facilities, they are constantly looking for new ingredients, techniques and processes for perfume production. Often, smaller fragrance manufacturers are their customers and buy ingredients from them that they cannot produce themselves for this price and in this quality.

With their market dominance, the "Big 4" often also regulate access to the raw materials needed for perfume production and thus directly or indirectly decide on the price of the ingredients. Entire cultivation areas for certain rose varieties work exclusively for them. The perfume creations based on about 30 ingredients are then offered for sale to fashion and lifestyle brands such as Hugo Boss or fragrance licensees as perfume oil. The price for 1 kg perfume oil for a premium fragrance starts at around 100 euros and can even reach 300 euros and more.

It is usually independent bottling companies that set the perfume oil in alcohol with some water to the desired concentration for fragrance production, fill it into bottles, pack it, and cellophane it. After that, the finished perfume is ready for pick-up in outer cartons in the warehouse.

6.2 How a New Perfume is Calculated in Industry and Trade

Typically, most perfume brands operating in the international market develop a launch and market strategy with different business scenarios for the launch of a new fragrance. These include a so-called "worst-case and break-even plan", but also various calculations of how a perfume could develop over three years in a market.

6

Even if passion for a certain fragrance may be in the foreground at the beginning of perfume development, there comes a moment when calculation and planning for different market situations is necessary. Perfume and perfumery are very strategic businesses, a fact that is often underestimated by outsiders and newcomers to this industry—in particular celebrities who want to bring their own fragrance to the market. On the other hand, without passion for perfumes and for this industry, there can only be accidental successes. But even if someone has been successful once, they will learn that the same strategy is very unlikely to work a second time.

The first phase of perfume development includes the creation of a concept with positioning and a profit and loss plan (P&L), which is based on administrative, marketing, and other costs as well as individual product costs (COGs), i.e. the production costs for the perfume itself.

The production costs include individual parts of the perfume such as the price for fragrance oil, pump, cap, bottle, decoration, and packaging as well as all costs from filling to the finished, repackaged perfume that is ready for pick-up in the warehouse. The correct calculation of the production costs of a perfume, which consists of numerous individual parts and employs different suppliers, is time-consuming, requires experience and also a certain amount of intuition—with the search for suitable suppliers being particularly complicated. They often have to be motivated for a project with initially smaller quantities of planned perfume pieces.

The initial order for the fragrance oil of "Cool Water", the eau de toilette classic, consisted of just 300 kg. Takasago, the fragrance house for which the perfumer and "Cool Water" creator Pierre Bourdon worked at the time, was not even sure whether its customer, the perfume brand Lancaster / Coty, had the potential to market the perfume. Without the intervention

of Pierre Bourdon and the requests of the Lancaster marketing responsible, the fragrance "Cool Water" would probably have gone to another manufacturer.

Other suppliers are also expecting larger orders. For example, bottle, i.e. glass, as well as pump and cap suppliers are happy to calculate with an order in the range of 30,000 to 50,000 pieces and upwards. Motivation is also needed when agreeing on delivery times for orders and reorders, as well as for the punctual delivery of perfume to the filler, often an independent company. The fragrance oil is delivered earlier, for example, because it still has to be diluted for the filling of an eau de parfum. However, negotiation skills are also required because not only do many suppliers require certain minimum quantities, but also larger quantities have to be bought in stock to get a good price. This often results in additional storage costs, which also flow into the calculation.

- **Service partners**

At the beginning of a new perfume development, a brand perfume manufacturer also has to decide whether to afford a new bottle of their own or to fall back on an existing standard bottle. If you decide on your own new bottle, not only design, but also tool costs for glass production will arise. The tool costs can quickly reach a five-digit amount, are charged by the glass manufacturer in addition to the bottle piece price and flow into the production price.

If you want to save time and money, you can choose a standard bottle that can be individually adapted to the new perfume, for example by color and other surface treatments. For this purpose, bottle manufacturers such as Heinz Glas in Bavarian Tettau, Verescence in France, Bormioli Luigi in Italy or Stoelzle Glas in Austria have their own catalogs. These bottle manufacturers, like Coverpla in France, also offer a complete service package, which also includes the procurement of perfume pumps and caps. In many cases, they work closely with packaging manufacturers such as Edelman and contract fillers for perfumes, so that the individual development steps can be efficiently coordinated for the client.

■ Measuring for Development and Trade

One of the best trade fairs in Europe for building your own supplier network for perfume production is the annual Luxe Pack Monaco, which takes place in the autumn. It aims to bring together manufacturers of packaging and packaging materials from all over the world, with a focus on the perfume and cosmetics industry. The fair usually takes place at the same time as the TFWA Cannes, a trade fair specializing in duty-free and travel retail. It attracts retailers from all over the world who are looking for new perfume and cosmetic products for their markets. So what is being developed as a perfume can also be offered for international sale right away. If time is too tight for a perfume brand to present its products to retailers in the same week, the Exsence in Milan and the Pitti Fraganze in Florence are recommended. Both trade fairs mainly attract retailers who are interested in the distribution of new niche perfumes.

■ Manufacturing costs

How do industry and trade calculate? The pure manufacturing costs of a perfume of large brand perfume manufacturers are 8 to 15% of the selling price to the end consumer or are planned accordingly. So if a perfume costs 100 euros in the store, its total manufacturing costs should not exceed 8 to 15 euros depending on the execution. The manufacturing costs for a perfume may not be more, since further costs for marketing, promotion, distribution, and other areas arise.

If you have determined the pure manufacturing costs, you also know in reverse what the minimum selling price of a perfume to the end consumer would have to be, and can calculate the amount of perfume to be sold in order to cover not only further costs but also to generate a profit. However, in order to be able to do this, a brand perfume manufacturer must know how the perfume trade, e.g. the stationary, owner-managed perfume store, calculates and what additional marketing and promotion support is expected.

Most brand perfume manufacturers, such as COTY, are in a middle position between suppliers and trading partners. COTY has its suppliers, such as the perfume houses or perfume manufacturers or glass manufacturers, to develop its products and then delivers its finished perfumes to the perfume trade, i.e. to the perfume stores that advise end consumers and sell fragrances and cosmetic products.

If the purchase prices of the suppliers change or the trade wants more support for the launch of a new perfume, the whole budget calculation can quickly get out of hand for a brand perfume manufacturer. There is certainly a growing trend for brand perfume manufacturers to sell their fragrance products directly to end consumers, for example via their own online shops. Some manufacturers also operate their own stationary perfume stores. An example of this is Estée Lauder with its perfume brand Jo Malone, which is offered in its own boutiques.

■ Distribution partners

In Germany, perfume is usually sold by independent retailers such as Douglas, Müller, dm, Rossmann, or department stores. But the traditional family-run perfume shop still plays a big role in Germany, Austria, and Switzerland, as well as in Italy. So there are different distribution channels through which a perfume can be marketed—provided that a perfume is not only offered through family-run perfume shops or their individual stores/doors. Many family-run perfume shops such as Wiede-mann, Stephan, Cebulla, Becker, Pieper, Al-brecht or Schuback, which together have about 2000 doors in Germany, are mem-bers of buying groups. The largest German group of this kind is the Beauty Alliance (YBPN) in Bielefeld with 1100 doors and 242 members (2020), followed by the Wir-für-Sie group in Mülheim-Kärlich.

Every family-run perfume shop, even if it belongs to a buying group, has its own expectations and experiences for the suc-cessful launch of a new perfume. Even if a buying group prescribes certain condi-tions uniformly for its members, perfume brands, and distributors have to negoti-ate individually with the family-run per-fume shops. This also applies to the Nobi-lis Group in Wiesbaden, Germany's largest perfume wholesaler, which offers perfume brands from different manufacturers. Only large groups like Douglas or Müller have a central purchasing department.

The perfume landscape in this country is somewhat reminiscent of the Holy Roman Empire in the 17th century, then a patch-work of 300 German states. For example, if a trader set out from Cologne to Königs-berg in the 19th century, he had to pass 80 customs stations. The approximately 2000 family-run perfume shops, which are un-fortunately becoming fewer and fewer and are increasingly being bought up by large groups for reasons of age, have contrib-uted a lot to the enrichment of the per-fume world. This includes, for example, the spread of niche perfumes. They also offer small perfume brands a chance to enter the market with their scents. The Bundesver-band Parfümerien e.V., trade association for the retail trade with perfumes and cosmet-ics, supports this additionally.

On the occasion of the annual perfume conference, which usually takes place in spring, a perfume niche trade fair opens its doors in Düsseldorf. The focus is on inno-vative perfume novelties. A group of fam-ily-run perfume shops that have joined forces under the name "first in beauty" of-ten offer a new perfume, from a small brand, a chance to enter the market.

- **Conditions, bonuses, discounts, advertising subsidies ...**

What conditions does an owner-operated perfume shop, an internationally operat-ing perfume chain such as Douglas, and a wholesaler, ie distributor, like the Nobi-lis Group, when a branded perfume man-ufacturer wants to distribute its products through them? An owner-operated per-fume shop and a perfume chain with cus-tomer advice are often offered a 40 per-cent margin for listing the products. In Ger-many, this corresponds to a factor of 1.99 at 19% VAT, which divides the selling price (SP) by the end consumer. So if a premium perfume costs 149.00 euros in the store, the perfume shop buys it from the manu-facturer for 74.87 euros. In order to make this margin attractive for the retailer, the industry often offers a sales-oriented an-nual bonus based on the purchase value (EK) to the retailer. Retailers like Douglas and large, well-known perfume shops then expect up to 25% bonus from the annual turnover without VAT, which the manufac-turer has generated through the retailer. In many cases, especially with smaller perfume shops, a bonus is only paid out from a min-imum turnover.

A typical bonus scale starts at an annual turnover of 7500 euros and is remunerated with 3 or 5% bonus. Some perfume shops like the Beauty Alliance (YBPN) also ex-pect a discount on the invoice value of the order. This can be up to 10% in addition to the usual 3% discount in Germany. In the so-called annual meetings between retail-ers and manufacturers, the entire package of conditions is re-established each time. It plays a role whether a perfume is newly launched or is already in the market. The following conditions are not uncommon:

- *Payment terms:* 14 days 2% discount, 28 days net
- *Invoice discount:* 5 to 10% of the order value
- *Test sample/free samples ("Vial on card"):* 5 to 15% of the order value
- *Decorations:* Twice a year per perfume shop/door with an invoicing rate depending on the size of the shop window between 70 and 120 euros
- *Advertising cost subsidy* (WKZ) for example for mailings or advertisements: 10% of the turnover with the dealer
- *Margin:* 40%
- *Bonus:* 20% on the achieved EK sales
- *Minimum order value:* 400 euros per order

If a perfume manufacturer can afford it—also to position itself more attractively vis-à-vis competitors—it offers the trade a 50 percent margin. Even this is not considered excessively high in comparison to other sectors such as fashion or accessories. That is why stationary perfume shops often also offer fashionable items and hair accessories. Here a margin of 100% and more is often calculated. Including value added tax, a 50 percent margin results in a factor of 2.38 in Germany.

Wholesalers, i.e. distributors, calculate with an even higher margin. Their customers are the individual perfume shops in a region or different countries with which they each make conditions agreements and then take over logistics, shipping, and invoicing for the products of a perfume manufacturer. They carry out various other services for the brand and products of a manufacturer. These include, for example, PR and Influencer Marketing as well as the development of advertising and promotion materials. In addition, wholesalers have their own sales team that can carry out assignments and training in individual perfume shops/doors.

As a rule, a distributor guarantees the manufacturer a minimum annual turnover. He picks up the products directly from the manufacturer or his company, bears the transport costs and keeps them in his own warehouse for resale to perfume shops. For this service, a factor of 5 from the retail price is not unusual.

The perfume trade and its industry act extremely strategically, because the success of a perfume also depends on the decision for the right distribution channel, partner and its conditions. Especially with new perfume introductions, the trade expects additional investments in the first year, a so-called overinvestment. This can quickly lead to a cost dynamics underestimated by new perfume manufacturers. Here is an example of some of the resulting costs:

If the retail price including value-added tax in Germany for a premium perfume is 149 euros, the stationary perfume shop buys it with a 50 margin, i.e. for a cost price of 62.61 euros (factor 2.38), from a brand perfume manufacturer. The perfume brand calculates 10% manufacturing costs itself, i.e. in our example with 14.90 euros from the retail price (149 euros) for the finished, filled and stored perfume ready for collection. The costs of the perfume oil of the fragrance supplier and the costs for the bottle with packaging are included in this price. One-time extra costs such as tools for bottle development are usually also included. In our case, the perfume brand has 47.71 euros (62.61 euros minus 14.90 euros) left per perfume to cover further costs for PR, advertising, administration, marketing and promotion materials such as free samples and testers.

Perfume shops expect a brand new perfume to deliver about 12 to 15% free samples from the cost price. These are delivered free of charge with the first order and then with each order or each order. If the perfume is introduced to the market, the free samples are certainly reduced. Usually about 5% of the cost price are delivered free of charge with each order. A perfume shop also always has the opportunity to buy additional samples and testers from the perfume brand at a special price (usually the cost price plus processing fee). In Germany, perfume shops expect a payment term of 28 days with 3% discount. However, large retailers such as Douglas often demand a payment term of 90 days, which puts pressure on the cash flow of a manufacturer.

Large perfume shops often expect discounts on each order, which are often 10% plus success bonus at the end of the financial year. The bonus demanded by the retailer can be up to 25% of the cost price after deduction of the discount. In other words: Of the cost price of our example of 62.61 euros, a total of 32.5% are deducted for a discount of 10% and a 25% bonus. Together with the 10% manufacturing costs, this makes a total of over 40% additional costs. Advertising subsidies (WKZ) are not yet included, for example for advertisements.

Some shopping communities like the Beauty Alliance conclude marketing agreements with the manufacturer for a perfume launch. They stipulate that advertisements must be placed in their own customer magazines and online portals for an amount oriented towards the sales target. In addition, the manufacturer must pay a basic bonus (usually 2 to 3% of annual sales), often combined with a bonus for the shopping community's service performance.

Window decorations are mainly decided by a perfume shop itself and carried out by independent decorators who invoice the manufacturer for their service. Most perfume shops want to decorate their shops at least twice a year with the brand and of course with a large-scale launch of a new perfume. For decoration materials such as exhibits, posters, i.e. window panels showing the perfume, an additional 40 to 60 euros can be estimated. With, for example, 300 perfume doors in which a perfume is launched, this adds up to a substantial amount. A perfume brand has little left

of the EK price to pay for its own, internal costs, let alone to generate a profit. As a rule of thumb, therefore, it can be said that in the first one to two years after a perfume launch, a black zero cannot be achieved.

Perfume brands calculate according to the above-mentioned costs how many perfumes they have to sell to the trade in order to achieve a black zero in the planning at least. In order to justify the costs for a new perfume, large perfume brands therefore expect a minimum order quantity if a perfume shop wants to carry this new perfume. Typical for an initial order or the first order are six and twelve perfumes for smaller perfume shops and upwards in a size, for example 50 ml for larger sales outlets.

- ■ **Number of sales outlets or doors**

An exclusive premium perfume is carried in Germany in about 500 to 1000 perfume shops. If it is an exclusive niche perfume, the distribution is even below 200 at the beginning, sometimes just 50 sales outlets in the first year. It is therefore almost impossible for a newly launched niche perfume to cover the operating costs of the brand perfume manufacturer in the first years. In my experience, in addition to luck, the right hand, good negotiation and of course an innovative perfume concept with the corresponding scent, a six-figure amount in the middle range is needed to be able to position oneself professionally in the market. Of course, you can build a perfume and cosmetics empire from your backpack like Estée Lauder once did. But for that you need friends, patrons, and benefactors who stand behind a new brand as business angels—and who actually exist in our fascinating industry.

Germany is a particularly interesting market for a new perfume, although not without risk. A total of about 2.2 billion euros are spent on scents every year through all channels—from mass market to premium perfume. At first glance this looks very good, but you have to keep in mind that about 2000 new perfumes fight for their survival every year. Only up to 5% make it to the next year. A large part of the new scents therefore, as *markt intern*, the trade information sheet for the perfume and cosmetics industry, reports, does not even cover the launch costs. These 95% of the new scents are not immediately taken off the market, but no further investment is made in them, they are hardly ever shown and therefore not reordered by the trade. They have lost the war of scents and wander around like perfume zombies without will. The perfume business is subject to a cruel survival of the fittest.

However, this does not mean that those scents that lose the fight are bad—on the contrary, often the opposite is the case. They are unique, but they do not find their way to their noses in the jungle of scents. It is therefore understandable that in the ranks of most perfume brands—such as COTY—there are no in-house perfumers for cost reasons alone and that they cannot afford a department in which perfumes are created from scratch. They usually brief— for example for their brands Jil Sander and Chopard—mostly three of the above-mentioned fragrance suppliers and let them submit perfume suggestions. The marketing departments of Jil Sander and Chopard, to stay with the example, usually ask for some revisions of the proposed fragrance candidates. After all, the final scent must please the licensee of the brand, fit the image, but also the portfolio of existing perfumes. After all, you don't want to create internal competition for the well-selling perfumes of your own brands.

■ **In-House Perfumers**

Of course, there are also brands that consciously afford their own perfumer—sometimes even several. Often these are the stars of the industry who have made a name for themselves with various perfume creations. This includes Thierry Wasser, who was appointed in-house perfumer of Guerlain in 2008. He has made a name for himself with "Hypnosis" by Lancôme and created the fragrance "La Petite Robe Noire" for Guerlain. In 2019, Olivier Polge manages to become the nose at Chanel. He has documented his experience with perfume poetically in the Youtube video "I Am A Nose".

Anyone employed as an in-house perfumer by a major brand such as Chanel,Dior,Guerlain,Hermès, or L'Oreal has won the big prize.

At Chanel the workplace is in Grasse, with regular visits to the Paris headquarters in Paris. The tasks also include managing the approximately 50 existing Chanel perfumes. Every other year, a new perfume or some additional products are added to the collection. Of course, it also enhances the image of the perfumer that the fragrance "Chanel No 5" is sold worldwide every 30 seconds.

In 1910, fashion designer Coco Chanel (1883–1971) founded the Chanel fashion empire. In 1921 she brought the perfume "Chanel No 5" to the market. Today (2023) the company Chanel S.A.S. is owned by brothers Alain and Gérard Wertheimer. They are the grandchildren of Pierre Wertheimer, a former business partner of Coco Chanel. According to the American *Forbes Magazine,* each of the two Wertheimer brothers has a private fortune of 12.7 billion US dollars (as of November 2017). Forbes estimates that Chanel CEO Alain Wertheimer alone now has a net worth of $30.5 billion as of May 2023. This makes them among the richest people in France.

The in-house perfumer of the brand Guerlain has his workplace in Paris. Guerlain is one of the oldest perfume houses in the world. From 1828 to 1994, the company was run by the family of the same name. After that it was sold to the LVMH Group (Moet Hennessy— Louis Vuitton SE), the world's largest luxury conglomerate, owned by the Frenchman Bernard Arnault. In the list of the world's richest people published by the economic journal *Forbes Magazine*, his net worth fortune is given as about 222.5 billion US dollars (2023), which puts him in a top place. The turnover of the corporate empire with around 196,000 employees is around 79 billion US dollars annually (2022). Since the company holds the rights to over 75 different brands, its in-house perfumer is generally also responsible for the scents of Christian Dior,Givenchy,Acqua di Parma,Kenzo and Fendi.

6.3 Areas of Perfumery: Haute Couture (Fine Perfumery), Functional Perfumery, Aromatherapy

The spectrum of perfumery comprises three large application areas. About 21% of all fragrance products sold fall into the category of haute couture/fine perfumery This includes perfumes, eau de parfums and eau de toilettes.

The classic claim of **haute couture** / fine perfumery is the creation of a perfume for the personal enjoyment of a wearer, but also for its positive effect on his or her social environment. One might think that fine perfumery is an olfactory "l'art pour l'art" and can only satisfy itself. But that is not the case. Fine perfumery is strategic perfume creation that must meet the needs of different individuals, self-claims, and market mentalities in order to be successful. Therefore, perfumers must not only be the much-praised noses, but they must also have a high degree of emotional and social intelligence combined with creativity, imagination, and sound knowledge of perfumery, trends, consumers and markets—even for different fragrance markets, if a perfumer wants to work internationally.

Certainly, most perfumes are created to accentuate the personality and individuality of the wearer. But in fine perfumery, wishes and moods that the wearer would like to experience also play a role. Psychologists speak in this context of the ideal self—just as one would like to experience oneself. The

perfumer must also take this into account. This is especially the case with the increasing number of perfumes for large and small celebrities. The fans of celebrities want to feel closer to their stars. In order for this to work olfactorily, perfumers must have a detailed briefing on how the respective celebrity is seen and experienced. So they really have to be able to immerse themselves in another person in order to olfactorily describe them attractively for themselves and the fans. This in turn requires knowledge of what the fans as target group find attractive as perfume themselves. After all, the new celebrity perfume is supposed to be a hit on the market.

Therefore, a perfumer is expected to have a lot of reciprocal empathy, but also strategic work, in order to meet different olfactory requirements. In addition, the perfumer must also take into account fragrance trends and typical fragrance preferences of individual markets in which the perfume is to be launched.

However, the biggest challenge for fine perfumery is to create a fragrance that arrives in different countries at the same time. This is becoming increasingly difficult. Within a few years, the market has changed fundamentally. Between 1999 and 2003, a movement took place that characterizes the fragrance markets today. In 1999 there were still global fragrance preferences. For each country there were four to five fragrances, such as Allure,Trésor, or CK one, which were presented as global perfumes in the top charts across the markets. But between 1999 and 2003, national fragrance individuality increased significantly. Since 2003, there are usually only one or two fragrances—such as Chanel No 5—that remain at the top in European countries for a longer period of time. Today one can say that the large countries can no longer be optimally served without their own specialized fine perfumery.

Within Europe, three very distinct national women's fragrance markets have developed with France, Spain, and Germany.

In Germany, "national" fragrance preferences have been ruling since 2003—at least with regard to new fragrance introductions. I will go into this in more detail in ▶ Chap. 15.

Thefunctional perfumery overlaps in many ways with fine perfumery, but has a larger market share. It includes the scenting of two large areas:

1. **Personal Care** or **beauty care** such as facial care and decorative cosmetics (make-up), hair care (e.g. shampoo), or body care (e.g. deodorant, shower gel), with e.g. shampoo and shower gel being the most used care products not only of German women.
2. **Home Care** or **household care** such as Fabric Care or detergents (e.g. fabric softener) or dishwashing detergent (e.g. dishwashing detergent). Furthermore, e.g. car care products, household cleaners, and room fragrances; even leather care products are listed as a separate subgroup. The industrial associations of the individual countries, such as the IKW in Germany, provide current information on the market importance of individual categories of beauty and household care at retail prices on their websites (▶ IKW).

From a fragrance psychological point of view, functional perfumery is more focused on psychophysical effects. Its main concern is to increase body feeling, self- and product experience, motivation in combination with simple/better/more satisfied application as well as higher acceptance and avoidance of displeasure, for example by masking or covering odors.

The aromatherapy with the smallest market share of the three application areas of perfumery is a sleeping giant. Insiders assume that the global market volume of US\$ 5.9 Billion in 2022 could be around 12 billion US dollars in 2032. In comparison, the total global essential oils market was already valued at US\$ 21.79 billion in 2022 and is projected to reach US\$ 27.5 billion by 2032. In its effect, aromatherapy aims at health, change or increase in well-being as well as therapy and healing of certain diseases and conditions. It is about the fragrance effect on the user, with psychosomatic factors in the foreground.

While research has traditionally dealt with the dimensions of "relaxed vs. active", new olfactory therapeutic goals have been added, as already mentioned. True to the motto "Everything seen begins in a place that was first unseen" (Ralph Waldo Emerson), aromatherapy is being researched as an adjuvant treatment for weight loss, depression, stress, burn-out, inner restlessness, and pain relief. In addition, the diagnosis and treatment of clinical symptoms such as Alzheimer's disease as well as sleep, libido, and concentration disorders are being investigated. Furthermore, aromatherapy is being used to increase and improve resilience, mindfulness, as well as happiness and creativity.

With all of its application goals, aromatherapy has not yet been able to break through in clinical everyday life. One reason for this may be that the topic of aromatherapy, especially the therapeutic approach, is only mentioned in passing in medical training. Although students of medicine and nursing trainees often have a positive attitude towards complementary and alternative medicine (CAM), there is still little information about aromatherapy in their curricula (Pearson et al. 2019). This is also due to the fact that, as we have already discussed, the effect has now been confirmed for many therapy goals, but the effectiveness of aromatherapy compared to classical medical interventions is still met with skepticism by most physicians. This also applies to other forms of therapy such as massage. They like to refer to comparative nursing studies that currently come to the conclusion: "In comparison to usual care … the evidence for the efficacy of massage and aromatherapy in reducing anxiety, pain and improving quality of life was not conclusive", for example Bridget Candy (Candy et al. 2019).

6

Aromatherapists will certainly read this with mixed feelings, because many studies that document the effect of aromatherapeutic treatments do not or not sufficiently orient themselves to scientific standards. And this is demanded in medicine. In scientific studies, test subjects are randomly assigned to so-called experimental and control groups, i.e. the subjects do not know whether they will receive the treatment or a placebo. Another step further are the so-called double-blind studies, in which the test leader also does not know during the study which test subject belongs to which group in order not to unintentionally influence the study result verbally or non-verbally and thus distort it.

Nevertheless, plants such as the real St. John's wort (St. John's Wort) have long been known in medicine for their antidepressant and mood-enhancing effects. It is recommended for use, among other things, as an essential oil for inhalation. The relief of depression is a focus of aromatherapy. Worldwide, depression is one of the largest health problems or the most common mental disorder. According to estimates, 350 million people are affected worldwide. About 25% of women and 12% of men suffer from at least one depressive phase during their lifetime (Haller et al. 2019). If aromatherapy succeeds in establishing itself as a treatment approach, this would also be its breakthrough in clinical everyday life. The chances of the essential oil industry, which is worth billions of dollars, are not bad for mild to moderate depression. This is mainly due to St. John's wort.

The ancient Greeks and Romans used St. John's wort as a remedy, and in the Middle Ages its positive effects on the psyche were already recognized. St. John's wort—the popular name was devil's or witch's herb—helped "against the dizziness and against the terrible melancholic thoughts" (Johann Hieronymus Kniphof—1704-1763—a German doctor and botanist, in Botanica in originali, 18th century). At the end of the 19th century it almost fell into oblivion. In 1930 it reappeared briefly, was even included in some pharmacopoeias, then disappeared again in many new editions until it reappeared again at the end of the 1970s and has been listed, for example, in the German Pharmacopoeia since then.

St. John's wort is now well documented as an antidepressant for mild and moderate depression. A recent comparative study by Heidemarie Haller from the University of Duisburg-Essen comes to the following conclusion: "In patients with mild to moderate depression, moderate evidence suggested that St. John's wort was more effective than placebo and its efficacy compared to standard antidepressants in terms of the severity and response rates of depression, while St. John's wort caused significantly fewer adverse events" (Haller et al. 2019). The study also suggests that the combination with a mindfulness-based cognitive therapy, which I will discuss later, is particularly effective in preventing a relapse of depression. The combination of mindfulness therapy with standard antidepressants seems to be very effective. We can look forward to further studies that combine cognitive therapy approaches with aromatherapy or supportive scent therapy. Certainly, the effect of talk therapy in combination with supportive scent therapy will have to be investigated. Cognitive therapies always run the risk of downplaying justified external conditions of psychological stress, as is often the case with depression. This can lead to external prevailing conditions being quickly legitimized, where actually personal outrage would be more appropriate.

This raises the question for aromatherapy or fragrance-supported therapy which health problem which psychotherapy or meditation is particularly effective in combination. A recent study, for example, finds that the combination of aromatherapy with music therapy is associated with a significant decrease in anxiety and stress (Son et al. 2019).

I have discussed further promising approaches in ▶ Chap. 1 (aromatherapy—how essential oils work). The confirmation of their effect will also decide which role aromatherapy or fragrance-supported therapy will play in psychotherapeutic practice (Zimmermann 2017).

In the past, it was small companies that supplied the essential oils. This has long since changed. Industrial giants like Cargill (USA), DSM (Holland), doTERRA (USA) and Young Living Essential Oils (USA) now dominate the market.

Even large fragrance suppliers like Givaudan (Switzerland) or Mane SA (France) are already playing a significant role in this market and have their own plantations, for example for lavender. With the increasing market importance of aromatherapy, the "Big 4" of fragrance suppliers like Givaudan, but also IFF (the American company has merged with the food division of US giant DuPont in 2019), Firmenich (Switzerland) and Symrise (Germany) will have a say in this application area of perfumery and take appropriate influence on trends. Firmenich and Givaudan are already among the top 5 manufacturers of essential oils. Here are a few examples that are not only used in aromatherapy, but also in fine perfumery, functional perfumery (cleaning and laundry products) as well as in food and beverage production:

- Orange oil
- Lemon oil
- Lime oil
- Peppermint oil
- Eucalyptus oil
- Jasmine oil
- Sage oil
- Tea tree oil
- Rosemary oil
- Lavender oil
- Rose oil

But anise, basil, bergamot, ginger, mint, pine, cardamom, cypress, chamomile, marjoram, mandarin, neroli, oregano, sandalwood, grapefruit, thyme, ylang-ylang, cedar, lemongrass, juniper, wintergreen, and cinnamon are also processed into essential oils.

Actually, one should still go into a neighboring area of the perfume industry, the application area **oral and dental care** and thus on **aromas** such as mint flavors, as they are used there. Because most flavorings or food flavor and additive flavors are also smelled. But that would go beyond the scope of this book. Just so much: The turnover of fragrances and flavorings is almost equally strong with only a slight product dominance of food flavors and additives over fragrances. Worldwide, the total market grows on average (2017 to 2020) by about 5% (to reach USD 37.3 billion by 2026), that of essential oils by about 8%.

Surprisingly, the growth in fragrances and flavors is almost five times greater than the annual increase in the world population. Scents like flavors arouse enormous desire and apparently have something of a drug. This means for the individual: Scent and aroma are becoming increasingly important.

6.4 On the Situation of the Perfume Specialty Trade

Actually, one would have to assume that all sales channels of perfumes to end consumers would benefit from the positive development of the large perfume suppliers. Let's take a closer look at this for Germany here, even though I will go into it in more detail in ▶ Sect. 13.1.

Perfume premium brands such as Dior are mainly offered in the perfume specialty trade, for example at Douglas. This also applies to cosmetics such as care and make-up. Within the large main sales channels for such products at end consumer prices, the perfume specialty trade (Parfümerie-Fachhandel) in Germany had a market share of around 19% in 2022. The Beauty Alliance Group, an association of independent perfume retailers with over 1100 sales outlets, is the market leader in Germany and one of the largest providers. Drugstores such as market leader dm or Rossmann recorded a market share of around 51% in the field of beauty care products (statista 2022). They have mainly specialized in the middle price segment in the lifestyle and celebrity perfume market. Here, small Eau-de-Parfum sizes from 10 to 40 ml are in the foreground. Consumer markets (Verbrauchermärkte) without elaborate store design and product presentation as well as with only minimal advice have a market share of around 7%. Discount stores with their relatively narrow and flat range of goods have a market share of around 8%, E-Commerce (typically without brick and mortar retail—at least one physical location that customers can visit) with 6% and pharmacies with a market share of around 4%, which are increasingly discovering aromatherapy brands and medical skin care for themselves. Department stores with around 5% and the food retail trade with only around 1%.

In the past 10 to 15 years, the perfume trade in Germany has changed more than many people would like. Many smaller, family-run perfume shops have disappeared or been taken over by large perfume chains. The fear of the future is spreading. The drugstore around the corner has long since overtaken many perfume shops in terms of range and is getting closer and closer to the classic perfume range.

But the fear has other reasons as well. Competence in skin care has shifted more and more from perfume specialty stores to dermatologists and plastic surgeons, pharmacies, drugstore chains, and health food stores, or has to be shared with them. Competence in fragrance went to niche perfume shops, perfume chains and above all to Internet providers. The latter pose a special problem for stationary perfume shops.

The customers are advised there, sniff the new perfumes, but then buy online—often even 30% cheaper. The same is true of the price-competitive make-up market, which the stationary perfume shops have to share with newly created make-up-only chains such as Inglot, Kiko or Mac Cosmetics. In addition, there are the other distribution channels already mentioned.

Smaller stationary perfume shops are only visited by 20 customers a day. If the average selling price for a perfume or beauty product is 70 euros, the business costs can no longer be covered. Even the typical perfume shop customers are not getting any younger. The classic perfume shop customer is over 45 years old, female and belongs to the upper middle class.

The situation is similar in other Central European countries. Here too, it is the customer in the "45+" age group who wants to invest in her own well-being. This benefits not only exclusive care brands, but also the fragrance market.

In France, around 7 million liters of fine fragrances are sold every year. Alone a third of this is consumed by the target group "45+". Among the classics of noble fragrances, their share is even much higher. Some of the top 10 prestige brands sell more than 65% of their fragrances to this age group—and this is true for both women and men. Statistics also show that this group perfumes more often, even several times a day. Actually, this should be a reason for joy for stationary perfume retailing. Because with an ever-aging population, the proportion of people over 45 also increases. The question is only whether stationary perfume retailing will succeed in winning the daughters of these women for their place of sale. I will come back to this later.

6.5 You as a Perfumer

If you were working as a perfumer, would you rather work in the so-called. Fine fragrance or in functional fragrance or aromatherapy?

If it's purely about revenue, functional fragrance offers the greatest income. If personal image and name recognition as well as the prestige of the company are important to you, your choice is likely to be the fine fragrance. But if you value independence, I have a completely different idea for you further down.

There are companies where a perfumer can work both in fine fragrance and in functional fragrance. This is the case with an in-house perfumer of the company L'Oréal in the north of Paris. Here, within fine fragrance, we are talking about brands such as Lancôme and within functional fragrance, for example, the fragrance of hair care products. Almost all beauty care products are perfumed. According to IKW 2022, consumers in Germany alone spend just over EUR 14.3 billion on this annually ▶ IKW. In the USA it is over 90 billion dollars (statista 2022)—and rising (after the Corona crisis). In Germany, skin, facial, and hair care products are at the top of the beauty care market. In 2022, consumers spent EUR 6.4 billion on this. In proportion to the market importance of these segments, the large fragrance suppliers have their own departments in which perfumers work on the fragrance of corresponding products. For decorative cosmetics, i.e. make-up, German consumers spent EUR 1.7 billion in 2022. The fragrance, for example in lipsticks, plays a big role here too.

In relation to this, the market share of women's and men's fragrances is relatively small in Germany. In 2022 it was just about EUR 1.7 billion, which is only just over 10%, with two thirds going to women's fragrances. In France, the share of women's and men's fragrances in the total beauty care market is around 20%. However, no further increase in the share of fine fragrance is to be expected in the next few years. It is the functional fragrance that works for the largest segments of beauty care. Of course it is the fine fragrance in which the most expensive perfume oils are used and sold. Functional fragrance, on the other hand, calculates with fragrance oil prices that are sometimes below EUR 10 per kilo. But quantity is what counts.

Let's take another quick look at L'Oréal, the world's leading beauty care manufacturer and provider. With fine fragrance brands such as YSL, Giorgio Armani and functional perfumed products such as hair care, the group has a net worth of well over 240 billion US dollars (2023). The group is still partly owned by the Bettencourt-Meyers family. Françoise Bettencourt-Meyers is the world's richest woman with a net worth of about 81 billion dollars (2023)

Another possibility is to work as an independent perfumer. The classic heart of French perfumery is still the small town of Grasse with its region. More than 60 perfume companies and about 120 professionally trained perfumers are based here, but I am not aware of a single unemployed perfumer.

An alternative to Grasse is the US state of Florida with numerous small fragrance companies that have partly specialized in the functional perfuming of other product categories and also supply to South America and the Caribbean. Perfumers who work from Florida for different markets must have a great deal of knowledge and empathy for different ethnic groups and market mentalities with their "likes" and "dislikes". This gives the opportunity to first internationalize gradually starting from the local market.

For example, there are companies in the Miami area that have specialized in the favorite fragrance memories of exiled Cubans. In 1993, especially many Cubans came to Florida—although not always entirely voluntarily. As a result, the nose was also afflicted with homesickness, and one longed for scents from one's own childhood. "Royal Violets" is such a fragrance

6

that has existed in Cuba since 1927. Reyes, Augustin, great-grandson of the then perfumer, has given this creation back to the exiled Cubans. "Royal Violets" has a special secret: it is almost a spiritual baby fragrance, with which even the laundry of the little ones is perfumed. In the tropical climate it is also said to keep vermin and misfortune at bay.

Often underestimated, but a very great opportunity for perfumers is the scenting of household products. These include, as mentioned above, first and foremost washing, cleaning and dishwashing detergents, but also room scents, fabric softeners, and even car and leather care products. In Europe there are over 36 billion washing machine loads annually, which corresponds to 1130 washes per second. In Germany alone, consumers spent around 5.1 billion euros on household care products in 2022. If you add beauty care, this market even achieved a turnover of over 19.5 billion euros.

Let's take a look at what options a perfumer would have if, for example, they wanted to enter this market in Florida. It should be noted that organic and vegan products are on the rise worldwide. A perfumer could therefore create scents for an organic detergent or, as a friend of mine who is a perfumer did, for vegan floor cleaners. They could smell like " Florida Water ", an eau de cologne created in 1808 by Robert I. Murray, which is shrouded in countless myths. For example, it is said that the scent goes back to a legendary fountain of youth in Florida. There would therefore be a lot of material for a so-called and particularly popular storytelling. It has been recognized that in our age of digital marketing, products with emotional, fascinating stories can be marketed and sold particularly well. "Florida Water" is a good example of this, because it also offers various additional benefits. The miracle stories that have been associated with the scent for years have almost equated it with holy wa-

ter. In Victorian etiquette books, "Florida Water" was recommended as a support for chastity, while at the same time warning against the dubious effects of heavy perfumes. In the end, it developed into a pleasantly protective mother-child scent. The attractive, slightly sweet smell of oranges, for example, combined with a shot of lavender—two fragrance notes that are known from aromatherapy—was helpful. The effect was supported by the transparent bottle, which resembled a vial more than a perfume container and through which the delicate aqua-green colored scent shimmered.

Household care products with a spiritual connotation, aromatherapeutic additional benefits and retro scents, supplemented by the telling of stories, can therefore be marketed in a modern and exciting way. Why do I mention this example? Often trends in fine perfumery come from functional perfumery, but also from other areas such as the flavor industry. Because perfumery loves trends and needs them like fashion. That's why there are always more than one trend in perfumery, and they also like to revive past trends. In addition, there are anti-trends that also have their target groups. Perfumery is also never boring because, almost naturally, new trends and fragrance preferences arise with the new seasons. In the next section I will go into more detail on the topic of "Trends in Perfumery".

However, one thing is certain: insiders of the fragrance industry are smiling. Perfumery has always been very creative, sometimes even a little cheeky. So she crowns as a trend also well-known, which she brings back into the spotlight as new. This only works because in perfumery you have a short memory (or want to have one) and quickly forgives pseudo-innovations and trends that were not. However, there are also permanent trends or, better said, permanent retro trends such as "Royal Violets". Perfumes of this trend remind of or refer to nostalgic fragrance formulations or aromas that are presented anew for each generation. Violet notes are a wonderful example.

Even if you, as mentioned at the beginning, do not intend to become a perfumer or bring your own perfume to market—the current insider information from the perfume industry and trade is still highly interesting, right? But if you do intend to, then ▶ Chap. 7 is exactly what you need, because here you will learn "step by step" how it works modern, i.e. digitally. In addition, there are plenty of tips from practice and perfume history on how to market yourself and your perfume wisely.

- the market dominance of the large perfume manufacturers and their consequences,
- how a new perfume is calculated in industry and trade,
- the areas of perfumery: fine perfumery, functional perfumery, and aromatherapy,
- the situation of the perfume specialist trade and
- you as a perfumer and the potential opportunities in this profession.

Summary

Even if one does not have the intention of becoming a perfumer or working in the fascinating but also very strategically oriented perfume industry and, for example, successfully launching one's own perfume on the market, this chapter has shared a lot of current insider information from the perfume industry and trade with the reader, which should be quite exciting for everyone who uses perfumes. In particular, how calculations and margins are made in the industry and trade and what one needs to be prepared for in negotiations with industry and trade is a well-kept secret. Specifically, this chapter dealt with five major topic areas:

References

Candy B et al (2019) The effectiveness of aromatherapy, massage and reflexology in people with palliative care needs: a systematic review. Palliat Med 4(2):179–194

Haller H et al (2019) Complementary therapies for clinical depression: an overview of systematic reviews. BMJ Open 9(8):1

Pearson ACS et al (2019) Perspectives on the use of aromatherapy from clinicians attending an integrative medicine continuing education event. BMC Complement Altern Med 19:174

Son HK et al (2019) Effects of aromatherapy combined with music therapy on anxiety, stress, and fundamental nursing skills in nursing students: a randomized controlled trial. Int J Environ Res Public Health 16(21):4185

Zimmermann E (2017) Aromapflege für Sie: Mit ätherischen Ölen begleiten, trösten und stärken. Thieme, Trias

Scent Online: Storytelling and Digital Marketing of Perfumes

Inspiration from the History of Perfumery for Fragrance Promotion and Modern Infuencer Marketing

Contents

7

The history of perfumery already knows the influencer marketing and good storytelling. In general, one can learn a lot from the history of perfumery. So it was not enough for the fragrance users of old days just to smell good and thus did not guarantee the success of perfumes. They expected more from their fragrance—at least one additional benefit. But it is also true that some perfumes, which were apparently created with quite normal and often inconspicuous plant fragrances, triggered great trends and brought their perfumers plenty of commercial success and luck. How was that possible? This chapter wants to show you that perfume topics of long past days not only address history-interested fragrance lovers, they also give suggestions for the digital marketing of perfumes today.

7.1 Retro-Fragrance Trends: Why Violets are Immortal

Perfumes and perfumery live from storytelling. In fact, they are, as said, emotional stories that are particularly inspiring. Ideal for storytelling and thus for perfume influencer marketing are ingredients of perfumes such as their trends, if they are communicated emotionally. The story of the introverted, innocent and shy-looking violet is a good example for this.

Plants seem more sympathetic to us when we anthropomorphize them, that is, attribute our feelings to them. The Swiss developmental psychologist Jean Piaget (1896–1980) referred to this as "animism". Young children often still believe until the age of seven that things like plants feel just like they do themselves. This perspective often evokes sympathy and smiles in adults. So when we talk about the introverted violet, we are assuming with a wink that this plant experiences human emotions and in this case even a personality trait. This makes it possible to establish a personal connection more quickly, and the plant is seen more emotionally. By the way: The distinction between extroverted and introverted people goes back to Carl Gustav Jung, a Swiss psychiatrist. Introverted people tend to focus more on a often rich inner life. They appear calm and reserved. They also prefer to spend their time alone in quiet environments. Extroverted people, on the other hand, are more outwardly oriented and usually do well in an active social environment.

When the rich inner life of introverts is still combined with "shy" and "innocent", as in our example with the violet, the combination triggers additional associations of need for protection, which is often still associated with involuntary isolated, lonely and needy. Vulnerable plants are the ideal emotional template for storytelling. The violet also surprises with various health-promoting effects, which also makes it the silent trend plant of the perfume industry.

Already in antiquity, wine was aromatized with violet root. As a flower, the violet was used very early on for various occasions as decoration in the form of garlands, wreaths or bouquets. Mostly female experience or the protection of women and femininity was in the foreground. Already in Greek mythology one meets the violet. Zeus, who was very inventive in his side steps, turned the beautiful nymph Io into a cow and laid her on a fragrant violet meadow. He turned another beauty, the daughter of the god Atlas, into a violet to hide her from the sun god Helios. Since then she has lived protected from his rays in the thicket of the forest. Also in poetry, the modest violet, often as a symbol of young love and femininity, was quoted and sung countless times, for example in the poem written in 1774 by Johann Wolfgang von Goethe:

The Violet

A violet on the meadow stood, bent in itself and unknown; it was a lovely violet. (Ein Veilchen auf der Wiese stand, gebückt in sich und unbekannt; es war ein herzigs Veilchen).

Botany lists over 500 species of violets, the most important with flowers of romantic-nostalgic, slightly bluish to intense purple. From the history of colors one knows that these were originally both masculine and feminine colors, which were later increasingly associated with femininity.

As a flower, the violet is only surpassed in popularity by the Rose. The most fragrant is the March violet, which after long winter days olfactorily prepares the soul for spring. The plant is also said to be good for the eyes, against bronchitis and stomach up-

set. The Greek doctor and father of medicine Hippocrates (460 BC to 370 BC) used violets to treat headaches and visual impairments. In addition, the sight of the first violet in spring was a sign of good luck. In honor of him, ancient Greeks and Romans celebrated wild parties. In pre-Christian times, the violet flower was associated with the awakening of nature and thus with fertility and femininity. Perhaps that is why the violet became the symbol of the city of Athens. A thousand years later, Hildegard von Bingen (1098–1179) also recommends the violet against visual impairments and cloudiness of the eyes. Another thousand years later, the plant received its last accolade: in 2007 it was named medicinal plant of the year.

Violets also achieved great things in the perfume industry. One could write an entire book about it, but I will limit myself to a few examples: In 1709, "Farina", an eau de cologne for women and men, came on the market. The scent, which apparently came from Italy, more precisely from Venice to Cologne, is said to have also inspired Goethe. This was already a very modern violet scent, as the plant was combined with others. "The first hint of refreshing bergamot is accompanied by a delicate jasmine and violet scent, rounded off by warm sandalwood and frankincense," was the scent description of "Farina". As a result, a veritable violet trend broke out. The main beneficiaries were soap and candle makers, the predecessors of perfumers and perfumeries. So Germany's probably oldest perfume shop, Boos in Andernach, was founded more than 300 years ago from a candle factory and soap factory. The legendary violet soap, which was developed into an independent violet perfume, was already famous at the beginning of the company. The famous American writer Charles Bukowski (1920–1994), born in Andernach, and his mother were among the lovers of this scent.

Another great violet trend goes back to the scent "Violetta di Parma". Violets were interpreted here—as was en vogue in the 19th century—still tender and lovely. The perfume by Borsari is said to have been on the market around 1870. The rediscovered fragrance formula is still considered a classic among violet scents today. That one can create great perfumes with violets not only for our great-grandmothers, but also for today's time, showed Guerlain. In 1906, "Après L'Ondée", an eau de toilette for women, came on the market. It interprets violets lightly floral-powdery and gives the scent impression "After the rain shower" again, as the name of the perfume says. "Après L'Ondée" inspired other perfumers to dedicate themselves to the topic "violet". This is how the above-mentioned scent "Royal Violets" came about.

Are violets immortal? Yes, absolutely! Now the scent of this plant is even being rediscovered by men. "A Kiss from Violet" is one of the latest examples. Gucci brought the perfume to the market for women and men in 2019. It was created by none other than Alberto Morillas, who had already hit the pulse of the times with "Acqua di Giò" by Armani. Another example is Hermessence Violette Volynka by Hermès a fragrance for women and men. It was launched in 2022. The nose behind this fragrance is Christine Nagel. Violets are, next to roses, the most celebrated retro scent or retro scent trend in the perfume industry. Because the seemingly innocent, cute violet always attracts the very big perfumers again. Goethe's poem describes a violet that wishes to be picked. A somewhat masochistic wish, one could think. But apparently the violet feels this way above all with perfumers. There it has the chance to become the trendiest plant again.

7.2 Beginning of the Perfume Industry: Scent With Additional Benefits

The history of the perfume industry is a goldmine for modern storytelling. Let's therefore take a closer look at individual areas of its history and see if we can gain insights from the early days of perfume creation for today's marketing of perfumes. At least I want to inspire you with stories from the beginnings of creating beautiful scents for your own storytelling. The amazing thing is, to say it right at this point: In its early days, perfume was important for beautiful smelling, but something else even more.

Certainly much from the beginnings of the perfume industry is in the dark, but archaeology has become increasingly interested in the topics of smell, scent and perfume in recent years and has collected some interesting findings that I would like to present here.

First of all: Whenever the perfumery started, it was probably interested in more than just an olfactory pleasure. In its early days, it was rather a form of functional perfumery, in which the beginnings of fine perfumery were integrated. Thanks to archaeology, we are finding more and more evidence that early perfumery already discovered multiple uses of fragrances for different needs of people and that this aspect was probably the most important for them in perfumes. This means that in early times as well as in antiquity, fragrances such as incense were probably never just limited to, for example, fumigations for the worship of the gods. People had quite different uses for their favorite fragrances—for example, as aromatic substances with which they perfumed dishes and drinks or different private and public spaces, but especially medical and hygienic applications. Incense is a good example of this. Even the Roman upper class used the expensive incense with its anti-inflammatory effect for dental care. It was chewed and was probably the first luxurious chewing gum that was then swallowed because it also had an anti-inflammatory and infection-reducing effect in the stomach area.

Of course, people have always loved good smells, but we can assume that the first perfumers let their perfumes promise additional effects that, as I said, also lay in the medical and hygienic application area. Perfumes like "Opium" or "Shalimar" of our days, which aim at pure olfactory pleasure, would have been possible as perfumes in the old days, but they would have only been sold well over a promised additional benefit such as against rheumatism and gout. Of course, perfumes, as I will discuss in ▶ Chap. 15, were a prestigious commodity, the enjoyment of which was mostly reserved for a selected circle of people who lived mainly in and around a palace or temple. Perfume or wearing such as well as fumigations were olfactory expression of power, divinity, wealth and luxury and therefore offered next to the pure olfactory pleasure already a social, but also a medical and hygienic multiple benefit.

In fact, functional perfumery is still often a real masterpiece today. An example would be a current hair perfume for women, which combines different requirements: a currently trendy fragrance direction in a light concentration from the environment of elegant ladies' notes, a natural protection against UV rays as well as a hair and scalp repair complex. The latter two functions should also be associated with the smell—all alcohol-free. Here, functional perfumery becomes an art because it is supposed to express olfactory different types of multiple benefits.

As already mentioned, since the beginning of perfumery, a multiple or additional benefit has been expected from a fragrance. So pure olfactory enjoyment, as we know it today, was not the only goal of the first noses. Because fine and functional perfumery as well as aromatherapy were still one back then. So it was quite understandable that fragrance creations had a broad-spectrum effect for as many areas of application as possible, even if specific fragrance notes were often used for specific places and occasions. Excavations by a team from

the German Archaeological Institute in Tayma, an oasis inhabited as early as 3000 BC at the old frankincense route in today's Saudi Arabia, provide interesting clues. Apparently, different areas such as temples, houses, public buildings, and graves were perfumed differently—in the spirit of modern room fragrance. Frankincense or frankincense was probably used more for the fragrance of luxurious living rooms and temples. Myrrh was probably the smell of public buildings and facilities. More and more clues are emerging from archaeology about how fragrances were used and what importance they had not only in ancient civilizations. For example, Dora Goldsmith, an Egyptologist at the Free University of Berlin, and her colleague, Robyn Price from the University of California, assume that the sense of smell was deliberately addressed not only to distinguish the king or queen from the rest of society and to emphasize their special position, but that specific areas and facilities such as the royal court were also given a specific aura by fragrance, even if not everyone in society had access to all fragrances such as the precious-divine frankincense.

So viewed, perfume was deliberately used functional perfume from the beginning. It probably started with the first approaches in the Stone Age from the knowledge of plants. As already mentioned, in South Africa 77,000 year old beds made of plants were found, on which aromatic leaves were strewn. On closer inspection, it turned out that these leaves were very rich in insecticidal chemicals, i.e. poisonous to insects and their larvae. For these beds, reed and leaves of the quince were used as basic materials. The latter belongs to the family of the laurels, with which one tried in the Middle Ages due to their wide range of medical applications to fight the plague. The beds were apparently maintained and renewed regularly, i.e. old beds were burned. Perfumery thus probably also started as a form of self-help to make living conditions more bearable, especially with regard to small pests such as fleas and lice. With the first approaches to functional perfumery, one probably wanted to relieve more or less intense and widespread itching, avoid scratching and skin rashes, and support the immune system.

▪ A first renaissance in art in pre-Christian times—triggered by perfume?

Hieroglyphs in Egyptian tombs and finds testify that Egyptians, but also Mesopotamians and Indians, began to create perfumes on a larger scale between 6000 and 3000 BC. One reason for this may have been the growing population in the cities. As mentioned above, archaeologists have discovered first settlements that date back to 10,000 BC. Living together in close quarters, medical-hygienic reasons, the formation of a self-contained and pleasure-seeking affluent ruling class as well as constantly growing knowledge through trade relations favored this development. As early as 3000 BC, there was a lively trade, above all, between Mesopotamia and Egypt. The land of two rivers had a great influence on Egyptian culture and certainly also on their taste for scent.

So far it has been assumed that the first perfumers were priests who used aromatic woods and especially resins for divine offerings or for fumigation or smoking during holy rituals. However, it is now more likely that women drove the perfume industry. These were often housewives who both gustatorically and olfactorily enriched dishes with aromas and used botany for medical treatment. In addition, aromatic substances from the environment were already used early on for smoke, rubs, oils, and ointments to protect against mosquitoes and other pests such as fleas and lice. In this context, spices and fragrances that came through expeditions and trade into their own region and smelled and worked even better than the local aromas exerted a special fascination. Since these were sometimes difficult and expensive to obtain, only a small part of the population such as kings, pharaohs, high-ranking officials, and priests could afford them.

An example of this is the costly traditional evening fumigation in ancient Egypt, called Kyphi. With this, the gods were honored with an offering incense consisting of 10 to 30 fragrant ingredients, which already comes very close to a modern perfume formulation. Frankincense is an air-dried resin from the tree of the same name and was difficult to obtain even for Egyptian aristocrats. The ancient Egyptians loved the resin beads of frankincense so much that they called them the "sweat of the gods".

Incense not only smelled good, they were also used by the residents of Mesopotamia, especially near the river, as cleansing, preservative, and health remedies. This way, insects such as mosquitoes, wasps, flies, and moths could be repelled. In addition, incense was supposed to support health, for example as dental care, and help against inflammatory conditions of all kinds and infections. It was not only valued by the ancient Egyptians, but equally by traditional Indian, Persian, and Chinese medicine. So the offering to the gods served an additional purpose. Probably the use of certain fragrances in holy rituals goes back to personally experienced fragrance enjoyment, but certainly to already known effects and benefits. The same applies to care products of the time, which, from today's point of view, creatively combined fragrance and care in ointments and oils.

A special lover of fine fragrances and care was the Egyptian Pharaoh Hatshepsut (around 1495–1459 BC), one of the most powerful women in world history. So after about 3500 years, a bottle that once belonged to her and was closed with a small lump of clay was found with remnants of a dried liquid. At first it was assumed that it was a perfume with a larger proportion of incense, the favorite fragrance of the ruler.

But it was a care product against skin diseases with fragrant palm oil, nutmeg oil, and tar as essential ingredients. However, one should have foregone tar. Because it contains benzopyrene, a highly carcinogenic substance. When the mummy of the Pharaoh was discovered in 1903, it was found that she had suffered from cancer—probably a side effect of her care product.

The Egyptians, also made possible by expeditions commissioned by Pharaoh Sahure (around 2500 BC), imported incense as early as the 3rd millennium BC. like other treasures for example from the Goldland Punt, which is probably located near the Horn of Africa. This was a fascinating trading place at the often branching incense route, one of the oldest trade routes in the world. Incense lovers like Hatshepsut even undertook their own expeditions to Punt. By ship and caravan, the Pharaoh brought back numerous incense trees from the journey, which she had planted around her temple. Heaven and earth should overflow with the fragrance of incense. Being able to produce one's own incense was not only a luxury, but also an expression of divinity, which Hatshepsut, like all Pharaohs, claimed for herself. How well she succeeded in producing her own incense perfume during her 20-year reign is unfortunately not handed down. The journey of the ruler is the first evidence in the history of perfumery of the efforts that fragrance lovers have to make to achieve their olfactory pleasures.

As said: The International Perfume Museum in Cannes dates the birth of perfume to 7000 BC based on found containers and names the Middle East as the place of birth. But one does not know exactly, since the containers could also be used for other things. The oldest containers were made of clay, with pottery originally coming from China. In the Chinese province of Hunan, a clay vessel was discovered that was fired 18,000 years ago. So it could well be that one will find the oldest perfume and cosmetic containers in China in the foreseeable future, which could give the history of perfume an interesting twist.

In contrast to clay, glass was discovered much later. The first glass beads found in Egypt and the eastern part of Mesopotamia are therefore assigned to the time around 3500 BC. Around 3000 BC, a glassy layer was developed in Central Mesopotamia for pots, vases and other containers. The first glass bottles appeared in Egypt around 1500 BC, which were possibly already used as flacons. The International Perfume Museum also assumes that around 4000 BC people began to use mainly resins for ritual incense in smoke barrels or censers. However, fragrances were always presented—after all, there is a hint of the favorite fragrance notes of the time, which still play a role in perfume today.

Even the first recorded perfumer Tapputi used resins. She created perfumes in Babylon around 1200 BC. Myrrh and calamus in combination with resins were her preferred ingredients. Tapputi was a lover, supervisor, chemist and perfumer of the harem at the royal palace. So she was already a modern woman with different tasks and duties back then. Her perfumes are said to have been true works of art. Of course, the techniques for capturing scents were still limited at the time, after all, one knew nothing about distillation.

At the beginning of perfume making, scents were probably won in fat or by means of cold maceration (maceration from Latin macerare = soaking) in water, but also by hot boiling, certainly at the beginning as essential oils by simply pressing plants for the production of scent. These olfactory impressions were very popular, even if they were far from the fragrance strength and brilliance of today's perfumes. In addition, the range of ingredients available for perfume creation at that time was very limited. Nevertheless, Tapputi—like her present-day colleagues—was always on the lookout for new olfactory impressions

7

or fragrance creations for different target groups and their needs.

According to today's understanding, Tapputi was already a master in the field of functional perfumery. Her perfumes had to smell good, but also work. Probably she can even be called the first fine perfumer. After all, she had an excellent nose and great imagination when it came to combining fragrance ingredients for her creations. She probably gave her clientele an extraordinary olfactory pleasure for the time being, otherwise she would not have been handed down to us as a perfumer. She must also have had the necessary strategic and reflexive thinking as well as enough empathy to be able to serve different individuals with their experience, desires, and needs.

This raises an exciting question of art history. In this way, a certain level of artistic achievement is also explained by the change in consciousness towards one's own individuality. This can also be transferred to fine perfumery. Perhaps the discovery of "individuality", which is often associated with the Renaissance in art history, took place in perfumery history as early as 1200 BC. After all, the creation of a successful perfume requires the corresponding level of reflection, self-confidence, knowledge, and interest in experimentation. Certainly, with the change in consciousness in the Renaissance, a certain degree of separation of man from the absolute claim of God was also associated. Nevertheless, if one were to set aside divine separation, as it could certainly not be thought of around 1200 BC, one would have to postulate: The hour of birth of perfumery, especially of fine perfumery, is at least since Tapputi also the hour of birth of the development of consciousness for one's own and other people's individuality and their representation. And that would mean that there was already a certain form of Renaissance in art through perfumery in pre-Christian times.

Maybe we humans were able to recognize, express, and experience individuality as art on an olfactory level much earlier than in painting. At least a first understanding of collective and personal individuality exists. You wanted to create something noble, valuable, and fragrant for the group, but also for very specific people, in order to express their aura. But, as mentioned above, one had to satisfy primary expectations of health and hygiene above all by means of perfumes. These functions were in the foreground during the creation. Nevertheless, when creating the fragrance, special account was taken of the king or the higher-ups as individuals in order to set them apart from others. Probably there were already fragrances that were created specifically with a view to identity, that is to say the uniqueness of a being. At least one can assume that the perfumes created for them had something special and were an expression of their person and position.

My thesis is therefore: The Renaissance in art, especially in painting from the 14th century onwards, had its precursor in the art of olfaction. Even contemporaries knew this. So, for example, Crusaders from the Second Crusade (1147–1149) brought perfumes, perfume ingredients, and creation techniques from Arabia to Europe, which were far superior to those used in Europe at the time and inspired them from the ground up.

The use of myrrh—not only one of Tapputi's favorite ingredients—reveals the secret and essence of the perfume of that time and why it became so attractive as it is today. Traditionally, myrrh was also used as an aphrodisiac. Women and men wore it as perfume, and beds were sprinkled with it before sexual intercourse. At the same time, myrrh offered another benefit: it disinfects the skin and prevents inflammation. calamus brings another twist to the perfume. It has a wonderfully refreshing, soft, spicy note. But Tapputi also appreciated it for another reason: its root extract was popular for its mood-enhancing effect, which in slightly higher doses even caused hallucinations. So perfumes were already the most beautiful of all drugs back then. And they combined multiple effects: they smelled good, served to increase sensual pleasure, were disinfectants, offered mood enhancement, and sensory intoxication and were used as tribute to the gods because of their value, often in a state of intoxication. This made them valuable gifts for everyone who was important to you and with whom you wanted to share the experience.

Perfume creations in the New World also had these reasons and also offered additional benefits. For example, the Indians healed with plant decoctions. Freshly crushed leaves served as wound dressings. Tobacco juice helped against ticks and was also used to disinfect wounds. Various aromas not only served medical healing, they were also used as an aphrodisiac.

Fragrant creams made of beeswax were used as ointment for burns and insect bites. Sweet Leaf, an Aztec sweet herb, served to increase sensual pleasure with its very sweet and slightly musky aroma and was at the same time the preferred universal remedy of many Indian tribes in Central America.

The indigenous people of America also practiced smoke-supported, aromatherapeutic applications with smoke bowl blessing rituals. The goal was not only regeneration, but also the cleansing of negative energy—both emotional and mental and physical. For this purpose, a bundle of special dried herbs, including sage, was set on fire, so that an aromatic smoke developed, which was blown into the area to be cleansed or to the person concerned.

All effects of a perfume, i.e. a fragrance, are based on the possibility of change, of transformation. This is the main reason for the success of the perfume industry. Even

in ancient times, perfume was an offer of transformation. Psychologically speaking, one tried to get closer to one's ideal self, one's ideal experience, or the desired experience with the perfume, i.e. how one would like to feel now, or to contact the divine or a higher self. And this must have succeeded subjectively quite often.

Thus, perfume has always had a psychotherapeutic function. Because psychology knows: People are happier the closer they experience the distance to their ideal self or between how they feel at the moment and how they would like to feel. With a greater perceived distance, people feel more stressed and less happy. That explains why I became a clinical psychologist and fragrance psychologist. It has always been the aim of the perfume industry that people feel better and more beautiful through fragrance.

Myrrh and Calamus have retained their attractiveness up to modern times. Calamus can be found, for example, in the unisex fragrances "Extrait d'Atelier Maître Chausseur" and "Comme des Garçons Series 1 Leaves: Calamus", myrrh in the unisex perfumes "La Myrrhe" by Serge Lutens and "Myrrhe Ardente" by Annick Goutal, which can still be found in the market here and there. All myrrh lovers can consider themselves lucky when they hold "Etro" "Messe de Minuit" in their hands. In the base note, myrrh, but also Frankincense and honey, the oldest perfume ingredients in the world, dominate.

The art of perfume in Grasse has been rightly declared a UNESCO World Heritage Site. Even if the preferences for fragrance have changed over the millennia, the ingredients have remained largely the same. That is why the protection of fields on which valuable fragrance plants or flowers and plants grow is important—even if some of them should not be so en vogue at the moment.

7

7.3 The First Fragrance Revolution: How Discoveries Changed the Perfume Industry

The perfume industry has so far experienced two revolutions, the third one is just beginning. In this section I would like to report on the first revolution: the discovery of alcoholic perfume.

The first creations of perfumery in the world consisted of beautiful smelling, but rather solid and resinous-waxy products. At the same time there were already fragrant essential oils, which were pressed from the oil of the plant, the essence. It was not until much later that it was extracted by distillation, as is often still done with steam today. Although archaeological research dates the discovery of distillation to pre-Christian times, the Andalusian botanist and physician Ibn al-Bayér (1197–1248 AD) is mentioned in documents as the first in connection with this method for the production of essential oils. Surely she was already used earlier for the distillation of rose petals by a Persian doctor. Meanwhile, the use of essential or essential oils has developed into a common therapeutic practice and production into an industry. Essential oils are used by both doctors and alternative practitioners to strengthen health (Zimmerman 2017). This requires a sound knowledge of application, for example with regard to the purity of the oils. Improper use can cause damage such as allergic reactions and skin irritation. In doubt, essential oils should therefore not be applied undiluted directly to the skin. I recommend applying them to a scent, i.e. paper strip (coffee filter paper also works) before touching the skin.

The first great olfactory revolution in perfumery took place through the combination of fragrance oils with alcohol and a little water. This was the hour of the discovery of the perfumes as we know them today. Alcoholic perfumery became a fascinating art, and perfumers became true artists. Because the effect of the fragrance molecules, which evaporate faster, more brilliantly, and finer through alcohol, was breathtakingly new for the sense of smell. Flowers and plants were never smelled so pure and pure before—like a mood-enhancing, olfactory symphony. From now on, perfume creation itself took precedence over all additional benefits.

Although the alcohol can rise unpleasantly in the nose and head too quickly, nothing else lets a flower smell so much. The Persian doctor Avicenna (980–1037) is said to have been the first to master the distillation of roses. He was thus probably the first true master of beautiful smell, who could enchant the souls of people with his creations. Surely there were already unknown Maîtres de Parfum before, but the apparently rediscovered alcoholic perfumery (the Egyptians already knew it in 400 BC) created a new dimension of smell, which was immediately experienced by everyone: the most fascinating of all olfactory pleasures, which one could have created as a personal favorite perfume—provided one could afford it.

It was Italians who experimented with different types of alcohol in the 12th to 14th centuries to create the most beautiful perfumes. It was quickly recognized that the type of alcohol plays a major role in the fragrance experience. In Germany, perfume makers had to work with alcohol made from sugar beets from 1807 onwards, which sometimes stung the nose. Although perfume makers here developed various techniques to make it softer, noses and perfume lovers still recognized which alcohol was used for a creation, which is not surprising given the large alcohol content in perfumes. In addition, the alcohol made from sugar beets did not evaporate so quickly. The reason for the triumph of the sugar beet in our latitudes was Napoleon's Continental Blockade from 1807 to 1813. It caused a boom in the sugar beet sugar industry, which after the lifting of the blockade through the import of cheaper cane sugar from the colonies first collapsed and then was supported by the state. This created the world's first sugar beet sugar factory in Silesia.

In Germany, agricultural alcohol has been produced under state supervision in agricultural distilleries for a long time, but initially without the idea of using it in perfumes. France, on the other hand, always had the better alcohol for perfumes. It came from cane sugar, wheat, or grapes, or was used in L'Eau de Vie quality from fruit. This alcohol smelled rounder and was more flattering to the nose.

Ethanol, that is, alcohol, obtained, for example, by distilling grapes, was probably already used in Arab perfumery between the 7th and 9th centuries. But in the eyes of some, alcohol is more of a curse than a blessing. So there are already very good alcohol-free perfume carriers, for example from the Swiss biochemistry giant Lonza. However, fine noses do not want to do without alcohol when perfume, even if it is said to be at risk of olfactory sensitivity or respiratory reaction. The Swedish scientist Eva Millqvist believes that ethanol as a solvent can amplify the respiratory reaction in sensitive people, which can lead to allergic reactions.

You should therefore always apply perfume to a scent strip and always sniff before you smell more intensely. Smelling also takes some patience and distance from the nose. Wait about 30 seconds to a minute before you smell a freshly applied scent on a scent strip so that the alcohol evaporates a little. Sniff the scent from a distance of 30 centimeters from your nose before you let it get closer to you. I also recommend that you only smell scents on scent strips during fragrance-supported therapies.

Back to the Italians experimenting with alcohol and other solutions. They wanted to find the most suitable perfume carriers and creation techniques for the constantly growing perfume industry. For them, the best fragrance experience was in the foreground. They were inspired by perfume creation techniques and ingredients that the crusaders brought back from Arabia after the Second Crusade (1147–1149).

In the 14th century, the time finally came. Thanks to an Italian invention for the production of perfume, a perfume came onto the market that dominated the market for over 400 years until the loss of its recipe and is still the longest-selling perfume today: "Aqua Reginae Hungariae", "Hungarian Queen's water of life" or in German "Ungarisches Wasser", as it is still called today (Strassmann 2012). It was the first perfume to be sold internationally that olfactorily dominated all royal houses in Europe. It had a particularly clear alcoholic solution of 95% alcohol, which was used specifically for the perfume. Today we would say that it was an eau de cologne or a very light eau de toilette. This fit well with the trend of the time because it could be given on handkerchiefs and pocket handkerchiefs and delighted both sexes with its scent. This creation is due to the main ingredient rosemary, the plant of eternal fidelity, and was launched by Elizabeth of Poland, Queen of Hungary (1326–1361). Unfortunately, the exact perfume recipe was lost in the 18th century.

7.4 Digital Perfume Marketing: Tips From the Most Successful Influencers

- **The first documented perfume influencer marketing**

Elizabeth of Poland is not only the most successful perfumer of all time with her long-lasting perfume, she was also a master of a different kind: she was the first to use influencer marketing as a conscious sales strategy in the perfume industry.

French kings were the official opinion leaders or influencers of their time. But true masters in this field were the ladies at and around the court who knew how to win over the respective king for something. Nowadays, almost all trends are launched and spread through digital word-of-mouth or opinion leaders in social media. In theory, everyone has the chance to become an influencer. Originally, however, this influence was reserved primarily for successful kings and queens, to which French monarchs in particular belonged. Among them was the influential Charles V (1338–1380), also called the Wise. He is considered the first known modern French fragrance lover. He also believed all his life that he had discovered a perfume that triggered a megatrend in the perfume industry. But it was probably something else: Elizabeth of Poland did everything she could to make Charles discover her perfume, and she succeeded so well that he believed he had discovered this scent himself.

Elisabeth's perfume was based on an unusually attractive perfume formulation. It not only smelled extraordinarily good, which is why everyone wanted it anyway, but it was also said to have rejuvenating and healing effects on pest and gout by a friendly monk. Nevertheless, the monarch was not able to successfully market her family perfume. Because there was a powerful competitor: the perfume "Carmelite Water" created especially for Charles V., which attracted all the attention at the French court. Elizabeth of Poland had to develop a plan to get her perfume into the orbit of the ruler and make him find her scent more suitable and attractive—preferably so that he would feel like the discoverer of the scent himself. For this purpose, her daughter Elizabeth of Pomerania (1347–1393), pretty, tall for her time, and well-proportioned, was to help. She had to get access to Charles V.'s immediate environment so that he would supposedly randomly smell the perfume and find it much more suitable and attractive. The main obstacle here were the ladies at court who were jealous of an-

yone coming too close to the king. A marriage opened the way to the monarch: Elizabeth of Poland married her daughter to the brother of Charles V.'s mother, the Roman-German Emperor Charles IV. Elizabeth of Pomerania thus came into the vicinity of Charles V., and the plan worked. The fragrance-loving monarch made the family creation "Aqua Reginae Hungariae" the biggest perfume success in Europe so far (Schwedt 2008). The scent was developed a little further from time to time—also by Charles V. himself. Although the original formulation was lost, the scent can still be bought today in different versions that refer to the original perfume, for example from Crabtree & Evelyn, where it is offered as "Hungary Water Eau de Cologne".

■ **Techniques and methods of modern influencers**

The history of today's successful influencers in the perfume and cosmetic industry began with Michelle Phan on Youtube in May 2007 (Phan 2014). She had applied for a position as an advisor at a Lancôme counter in Florida and was rejected. After that she started to share her love and passion for make-up with others on Youtube. As a so-called "beauty vlogger" (combination of video and blogging) she reached 8 million beauty fans in a few years. Michelle was sponsored by the French L'Oreal group very early on and, like all beauty and fashion bloggers, presented her latest purchases in so-called hauls, advised her followers and crowned her personal favorite products. This developed into a profession that combines journalism and marketing.

A person who is referred to as an influencer today is someone who, due to their trustworthiness, knowledge, and media presence, acts as an opinion leader, primarily via the Internet or digital marketing, and thus influences other people's purchase decisions. This helps brands, products, and services to break through more quickly. In addition, they reach multipliers and achieve greater willingness to test offers (Backaler 2018; Prahl 2019).

The first generation of perfume influencers who made a name for themselves as beauty vloggers, such as Bethany Mota, Zoe Sugg, and Ingrid Nilson, primarily recommended their favorite make-up products. But they also made discoveries that could be applied immediately in perfume advice. Top beauty vloggers publish an average of two posts per week on Youtube. They recognized—perhaps initially purely intuitively—the DNA of the perfume industry. Here, in the first instance, it is all about the topics of beauty, attractiveness, and increased well-being. In the second step, they realized that they had to address their followers differently depending on the day of the week, because, for example, the desire for well-being looks different at the beginning of the week than at the weekend. In other words, depending on the day of the week, you need different things to feel good, because you expect a different benefit from beauty care more or less consciously.

- Feeling good on Monday and Tuesday means having more security and knowing that you have the right do-it-yourself beauty tools to start the week off right. In addition, the follower is built up and encouraged, for example, with the following statement: "I want you all to see how beautiful you are—with and without make-up."

It is known from perfume practice that the day of the week also influences the choice of fragrance. Customers are more open to fine-herbal chypre or leather notes at the beginning of the week. They act like a shield, convey an aura of self-confidence and competence, exactly what you need for employee meetings that take place in many companies on Mondays.

- Feeling good on Thursday and Friday means having do-it-yourself tools with which you can try something new or transform yourself to indulge in promising fantasies. The typical message of the first influencers was accordingly: "Become the person you want to be. Life is too short to have just one look, experiment! Find all sides of yourself!" On Thursdays and Fridays, perfume customers are much more likely to try new fragrances. It is no coincidence that many perfume launches take place on Thursdays, the day when new films usually come to the cinema.

The first modern influencers made another discovery with their mostly three-minute short videos: You can fascinate others best when you turn the advice upside down, or start from the back to advise customers. Starting from the desired use and anticipated experience, the conclusion is drawn to the ideal brand or the ideal product. In other words, it is advised from the future into the present, or, psychologically speaking, from the ideal self to the current self. In the concrete case, for example, the skin feeling on the upcoming vacation is concluded from the current skin feeling. This would be the case if someone is interested in sun protection or a sun fragrance, but actually looks forward to a party with friends under the open sky at the weekend. This anticipated well-being is initially in the foreground—a finding that has changed advertising. While the classic brand video still focused on the product, is now the desired experience. Beauty vlogger videos are also not based on the usual commercials, but rather look like recommendations from friends who can be bought on linked websites as "must-haves".

The first wave of perfume influencers, who mainly came from the USA and Great Britain, only reached Germany a little later. The best-known influencers of the early hours included Daaruum, Ischtar Isik, Bianca "Bibi" Heinicke, and Sami Slimani, one of the first men in the industry.

More and more influencers like Christophe Cervasel,Sylvie Ganter from Atelier Cologne or Bibi then created their own

7

perfume and beauty brands. Many have also become sought-after brand makers because they reach specific target groups, especially new customers, with a relatively small number of followers. Influencer fame Bean (@beansiie), a teenager from San Diego with 75,000 followers on Instagram and Tumblr, managed to turn Willa, a social commerce start-up, into a sought-after cosmetics brand in 2015.

- **Trends in influencer marketing in the perfume sector**

At first, brands, companies, and perfume stores such as Sephora or the Wiedemann perfume store in Bavaria used their own employees on Instagram or Youtube as influencers as part of "Instore Beauty Vlogging", for example under the name "Bavarian Beauty Gang". At the same time, external influencers who were employed by advertising agencies were obliged. Meanwhile, influencer marketing and the number of influencers have increased explosively, and mostly influencers are paid by companies to present their products (Knost & Seeger 2020). The compensation was particularly good if they had a large number of followers. As a result, the so-called Micro-Influencer-Marketing was supposed to counteract the suspicion of purchase. Here, the focus is no longer on the number of followers, but on authentic expertise in a certain area (Schach & Lommatzsch 2018).

It is advantageous if the opinion maker is locally networked. Stationary perfume stores count between 20 and 200 customers daily, small perfume stores often only around 20—trend falling. Every customer who can be advertised by an influencer helps the business. In order to generate a larger number of customers, many stationary perfume stores now rely on an additional online shop. For many of them, this now accounts for a third of sales.

For some years now, more and more brands and retailers have been offering partnerships to customers they consider to be

opinion leaders. They should act as local influencers outside the business, initially digitally, later in personal conversations, for example in a café. This idea is not entirely new. Just think of the many Avon consultants of yesteryear who invited themselves to house parties and thus made direct sales a success. Direct selling has now developed into a social commerce or social sales network with Influencer-Marketing in which the influencer acts as an independent business partner of a provider and his environment or his followers benefit from his compensation, in line with the motto "Who shares or recommends, earns with".

- **How to market yourself as a brand: Tips for perfume vlogging**

Short videos on Youtube, for example, can quickly and relatively easily achieve a certain degree of popularity as a beauty vlogger. Short videos of three to five minutes are ideal for introducing perfumes. An exciting introduction helps to keep even impatient viewers longer. In 2019, around 30,000 new beauty videos were uploaded monthly, with a trend towards increasing. 98% of them come from independent filmmakers. It is estimated that by 2022 there will be over 1,000 beauty influencers on Instagram and YouTube. As of February 2023, Latina influencer Sandra Cires Art was the most subscribed beauty content creator on the video platform, with over 16 million subscribers to the channel. One of the most popular perfume influencers is 33-year-old German cologne connoisseur Jeremy Fragrance. He has almost 2 million YouTube subscribers and almost 6 million TikTok followers (2022).

A picture is worth a thousand words, even in a perfume video. Pictures are snapshots for the brain, they overcome language barriers, create greater virality, are commented on and shared more often. Videos are even shared twelve times more often than pure text posts in connection with a link. That is why influencer marketing with beauty vlogging is ideal for the online presentation of perfumes.

Brand managers and advertising agencies are often amazed at how simple it is to produce videos for YouTube, for example, and become a beauty vlogger. Facebook, Pinterest, and Instagram are suitable for announcing and promoting videos.

Basically, all you need to create a video is

- a tablet or smartphone that also serves as a video camera,
- a YouTube account,
- Capture, a software program for video production and
- basic knowledge of the programs iMovie (Apple) or Movie Maker (Windows) for video editing.

A video about perfumes should not look as perfect as a big brand's commercial. It is important that the vlogger is not only seen, but also heard well. In addition, a perfume vlogger must appear authentic and competent and surprise the follower.

To start as a perfume vlogger, you should first produce twelve videos of no more than three to five minutes, which are posted twice a week as a continuing series on the net. After six weeks, the feedback usually shows how the new influencer is received by the audience.

A short script should be recorded for each video, which specifies which perfumes are shown and which explanations are given. In addition, the background decoration and the clothing of the vlogger should be determined. Often, a family member or someone from the circle of friends can be won over for the camera work and the lighting. Perfect for this are a digital camera with HD video recording and a tripod. You should definitely invest in a good external microphone. The cameraman or camerawoman should also always check whether the vlogger is looking into the camera and making eye contact with the viewers at the decisive moment. This is especially important when something positive is said about a perfume. It increases credibility.

All episodes should be filmed directly one after the other—preferably at the week-end, as the noise level from the street is then lower. Also think about different formats. This means that your video should look and work equally well on all devices of all sizes and types. Therefore, you should choose a camera setting that is more focused on close-ups, but still shows some background/depth. This also makes the perfume you are talking about each time clearly visible. When recording short videos, the old rule also applies: "Don't zoom, don't shake, don't swing". Suitable music can be found under ▶ www.incompetech.com and ▶ www.danosongs.com.

The finished videos can be shared with friends using corresponding links, e.g. on Facebook. Together you can quickly get to 1000 contacts. Short versions of the videos up to 15 seconds long are also suitable as announcements. Remember that when filming the videos, pictures should also be taken behind the scenes and of the perfumes with decoration—ideally 30–50 pictures with which you can then generate additional interest in your perfume vlogging on Instagram, Facebook, & Co.

The title of a video is also important for its success. He may be emotional—provided he has a relationship with the vlogger himself. So a series of short videos could be titled "Perfumes for Introverts". One of the individual episodes might then have the subtitle "Welcome to the innocent world of violets".

Tips

How to successfully develop into a brand and create your own perfume video so that it is exciting for viewers

Here are five essential components that I would like to mention to you for the success of your beauty vlogging:

1. First of all, after the welcome, address the current occasion, the reason or the why for your video. This can be aligned with a certain day (Monday /

7

Tuesday vs. Thursday / Friday) that immediately reveals the connection to the topic. Ideally, you do this in the first 10 to 20 seconds, so that everyone stays tuned to see how your video continues. That is, in this time span, your viewers should already have an idea of what will happen in the next few minutes and why it is told. Eye contact is very important in the welcome.

As an occasion or why, for example, you could take the week in which the "Plant a Flower Day" is celebrated. It always takes place in March, when the violet also blooms. Ideally, you would then also be surrounded by blue-violet violets, which, by the way, comes across very well on the web as a color, like all reddish tones. If the video is posted on Monday / Tuesday, you could explain what the violet scent can do at the beginning of the week. If the video is posted on Thursday / Friday, you could, for example, talk about the transformation that can be experienced with a violet scent.

Ideally, you would show a total of two to three perfumes from different brands (you have them in front of you). But you can also only introduce one perfume, but immediately say that you will also talk about other perfumes in the next video.

2. To convey authenticity and trust in the viewer, it is recommended to include your own personal "why" in the address: how you came to the scent and why you love it so much. Ideally, if you communicate friendliness, eye contact, and personal experience, you will also communicate competence and trust. For example, you could report how you found personal happiness with this violet scent and why you want to share this experience with others. By the way: Mistakes in the

video are no problem. That makes you sympathetic and authentic.

3. Now is the moment when you passionately, but not excessively, demonstrate your perfume as a personal gain and enjoyment to your viewers. For this, you smell, on behalf of your viewers, the perfume that you spray on a scent strip. Discuss the tactile experience by touching the bottle or taking it in your hand. Furthermore, show the name and packaging of the perfume. Now you can also name a few other ingredients of the perfume.

4. Now comes perhaps the most important scene in your perfume video: If the viewer begins to anticipate how the video might continue, make and show something unexpected. You integrate a surprise into the video that remains in memory and thus promotes the digital mouth-to-mouth propaganda for your video, so that the consulting experience is verbally shared with others. For example, let something or someone else decide on the scent that has not been in the video so far. It could be your cat or another quadruped.

5. Of course, everyone who likes your perfume video wants to see you again. Therefore, announce your next show when you post the video online and what you will report.

If you can now implement your script as a video in three to five minutes, optimally in two minutes, and also make eye contact with your audience, you have nothing to stop you from being a perfume vlogger.

One last thing: Choose the perfumes you want to show wisely. For influencer marketing, for example, these are ideal:

- Trendy perfumes
- Innovative perfumes

- Fragrant discoveries, e.g. on a trip
- Rediscoveries or fragrances that were long thought to be forgotten
- No mainstream products
- Private labels, exclusive brands that are not found in every perfume shop
- Good entry price. Premium perfumes are naturally not cheap. Experience shows that for these perfumes you need special or travel sizes for less than 50 euros
- The availability of samples and samples (for the products promoted by the influencer), which can be picked up by new customers in the store, but which can be ordered like the perfume itself on a website
- Finally, the most important thing. Make sure that the perfume you are showing also has a good "conversion rate" with your target group. This is the number of samples, for example, that you have to give for the perfume before a customer decides on the scent. A perfume sample costs the trade about 0.35 euros. Now you can calculate for yourself how many free samples you can afford at what margin so that it still pays off. Probably you will then come to the same result as many in fragrance marketing: Unless you want to make an "overinvestment" for a perfume to make it known, it is about twelve given free samples per sold perfume, which you can still afford.

Have I sparked your curiosity to explore the possibility of trying it out yourself as an influencer or perfume vlogger? I would be very happy about that, because our industry lives from new topics and views. And to all newcomers: Just be brave! Often knowledge from different industries can be combined creatively.

Summary

The dead live longer, that also applies in the perfume industry.

In this chapter, the dynamics of trends, in particular the retro trends in perfume ingredients were presented using the example of the violet. Then it was about the first revolution in the perfume industry: the discovery of alcoholic perfume—all with the aim of collecting a lot of interesting material for one's own storytelling or to fascinate as a perfume influencer.

In the second part of the chapter we then discussed how the perfume industry has developed in digital marketing of perfumes in recent years and reported on remarkable findings of today's influencers on how to successfully market a brand or its perfume especially through short videos played on social media channels. For this, very concrete ideas and tips for the creation and dramaturgy of short videos were given. Furthermore, the chapter gave indications of which perfumes are particularly suitable for online marketing.

References

Backaler J (2018) Digital influence: unleash the power of influencer marketing to accelerate your global business. Springer, Cham

Knost J, Seeger C (2020) Influencer-Marketing – Grundlagen, Strategie und Management. UVK, München

Phan M (2014) Make up: your life guide to beauty, style, and success – online and off. Harmony Books, New York

Prahl J (2019) Influencer – wie Online-Verkäufer das Internetshopping beeinflussen. Das Erste, (w) wie Wissen. Das Erste (NDR). Video verfügbar bis 11.11.2022

Schach A, Lommatzsch T (2018) Influencer Relations: Marketing und PR mit digitalen Meinungsführern. Springer Gabler, Wiesbaden

Schwedt G (2008) Betörende Düfte, sinnliche Aromen. Wiley, Weinheim

Strassmann R (2012) Duftheilkunde—Der Weg, Düften zu begegnen. Kräuterwissen—Klangreise—Wegzeichen. Hofstetten, St. Peterzell/SG

Zimmermann E (2017) Aromapflege für Sie: Mit ätherischen Ölen begleiten, trösten und stärken. Thieme, Trias

Big and Small Moments of Modern Perfumery

From the Search for the Blue Flower and the Atlantis of Perfumery—How Perfumery Became What It is Today

Contents

J. Mensing, *Beautiful SCENT*,
https://doi.org/10.1007/978-3-662-67259-4_8

8

I must begin this chapter with a thank you to Charles Baudelaire (1821–1867). He was not only one of the greatest French lyricists, he also gave perfume and scent in literature a completely different value from the middle of the 19th century. With him, after a long era of olfactory chastity, smell was again sought, experienced and enjoyed in multifaceted and pure eroticism. But with Baudelaire and his sinful-immoral and sensual-provocative imaginative poems, perfumers were also inspired to create new scents that led to the second revolution in perfumery. People began to experiment with artificial, i.e. synthetic fragrances to create innovative scent impressions that were previously unknown in perfumery.

8.1 Scent & Eroticism with Charles Baudelaire or: The Change in Thinking

It is known about Charles Baudelaire, who lived in Paris and was very much influenced by the urban life in his city in his poetry, that he was also a fanatic of smells. Body odors and sexual odors often played a central role in his poems. His passion for scent went so far that he wrote his famous poem collection "Les Fleurs du Mal" (The Flowers of Evil) on perfumed paper.

Many of his poems, scent impressions and memories were inspired by Jeanne Duval, Baudelaire's exotic lover and muse. She was an actress, dancer and a Caribbean beauty from Haiti, whom he called "Vénus Noire". Her influence can be seen, for example, in the poem "Le Serpent qui danse" (The Snake that Dances), which is part of the poem collection "Les Fleurs du Mal". Of the over 100 poems, six were banned for endangering morality with the publication of the collection (around 1857).

At that time, few people suspected that Charles Baudelaire with the "Flowers of Evil" or his imaginations, which were not a copy of reality, but transformations of appearances—based also on perverse scenes and provocative images of human imagination—would become one of the most important lyricists not only of the French language (Miertsch 2005). Victor Hugo, who wrote to Baudelaire in 1857, had an inkling of the unusual rank of the work: "Your Fleurs du Mal shine and shimmer like stars."

Even in our days, perfumers are still stimulated by Baudelaire's poetry for their creations. So the olfactory interpretation of his poem "Parfum Exotique" gave the Eau de Parfum "Baudelaire" by BYREDO a fragrance character of wild-exotic and mysteriously-bewitching attraction. Leather notes and black pepper drive each other on. Even with the various settings of Baudelaire's poems, there is still a strange eroticism in the air. Here are two examples by Georges Chelon and Michael Mansour: ▶ https://www.youtube.com/watch?-v=BO-Pya5G3P4, ▶ https://www.youtube.com/watch?v=5yRyMUSA8qs.

> **Parfum Exotique**
> When, with my eyes closed, on a warm autumn evening, I breathe in the scent of your warm breast, I see happy shores unfold, which are dazzlingly lit by the fires of a monotonous sun …
> Quand, les deux yeux fermés, en un soir chaud d'automne, Je respire l'odeur de ton sein chaleureux, Je vois se dérouler des rivages heureux, Qu'éblouissent les feux d'un soleil monotone …[1]

With Baudelaire, scent and perfume became much more important in literature. Both were important, could be complex and also clearly erotic and sinful and provoking for the senses. Starting in the second half of the 19th century, the topic was then also picked up literarily by others in all its facets. Until then, according to the French anthropologist Annick Le Guérer, scent and perfume played no special role in literature (Le Guérer 1992). This, to say it at this point, has changed fundamentally since then. Hans Rindisbacher offers a good overview of the different approaches in literature to the world of smell (Rindisbacher 2015).

The seductive power of certain scents and perfumes was certainly already known in ancient times and antiquity (Hurton 1994). As reported in ▶ Chap. 1, the promised effects of myrrh as a fragrance were so great that the attraction of the sexes was even said to increase to incest. In the literature of the modern era, especially in the "early romantic period", that is, from 1789 with the beginning of romanticism, the personal experience of eroticism in connection with fragrance and perfume was only vaguely hinted at in literature. In the natural mysticism of the Romantics, as we will see below, fragrance and perfume are also only side effects. Scents are part of a rather almost innocently experienced nature in romantic literature (Le Guérer 1992), whereas Baudelaire preferred the artificial, which only existed in his imagination, to nature (Flammersberger 2016).

With his break from a natural mysticism of beauty and harmony as the highest goal towards the imaginative artificial, which can also play with dissonance and provocation, Baudelaire was a pioneer of his time with his art. The decline of romanticism was in the air, which was also supposed to lead to the break of modernity with the, one could say: romantic, nature-oriented perfumery. Baudelaire was already part of a movement that now allowed other arts to rethink.

About ten years after the publication of "Les Fleurs du Mal", coumarin and then vanillin were discovered and developed as the first synthetic fragrances in 1868. It should first be the end of the romantic perfumery, which until then had been oriented towards the good smells of nature with its fragrances and had to. The use of synthetic

---

1   Translated by the author from Guernes, 7 ▶ https://christian-0-guernes.blogspot.com.

fragrance ingredients in perfumes began in 1881 and, as I will show, revolutionized perfumery. It was triggered by the perfumer Paul Parquet (1856–1916). He created a completely new men's perfume: "Fougère Royale" (Royal Fern). For this perfume he used a synthetic fragrance ingredient, namely coumarin, for the first time. This created a scent impression that did not exist in nature, but was created solely by his imagination. Paul Parquet knew what he was doing. So he is said to have said: "If God had given ferns a smell, they would smell like Fougère Royale."

Even if the first artificial perfume or the first scent with synthetic fragrance components, where also natural ones were used in the creation, had nothing sinful-provocative or even fantastic-erotic, the claim of modernity was also postulated for the perfume industry. With artificial means, one wanted to create a new olfactory stimulant and thus free oneself from the naturalness. This was a revolutionary step, but actually also an old human desire in search of more self-determination—for example, to test a new sensuality, eroticism, attractiveness or simply a different experience that came from the free thought or imagination. Through the changed thinking of the perfumers and the possibilities of the new ingredients, gender stereotypes could now also be overcome and fragrance, sensuality and eroticism could be redefined accordingly for both genders.

In 1889, the perfume industry was then ready, even if Baudelaire did not live to see it. A perfume rich in artificial fragrance substances Coumarin and Vanillin was brought to the market under the name "Jicky" by Guerlain. The fragrance, which was probably initially intended for modern and experimental young men, remained unchanged in the recipe for the perfume of the "new woman". A new female consciousness, which fought for its emancipation in the first large demonstrations from the second half of the 19th century (I will

talk about it later), now had its own fragrance for the first time. What was special was less that this perfume came from men's perfume, but rather that it had a fragrance impression that existed in nature, but not to this extent and with this vehemence in expression. Just the awareness of wearing a perfume with synthetic fragrance components made the wearer feel modern. This fit the zeitgeist and thus a woman who could now also olfactorily shed her previously ascribed naturally feminine role.

Other perfumes with the same or similar philosophy should follow "Jicky", especially with the discovery of synthetic aldehydes. The fascination at the beginning of the 20th century for aldehydes then led to its own "artificial floral fragrance direction" (floral-aldehydic), for which the already mentioned Chanel No 5 is an example, even if it contains many natural ingredients. But then there were also purely synthetic fragrances that would certainly have pleased Charles Baudelaire. This is probably not easy to understand for some contemporary fragrance lovers. At his time, however, Baudelaire believed more in the modern and therefore preferred the artificial to nature. He found nature rather ugly with its positive triviality, which he preferred to the monster of his imagination, as he said. His inspiration was the world of plantless metropolises, which were ugly for him too, but with their asphalt, their artificial lighting, their stone canyons, their loneliness in the hustle and bustle of people and above all with their sins, the template for his art.

■ **The relatively short triumph of modern perfumery**

The triumph of the artificial, that is to say synthetic fragrances, in perfumery, should then go unquestioned for just over a hundred years. Meanwhile, as we all know, a change has again taken place in the perfumer's way of thinking. Postmodern perfumery is again oriented towards nature, but now in its pure and unadulterated form.

Even Romanticism had an influence on today's postmodern experience of scent, which I will show using the example of the influence of German Romanticism on the current German experience of scent. Let me say at this point: Even if Romanticism did not yet know the olfactory eroticism of a Baudelaire with his provocative imaginations, it was not asexual, because innocence also has its attraction. But Romanticism was more of an intermediary stage for the experience and personal encounter with scent and perfume. First one had to break away from the influence of many classical philosophers and the strict moral guidelines of the church in order to be able to experience the full range of olfactory enjoyment. The credit for this goes above all to Charles Baudelaire, because for a long time the sense of smell was considered the "lowest" and "most archaic" of the five human senses and was therefore suppressed and demonized like many things that did not conform to custom. Above all, the church has held since the Middle Ages that smells and fragrances can quickly serve frivolity—although there has always been a double standard, because scents and perfumes were often useful means of combating disease, and the despised sense of smell revealed the state of health or illness.

Only a few people broke with antiquated ideas in public around the middle of the 19th century and took disadvantages for their art in the face of the prevailing opinion and custom. The bourgeoisie, accustomed until then to the positive-idealized poems of Romanticism, even though the aesthetics of ugliness was already represented in this epoch, reacted angrily to the break with the tradition of lyricism that Baudelaire initiated with his provocative imaginations (Flammersberger 2016). As I said: On July 7, 1857, the French prosecutor's office initiated a criminal prosecution for insulting public morality against Baudelaire, and he was sentenced to a hefty fine. Certainly other poems like "Lesbos" or "Femmes damnées" in the collection "Les Fleurs du Mal" contributed to this. Even his publisher was not spared by the prosecutor's office and was forced to flee to Belgium.

8.2 The Godfather of Perfumery Or: Are Perfumes Immoral?

A number of famous philosophers have expressed themselves—albeit mostly disparagingly—about the effect of scents. I am thinking primarily of Plato, Aristotle, Kant and Hegel. Their deliberations have led to three important questions that were discussed right up into modern times:

- Are perfumes immoral?
- Are scents coercion?
- Is perfumery an art?

Even 30 years ago, a job in the fragrance industry did not receive the recognition it deserved. Scent, beauty and intelligence seemed incompatible. Even my doctoral supervisor was surprised when I told him that I was interested in perfumes and perfumery. While he too was fascinated by the world of beautiful scents, at the same time he considered a job in this area to be little scientific. Probably he too still had the views of the great philosophers in his head.

Since Plato (428/427–348/347 BC) saw the exact classification of odors as impossible, he divided them into the categories "pleasant" and "unpleasant". The latter included scents and odors that aroused fleshly desires or animal instincts, and so perfumes were often mistrusted. With this boundary to the immoral, which certainly does not apply to all perfumes, advertising still plays today.

The mistrust of everything animalistic was reflected in the hierarchy of the senses. Aristotle (384–322 BC) assessed the five senses as essential for gaining knowledge, giving sight and hearing priority over smell. This view was long transferred to the assessment of scents. The contempt for the sense of smell was further reinforced by the plague, which was experienced as a divine punishment. In the Middle Ages it was believed that it would also be transmitted by smell.

With the beginning of the Enlightenment in the 17th century, philosophical preoccupation with the senses gradually lost its religious, moral and ethical impetus. But the reservations about smell and scents remained. This went so far that, for example, in the 18th century in England the wearing of perfume was temporarily banned. After all, the perfume was indirectly said to have the ability to work.

The English mathematician, state theorist and philosopher Thomas Hobbes (1588–1679) was the first to plead for the equalization of the senses. "We cannot think of anything if it was not created wholly or partly in one of our senses beforehand" (*Leviathan*, 1st chap., p. 11). Nevertheless, from a philosophical perspective, even fine scents were largely excluded.

It is said of Kant (1724–1804) that for him the beautiful had no smell, and in his aesthetics the smell—like with Hegel—is not discussed. Scent is for him a symbol of unfreedom. Thus he wrote: "Smell is like a taste at a distance, and others are forced to enjoy it, whether they want to or not, and therefore it is less sociable than taste, where the guest can choose from among many dishes and bottles according to his comfort, without others being forced to enjoy it" (*Anthropology in a Pragmatic Sense*, p. 158).

Hegel's (1770–1831) famous saying that the sense of smell is incompatible with the interests of art and spirit then hit the perfume industry right in the core. As a man of his time, he thereby denied it the possibility of being an art. Nevertheless, he left a little door open with the statement that the nose could only adopt an aesthetic attitude if it was guided by the mind and thereby relieved itself of its primary and natural function—that of smelling.

■ **Nietzsche—the godfather of perfumery**

With Friedrich Nietzsche (1844–1900) the turnaround came. Posthumously, one would have to name him the "godfather of perfumery". The indirect question of philosophy, whether perfumery can be scientific at all, receives more than one new direction through him. The nose is discovered as the most delicate scientific observation instrument, about which no philosopher has spoken with such veneration and gratitude until Nietzsche. He even chooses the nose as his sharpest scientific observation instrument: "It is still able to detect minimal differences in movement that even the spectroscope does not detect" (*The "Reason" in Philosophy* 2, KSA 6, p. 75 f.). The sociologist Jürgen Raab summarises Nietzsche's understanding of smell as follows: "For Nietzsche, the sense of smell is the sense of truth because it does not draw from the intellect, which is detached from the physical affects and therefore deceptive, but from the secure sources of animal instinct" (Raab 1998).

Nietzsche endows the nose as an organ for intuitive knowledge of reality with its own power and magic, even the ability to create elementary, non-spiritual, true art. For Nietzsche, this art is far superior to science. It is free and not controlled by the establishment. In this understanding of art, there is actually no mistake, only a side by side of the different works (Leistikow 2019). This can also be applied to perfumes. Since they are created by noses and thus by reliable and free sources without influence, they have the right to be recognized as art.

Accordingly, the perfume industry and its art, the perfume, would never have to justify itself according to Nietzsche. Even if there is sometimes passionate debate about what is beautiful or what smells good and what does not, in art as well as in the perfume industry it is not a question of being right or wrong (Leistikow 2019). The free perfume industry is thus the counterpart to the deceptive truth because it does not claim science, in this case chemistry, for itself. What smells good has been in the nose of the observer since Nietzsche. Even semblance, illusion and ambiguity are therefore not negative things for him if one wanted to attribute them to a perfume creation. These can also be expressions of beauty (Leistikow 2019). Of course, Nietzsche makes a complete sweep and provokes the sciences with his yearning for the "scrapping of the establishment" (Leistikow 2019). But after more than 2000 years of suppression, which began in antiquity, smell, perfume and the entire perfume industry needed a powerful voice through various male opinion leaders.

There were voices for the recognition of the nose and fascination for olfactory worlds even before Nietzsche, only they were less audible. A real image upgrade came through romanticism and its research on smells. It was the mystical in scents that attracted people. With them you could create your own world. That fit well with an era that was attracted to world of sagas and myths and sought sensual experiences in the intuitive, mysterious and divine. From the end of the 18th century to well into the 19th century, this was a central theme in literature, especially in prose and poetry. Above all, "lonely wanderers" such as Byron, Goethe, Shelley and Joseph Conrad fascinated. Directly or indirectly, they spoke of scents and odors that could revive memories of past events, certain locations, people and the associated moods and emotions. In turn, this aroused the interest of olfactory research.

8.3 Perfume & Poetry: The German Soul of Scent

It is little known what special meaning fragrance and smell have had in German-language literature, especially poetry, for centuries. German Romanticism of the 18th and 19th centuries, with its focus on nature and longing, continues to influence our olfactory sense directly or indirectly to this day. The postmodern soul of scent, especially that of the Germans, wants to be as close to and pure of nature as possible, ideally without any synthetic fragrances. The current big trend in perfumery, even though the next trend is already emerging through neuroperfumery, is in a way the rebirth of the nature romanticism of the 18th and 19th centuries and its further development. Postmodern perfumery is based on an increased awareness that nature must be protected and preserved by humans in order to secure its future and thus the future of all living beings. This includes resource-saving production with economic, efficient and responsible natural management. A demanding goal that must be constantly worked on, also because the global demand for fragrances is increasing faster than the world population is growing. So the annual global harvest of about 1000 tons of real vanilla can no longer cover the increasing demand. Postmodern perfumery has therefore long since reached its limits and will continue to develop into "plant peptide perfumery" and neuroperfumery, which are based primarily on biologically and medically important and effective molecules. I will discuss findings from peptide research with examples from perfumery at a later stage.

Romanticism with its works has significantly contributed to the image enhancement of perfume. This is pointed out by the Italian Germanist Massimiliano De Villa (De Villa 2017) and in this context he mentions

8

Friedrich Gottlieb Klopstock's (1724–1803) poem "The Summer Night" (*Odes*, Volume 1, Leipzig 1798, pp. 233–234.) a nature-praising, but in the mood melancholic, fragrance poem, written even before the epoch of Romanticism really began for the Germans. But it already touched the nature-romantic, slightly melancholy soul of scent of the Germans at that time:

The summer night

When the shimmer from the moon now pours down
Into the woods, and smells
With the scents of the linden tree
In the coolings waft;
So I am overshadowed by thoughts of the grave
Of the beloved, and I see in the forest
Only it dawn, and it blows me
From the blossom not forth.
I once enjoyed, o ye dead, it with you!
How the fragrance and the cooling wafted us around,
How you were adorned by the moon,
You o beautiful nature!

Goethe's much quoted beginning of the Mignon song from his novel *Wilhelm Meister's Apprenticeship* (1795/96), "Do you know the land where the lemon trees bloom …," is a call from the south dreaming of this epoch. The longing for the region beyond the Alps with its botany became the great basic theme in the German fragrance market. Over the years, the longing of the Germans has shifted even further south into the tropics, reflected in the great love for the fragrance direction "Fruity". Not surprisingly, "Fruity" has been shown over the years in different trends, mostly playing with the dream of the south or the tropics, which can be seen particularly well in the so-called mass-market (mass and prestige).

Heinrich von Ofterdingen (1772–1801) dreams in Novalis' novel of a blue flower, the Sehnsuchtssymbol of the Romantics par excellence, and sets out in search of it:

» "But what pulled him with full force was a tall, light blue flower that first stood at the source and touched him with its broad, shiny leaves. Countless flowers of all colors surrounded it, and the most delicious scent filled the air" (Novalis 1978).

Romantic fantasy flower notes have remained a popular theme to this day. They usually fall into the fragrance direction "Floriental" (Floral-amber or floral sensual), second favorite fragrance direction of German women. The already mentioned Florientals are a permanent trend in Germany and keep coming back to the fragrance market in new transformation.

"O fresh scent, o new sound!" wrote Ludwig Uhland (1787–1862) in his poem "Frühlingsglaube" from the year 1812, which sounds almost like a slogan of the German perfume market. As I said: Around 2000 new fragrances come onto the German market every year, which loves the new. About 30% of fragrance users discover a new fragrance every year, mainly during the spring and autumn months.

In the book *Des Knaben Wunderhorn* published between 1805 and 1808 by Clemens Brentano and Achim von Arnim, flowers are considered which will be mown away tomorrow—wonderful nature romanticism with an indirect steeplechase for the preservation of nature and for the value and recognition of natural scents. The search for an idyllic refuge, which quickly becomes the endangered habitat of plants in nature poetry, is today, especially in Germany and more than ever, a permanent theme, not only in aromatherapy.

The motif of the forest and the trees is in the center of nature poetry. The poem by Joseph von Eichendorff (1788–1857) "Hörst du nicht die Bäume rauschen" is one of the more famous ones. With Frie-drich Hölderlin (1770–1843) it are the oaks, which are mystified as sovereign individuals:

> **The Oak Trees**
> … But you, you magnificent ones! stand, like a people of Titans
> In the tamer world and only belong to you and the sky,
> Which nourishes and raised you, and to the earth, which bore you.
> None of you has gone to the school of humans yet,
> And you crowd happily and freely, from the strong root,
> Under each other up and seize, like the eagle the prey,
> With a powerful arm the space, and against the clouds …

The forest with the heartbeat of the trees has particularly appealed to the Germans. Bestselling author Peter Wohlleben (Wohlleben 2019) speaks in his book *The Hidden Life of Trees* of the hidden bond between man and nature. The writer Elias Canetti (1905–1994) once explained: "In no modern country in the world has the feeling of the forest remained as alive as in Germany" (Canetti 1992, p. 202).

And how is it in German perfumery? Is the scent of resin and forest in the air? Yes, you have to say. He has become a trend and avant-garde through niche perfumery in recent years with the "woody" fragrance direction. With niche perfumes it is about creating fragrance personalities that are not intended for wide distribution, but are aimed at a few connoisseurs. They are not advertised extensively in the media, but are introduced as part of a fragrance consultation. This "resin-wood-forest trend" was first picked up by small perfume boutiques and specialist perfumeries. In the mean-

time, it has developed into a silent power in the German perfume landscape. Psychologically speaking, this fragrance direction is about cleansing the senses, starting anew and focusing. Their fascination lies in a subtle strengthening, behind which more or less the conscious desire stands to find his power place ("Kraftort"). For the Germans this is in the forest.

The success of niche perfumes with woody notes has now led to larger providers such as Penhaligon's, Jo Malone, Montale, Credo, Le Labo, M. Micallef, Amouage, Parfums de Rosine, Byredo, Frederic Malle, Kilian and Serge Lutens. All offer the woody theme as modern fragrance enjoyment.

8.4 Perfume & Image: The Ups and Downs of Perfumery

Different disciplines devoted themselves to the research of fragrance in the 17th and 18th centuries, partly with an almost police-scientific interest, without smell being able to shake off its ambivalent nature and being considered dangerous by many. So it was still discussed and researched in the 19th century whether and how smells would transmit diseases. Above all, a connection was seen between smell and cholera. Although doctors finally ruled out a connection between smells and diseases, the majority of the population held on to the now discarded view (Tullett 2019).

In addition, when assessing a smell, a group-dynamic association could be observed. Those who did not belong to the group supposedly stank. Medicine, psychology, anthropology, ethnology, and history know numerous examples of smell stigmatization. Smells serve the classification and definition of relationships. From the perspective of 18th-century bourgeois society, the rabble stank, while the bourgeoisie perfumed itself (Corbin 1984).

8

The nose has always been used for interpersonal espionage. For example, historian Jonathan Reinharz (Reinharz 2014) reports that European Christians once relied on their noses to identify pagans. Odor stigmatization also took place on a political level. For example, in the 18th century, Anglican Englishmen accused their Catholic French enemies of smelling like garlic (Tullett 2019).

With the beginning of the 20th century, the then still young sociology also dealt with the topic of smell in a sporadic study. The thoughts of the philosopher and sociologist Georg Simmel (1858–1918) from the year 1908 are to be understood as part of this movement (Simmel 1992). After that, the interest of research in smell, which was always unstable and hygiene-oriented, declined due to the outbreak of the First World War. It was not until 1968 that systematic, independent and interdisciplinary research was carried out with the founding of the Monell Center in Philadelphia. It took another half century for Simmel's thoughts to be taken up in the essay *The Sociology of Odors* by Largey and Watson (Largey und Watson 1972) in sociology. More recent contributions to sociological olfactory research are the work of a group of American and Canadian anthropologists and sociologists: Constance Classen, David Howes and Anthony Synnott (Classen et al. 1994).

- **The image change for the perfume industry**

It is hardly possible to determine retrospectively what exactly led to the image change for smell, perfume and perfumery in our cultural circle and when this change took place among the majority of the population. Many things came together. Certainly, publications and especially new scents of the second perfume revolution in the 19th and early 20th centuries caused a rethink.

Only in the 20th century were scents considered chic in our latitudes. They fit the modern emancipation movement of women

and thus also the discussions of intellectuals. After the First World War, women preferred innovative men's perfumes, they did not want to smell like "flowers" anymore. So, in line with the zeitgeist, the scent "Chypre" by Coty came on the market in 1917: a lively, powerful perfume with an aromatic, fine-herbal green head note and almost rough, today no longer used oak moss for allergy reasons—a women's perfume with clear references to men's perfume.

The increased interest in smell and perfume was finally fertilized by discoveries in neurobiological olfactory research in the 20th and 21st centuries. In addition to purely professional books on applications and methods of perfumery, popular cultural-historical treatises on the use of scents and odors increasingly found their way around. In this context, great works were rediscovered, which contributed significantly to the image upgrade of the perfume. This includes, for example, *The Book of Perfume*, a bibliophile work on the use of different cultures with body cosmetics by Eugene Rimmel (1820–1887), a British perfumer and co-founder of the cosmetics company of the same name. The book virtually resurfaced and was reissued 120 years after its first appearance in 1985 several times (Rimmel 1988).

Also required reading is the work The Foul and the Fragrant: Odour and the French Social Imagination (Le miasme et la jonquille. L'odorat et l'imaginaire social. 18e - 19e siècles / Pesthauch und Blütenduft. Eine Geschichte des Geruchs) by the French historian Alain Corbin, a both meticulous and well-documented analysis and description of the hygienic and olfactory situation and its change in Paris from the middle of the 18th to the end of the 19th century (Corbin 1984). Both Corbin's much-noticed treatise and Simmel's work served the German writer Patrik Süskind as a model for his historical novel *The Perfume* (Das Parfum), the story of a murderer from 1985 (Süskind 1985). This bestseller with about

20 million copies sold worldwide has been translated into 49 languages. The book is about the crowning perfume that makes its wearer look so attractive that everyone is blinded by him and he gets unlimited power over people. Like no other perfume book, the novel was discussed in public and raised many old questions of the perfume again:

— Do perfumes make us more attractive?
— Is a perfume deceiving?
— Does the use of perfume have advantages?
— What does a scent have to be like thatalmost everyone lets you get very close?
— How does a fragrance affect the beauty and attractiveness of a person?

The last question falls into the area of psychological aesthetics research. Beauty ideals are subject to the change of taste and fashion. For about 50 years, human beauty has been systematically researched. At the beginning it was believed that beauty lies in the eye of the beholder, but soon studies showed that different people do indeed resemble each other in their sense of beauty. In addition to attractiveness of face, body and movement, social status, personality and voice, body odor and perfume have also been research topics in recent years.

Two discoveries are of particular interest:
— True, as long as it is not in the nosestings and there is no previous experience, it is almost exclusively learned what smells bad. Disgust is learned, but the basis for disgust is innate in every human being. There are, on the other hand, plenty of innate aesthetic preferences for smells. For example, it is genetically determined what is judged to be an attractive skin smell. Women naturally prefer the skin smell of men whose immune system complements their own and thus differs from it. I will discuss it later. Since scents smell differently on every skin, it is the combination of skin smell and scent that makes them "smell good". An ideally beautiful smelling men's perfume ei-

ther merges with the skin smell of the wearer or masks it perfectly. Researchers have calculated it: such an attractiveness scent would have to complement or mask about 120 different skin smells so that the wearer would have a safe advantage—for Patrick Süskind's novel hero Grenouille a probably solvable task. The calculation is based on MHC complementarity (Major Histocompatibility Complex) and includes a group of genes responsible for self-antigens on the surface of body cells and influencing immunological characteristics, which in turn are reflected in body odor.

— As already mentioned, the sense of smell is the only sense directly connected to the emotional center and there especially with the amygdala. It plays a central role in the emotional evaluation and recognition of situations, but also of faces and the interpretation of facial expressions. This gives it a decisive importance in the visual assessment and perception of beauty. It is also indispensable for the perception of any form of arousal, that is, affect- or lust-related sensations. If the area is additionally stimulated by scents, it can act as a real emotional amplifier and decide the degree of attractiveness, love, affection and happiness. In combination with the Hypothalamus and its network, scent can then additionally lead to the release of Dopamine and thus become the "smelling happiness". Which reaction a stimulation leads to depends on the mood one is in and how one would like to experience oneself. Since women's amygdalas are more strongly interconnected than men's, they usually smell better and a pleasant-attractive smell can be linked to a feeling of well-being that reaches up to happiness, scents should actually make them beautiful—at least for themselves. The actress Audrey Hepburn (1929–1993) is said to have once said: "I believe that happy girls are the prettiest girls" ("Ich glaube, dass glückliche Mädchen die schönsten Mädchen sind"). She certainly did not have the media scent and perfume in mind.

Younger men in particular are interested in the answer to the question: "What does a scent have to be like thatalmost everyone lets you get very close?" There are many myths surrounding this topic, which have contributed to the increased interest in smell and perfume.

8.5 Pheromones Or: The Current State of the Search for the Atlantis of Perfumery

In this section we set out in search of the Atlantis of perfumery. We discuss the often controversial topic of "pheromones in humans". I would like to ask you right at the beginning:

- **Do you belong to the believers or unbelievers in pheromones?**

Smell affects our choice of partner, that is undisputed. I will discuss it later. Not only body odor is discussed in this context, but also chemical signals that surprisingly reduce sexual arousal in men. For example, it has been shown that the smell of tears shed by women in sadness triggers this process in men (Berger et al. 2017)—a reaction that is also reflected on a physiological level, e.g. in a lower testosterone level, but also in a reduction of the brain activity that controls sexual stimulation (Gelstein et al. 2011). Cell physiologist Hanns Hatt from the Ruhr-Universität Bochum asks in this context whether one can conclude from such studies that there could also be a pheromonal effect in humans that differs from classical olfaction.

In this context, Pheromones are also understood to be messenger substances for information transfer that are perceived unconsciously and influence reproduction-related physiological processes or corresponding behavior. But this definition already begins the problem. Many researchers avoid the term "pheromone" in favor of "chemosignal" (chemical signal) with the argument that no general consensus can be reached as to what the term pheromone stands for in humans. This problem does not exist in popular science articles. Pheromones are discussed here as triggers for sexual readiness, as can be observed in the animal world. Whether we humans also communicate this readiness with pheromones has not yet been proven, but has triggered a feverish search with a large number of studies and publications. Although the existence of pheromone receptors has now been discovered in humans, which can be activated by the fragrance Hedion, a synthetic fragrance with a floral scent, it is not really certain whether the receptors are involved in sexual readiness. At least it was found that Hedion has an influence on human sexual behavior, if not a direct, then perhaps an indirect one. In the experiments on this, people reacted with increased trustworthiness under the influence of the fragrance Hedion when others first trusted them (Berger et al. 2017), according to the motto: "As you do to me, so I do to you".

In the animal world, pheromone relationships are often more direct. Often an animal cannot help but agree to mate when it perceives the dispatched messenger substance. Of course, this behavior cannot be transferred to humans, and for this reason, according to the current state of research, the existence of pheromones in humans is doubted (Wyatt 2015). Pheromone believers don't like to hear that, because they argue that humans are mammals and it is therefore possible, perhaps even likely, that we have pheromones. They don't run as directly as often in the animal kingdom, but chemical signals also have an effect on humans, for example, as we will see later, that women are perceived by men as happier and more relaxed and therefore more attractive.

The term "pheromone" has been around since 1959. It was introduced by a German-Swiss biochemist team. Of the three skin glands (sweat, scent and sebaceous glands), it is the scent glands, the so-called apocrine glands, that the pheromone research focuses on. Scent glands are not active from birth, they develop with puberty. They are located in the area of the armpits, the nipples and in the groin area and secrete fragrance substances with pheromone-like effect, especially in reaction to emotional stimuli such as desire or fear, but also in reaction to sexual hormones.

The proof of the widely held belief that there are human pheromones is mainly based on four steroid molecules (biological substances that convey a certain message or information to certain body cells): Androstenone, Androstenol, Androstadienone (AND) and Estratetraenol (EST). Whether and how these four steroid molecules affect the sexual readiness of humans is a much-discussed question. I would like to briefly discuss each substance before we turn to pheromone research.

Androstenone is an androgen (male sex hormone) and is formed as a chemical by-product during the breakdown of androgens. It is said to have a masculinizing effect and a masculine attractiveness associated with it. Androstenone is secreted mainly by the sweat glands of the armpits, especially in men. It is directly related to the sex hormone testosterone. Pheromone believers assume that research will identify Androstenone as one of the strongest pheromones for "sexual attraction".

Androstenol was first extracted from boar testicles before it could be synthesized. It occurs in humans, like Androstenone, in the sweat glands, but also in urine and other body fluids. Also, Androstenol resembles the male sex hormone testosterone in its chemical structure. Men produce it on average two to three times more than women. Androstenol has a slightly musky smell. As "alpha-Androstenol" it is a substance that smells faintly of sandalwood. Pheromone believers assume that research will identify Androstenol as a pheromone that promotes "friendliness" or "trust".

Androstadienone (AND) is a breakdown product of testosterone and is found in male sweat, but also in semen. So far, no receptors have been discovered in humans that can recognize AND. But pheromone believers do not give up hope because AND probably affects brain processes that, among other things, control the perception of faces. The influence on the amygdala is discussed.

Estratetraenol (EST) was first discovered in the urine of pregnant women in the late stages in the 1960s. EST as a compound resembles estrogens. Probably it is also present in armpit sweat. A whole range of effects are attributed to the alleged hormone, which are also shown in women themselves, including influences on psychological-emotional reactions. For example, it has been found that EST increases the mood in women.

- **The mysterious organ**

Originally it was thought that the potential effect of the chemical signals that are supposed to communicate sexual readiness, begins with a almost mysterious organ, the vomeronasal organ (VON) or the Jacobson organ, which smells the messenger substance. Even today, however, its function is often questioned for humans. Nevertheless, it exists in the form of tiny indentations on both sides of the nasal septum; more on that in a moment. Meanwhile, it is becoming clear that sexual readiness can also be communicated in the animal world without VON by chemical signals. You can see that the topic of "pheromones" is more complex in the animal world than previously thought.

For a long time it was believed that humans lack the vomeronasal organ. In the scientific literature, this was therefore referred to as a "olfactory devolution". The current research results contradict this. They do not assume that the sense of smell has decreased during human evolution (Shepherd 2013). Ethnologists were among the first to notice the importance of the sense of smell in interpersonal relationships that is still valid today. They pointed out, for example, that in some cultures, such as among the Inuit, people still smell each other's hands, which has developed into a handshake in Western civilizations.

In the mid-1980s, olfactory research was in for a surprise. A team from the University of Utah in the USA proved the existence of the vomeronasal organ in adults and documented its function. Up to this point, it had been assumed that it only existed in infants and regressed during development. The stimulation of the VNO now showed an interesting effect: adult test subjects reported not smelling anything, but feeling a "warm, vague sense of well-being." This discovery spurred the commercialization of pheromone perfumes, which did not necessarily contribute to the positive image of the perfume industry, while pheromones became a research topic. Here, attention was initially focused on the effect of androsterone (not to be confused with androstenone) on women in particular. In the animal kingdom, the steroid androsterone is known for its influence on sexual behavior. Humans also excrete androsterone, men much more than women. It is similar to the male hormone testosterone and should work on women unconsciously like an aphrodisiac, at least that was the hypothesis in the 1980s. A worldwide olfactory study conducted by the National Geographic Society in 1986 provided a first clarification of its effect on humans. Over a million scent cards ("Scratch and Sniff," which release the scent when rubbed) were distributed for evaluation in order to test the scent of human androsterone in a blind test. The results were sobering. Androsterone was rated as one of the least attractive scents, especially by women. This unexpectedly negative result of the pheromone effect raised other fundamental questions: Can there be any objective studies on the topic of pheromones or olfactory effects that aim at sexual readiness at all? Do commercial interests or personal fascinations play such a big role in this topic that research is continued until the desired data is available? With pheromone research, one can actually get this impression. Of course, the idea of finding scents or odors that can actually manipulate one's counterpart unconsciously is very tempting—to like someone, to respect them, to do something for them or to think something specific about them.

Despite the negative results of the National Geographic Society study, pheromone believers did not want to give up. They continued to research, and indeed certain effects of androstene molecules could be shown—for example, on mood, physiological arousal, visual perception, and brain activity (Savic und Berglund 2010).

As early as the early 1970s, Martha McClintock, an American psychologist, published an observation that was repeatedly referred to. It is the description of the amazing phenomenon that, in women of childbearing age living together, the menstrual cycle approaches. The pheromone unbelievers criticized these studies for statistical and methodological weaknesses. Even a repeat of the 2009 study in a Polish student dormitory could not confirm the McClintock results of Wellesley College in Massachusetts. But the pheromone believers referred to a study by Geoffrey Miller from 2007. In this study it was shown that "lap dancers" (women who dance in front of customers in close proximity and on their laps in a bar), who did not use the pill, were given 50% more money just before ovulation than a control group. Miller himself admitted that the results could be explained in different ways: for example, by changes in their personality, their tone of voice or other physical characteristics. For the pheromone believers, this study was at least the confirmation of an ancient knowledge: men judge the self-smell of women most attractive shortly before ovulation.

- **Why the pheromone believers also like to refer to Napoleon Bonaparte**

The research interest then focused, inter alia, on so-called copulins, which are actually not pheromones, but which women emit in phases of the cycle. Their smell is said to have attracted men at all times, probably also Napoleon Bonaparte, who wrote to his Joséphine: "Do not wash—come in three days."

Chemically, copulins are a mixture of short-chain fatty acids. They are found in female vaginal secretions. There they occur more during ovulation. In a study already carried out in the mid-1990s at the Vienna Ludwig Boltzmann Institute, whose methodology and results are now being questioned, young men were asked to sniff copulins. However, the mixture was so diluted that the test participants had no conscious olfactory perception. During the smelling, the men were asked to judge portrait photos of different women. Before and after smelling, the test participants had to provide a saliva sample for the measurement of the testosterone level. An increase in testosterone was recorded in all 66 test persons, regardless of how attractive the respective woman was rated on the photo. The men also perceived the attractiveness of the women to be equalized. Less attractive women now appeared more attractive. The influence of copulins was thus attributed to a kind of chemical change of consciousness. It was not clarified how copulins are actually perceived and what role the vomeronasal organ plays in this. Because according to recent research at the University of Dresden, only two thirds of all people have this relic from the early days. Accordingly, one would have expected that not all men from the Vienna study could be stimulated by copulins—at least not if the stimulation

goes through the vomeronasal organ. Accordingly, the Vienna study, which is based on a diploma thesis (*Female Pheromones—Effect and role of synthetic copulins in the hidden ovulation of humans*), which was submitted by Astrid Jütte in 1995, was received with mixed feelings in the academic world. The results, which were published in the same year in the journal *Gynäkologische Geburtshilfliche Rundschau* (Grammer und Jütte 1995), for example, were then so exciting for the press that they were taken over without much questioning and are still often quoted today. However, many still overlook a study from 2017 by Megan Williams and her colleague Coren Apicella, both from the University of Pennsylvania, which is based on Jütte's study. Based on the study of 243 men, she comes to the conclusion that copulins do not influence their sexual behavior: "We found no evidence that copulins influenced men's sexual motivation, the willingness to engage in sexual adventures, … [or] the perception of female attractiveness …" (Williams und Apicella 2017; Williams und Jacobson 2016). In other words, the question of whether copulins can act like pheromones is—to put it cautiously—also postponed here.

Similarly, investigations into the effects of two other potential human pheromones mentioned earlier, androstadienone (AND) and estratetraenol (EST), have so far not yielded consistent results (Chakkarath und Weidemann 2018). AND is discussed as a potential male pheromone, EST as a female one. AND and EST are said to at least raise the mood in the opposite sex. However, studies have not yet yielded a clear result because the influence of the test leader seemed to play a role. However, recent studies report that EST has an influence on the perception of the emotional state of women by men. Women are perceived as happier and more relaxed (Ye et al. 2019). Furthermore, it was observed in one study that EST increases the perception of emotional touch (Oren et al. 2019). In other words, the search for the Atlantis of perfumery will continue.

■ A Swiss scientist sheds light on the Atlantis of perfumery

The use of sex pheromones as ingredients in pheromone perfumes has, as said, damaged the image of modern perfumery. In particular, the promise of promised effects has always given the industry a touch of charlatanism, especially because scents came on the market that did not hold up in independent tests what they promised. Therefore, one cannot thank the biologist and Swiss immunologist Claus Wedekind enough that the topic has been discussed academically more complex since the turn of the millennium. The confirmation of the effect of scents on human sexual attraction has so far only been reliably provided by immunology, more precisely by organ transplantation research. Our cells are equipped with proteins that help the immune system to identify itself as one self. By chance, the following was discovered: Female laboratory staff were able to assign mice to different immune system groups based on their smell.

Claus Wedekind from the University of Bern and Lausanne went to the bottom of this in 1995 and 2005. If women could smell differences in the immune system in mice, shouldn't that be possible in men too? This was an exciting question because it was already known for a long time that fertility increases with different immune systems of man and woman. Conversely, the risk of pregnancy complications increases with similar immune systems of the partners.

Wedekind applied the typical T-shirt sniff test. The male test subjects were asked not to change a neutral white T-shirt for two days, to sleep in it, not to use perfume, aftershave or deodorant, not to eat anything sharp, not to smoke, to drink only water and to abstain from sex. After that, the T-shirt was sealed and presented to

women (100 students) in a blind test. The results spread quickly. Women judge the body odor of men with different immune systems to be more attractive ("smells like the boyfriend or the ex-lover"). However, if the body odor was too strong, the effect was lost.

In summary, one can say about pheromones: The connection between body odor and sexual attraction is now undisputed. Whether one can also use them as corresponding fragrance ingredients in perfumes will remain a much-discussed and controversial topic. This takes me back to the question I asked at the beginning: After reading this section, do you now belong to the pheromone believers or unbelievers?

8.6 The Second Fragrance Revolution: How Discoveries Led to a Better Smell

Increasingly more fragrance users wish for perfumes made from purely natural instead of synthetic ingredients. This trend has manifested itself in organic or vegan as well as in bio-perfumes. From the end of the 19th to well into the 20th century, it was exactly the opposite: synthetic stood for progress, innovation and future. One reason for this were the numerous chemical discoveries in the second half of the 19th century, which changed world trade.

For centuries, access to and trade in aromas and colors were strictly controlled. Too much was at stake for the trading houses of the countries that dominated the sea like Spain, Portugal and England. They had made a killing with the import monopoly. No wonder that solutions were sought in the interior. Vanilla is a good example of this.

Originally, the spice vanilla came from Mexico and Central America. It was already very popular with the native Mexicans, the Aztecs, as a spice because of its aromatic ingredient vanillin. Eventually, the Spanish conquerors brought it to Europe. Since it only grew in Mexico at first, the Spaniards held the monopoly on vanilla for a long time, which became the second most expensive spice in the world after saffron. Although there was a great demand from the beginning, but time-consuming cultivation and processing stood in the way of a quick delivery. In addition, it was not possible to grow vanilla in other areas, because a special bee and hummingbird species that only occurred in Mexico and Central America was needed for pollination. Not until 1837 did a Belgian botanist and 1841 a plantation slave succeed in artificial pollination.

But artificial pollination did not solve all problems. The long processing of vanilla remained. Its fruits had to be blanched, i.e. heated in water and then dried in the sun for weeks. After that they came to ripen in boxes before they acquired their characteristic aroma.

- **Sensational news for the fragrance world from a German small town—not from Paris**

In 1874, the sensation came from a place the world did not know. In the then only 6000-soul small town of Holzminden in southern Lower Saxony, the German chemist team of Wilhelm Haarmann and Ferdinand Tiemann succeeded in the synthetic production of Vanilla. In the same year, the two founded Haarmann's Vanillinfabrik. In 1876, Karl Reimer joined as a third owner of the company. He had discovered how to produce vanillin even more efficiently and cheaply. The company was now called Haarmann & Reimer. Another discovery by Tiemann increased the profitability of vanilla production. Two years later, it was even possible to produce violet aroma artificially. This gave them a real trend-fragrance building block in their hands, because violet is rediscovered by almost every generation. By the way, Haarmann & Reimer was bought by Bayer AG in 1953 and merged into the company Symrise in 2003.

The second half of the 19th century also brought breakthroughs in other areas of synthetic research. In 1856, the first synthetic dye, Mauvein or Mauve, a weak reddish violet, was discovered. In 1878, perhaps the greatest success was achieved in the field of color: the first full synthesis of Indigo, the deep blue. This was followed by a bitter competition between manufacturers of natural and synthetic Indigo. But thanks to industrial synthesis processes, Indigo could be produced more cheaply and in even more spectacular blue, so that the market for natural Indigo finally collapsed.

The first synthetic production of a plant substance that was to revolutionize the perfume industry took place in 1868 with Coumarin. Like synthetic vanilla (vanillin), coumarin could be marketed industrially from 1876. The plant substance obtained from tonka beans smells highly aromatic and sweet and tastes like vanilla and woodruff in flavor. Synthetic coumarin and vanilla were a wonder of modern chemistry and were actually just waiting to be discovered in a perfume of the future.

Nobody asked if the use of Coumarin and Vanilla—in whatever form they were obtained—would entail any health risks. Not until 1954 was coumarin banned as a flavor in the USA because of toxic reactions and later also restricted on the other side of the Atlantic in its maximum content as an ingredient. But even the consumption of vanilla, the world's favorite flavor No. 1, is now discouraged in large quantities. Paracelsus already put it in the early 16th century: "All things are poison, and nothing is without poison; only the dose makes it that a thing is not poison."

- **The second revolution in the perfume industry by Paul Parquet**

The second perfume revolution began in 1881 with the use of synthetic fragrance molecules in perfumes. It was triggered by the perfumer Paul Parquet (1856–1916), co-owner of the perfume house Houbigant founded in Paris in 1775, which still delights olfactorily today with its perfumes. Monsieur Parquet was committed to the long tradition of his perfume house and the French perfume industry. Already during a whole century before Parquet's time, Houbigant had achieved everything that a perfume house could wish for. His customers are said to have included Marie-Antoinette (1755–1793), the later queen of France executed on the guillotine. Napoleon is also said to have had a special perfume created for Empress Joséphine at Houbigant.

At the time Parquets there was a great competition among the perfumers and fragrance houses. Especially the relatively young, 1828 founded up-and-coming fragrance house Guerlain developed more and more into a dynasty and threatened to overtake Houbigant. Parquet had to act. He could no longer rely on classic French perfumery, but had to take a chance on something new. He was aware that only a revolutionary leap into the future of perfumery could eliminate the dusty image of his fragrance house. Parquet was not alone in these considerations. For a long time, perfumers have been looking for new ingredients to create a completely new fragrance experience.

Parquets decision in this situation should make him one of the greatest in the art of perfumery and a pioneer of modern perfumery in many ways. He created a completely new men's perfume: "Fougère Royale" (royal fern). For this perfume, he was the first to use a synthetic fragrance ingredient, namely Cumarin, one. Actually, one would have expected this fragrance ingredient—if at all—in a women's perfume. But Parquet went one step further: He gave the almost odorless fern a smell. "Fougère Royale" thus broke with the dominance of natural scents as the first fragrance and introduced Fantasy Notes one. That was new too. So far, perfumers had adhered to the world of smells of nature. Parquet invented a smell for "non-smelling nature"—not just

any, but a fragrance that was to be imitated most often and represented a whole new fragrance family. As mentioned above in connection with Charles Baudelaire: Paul Parquet knew what he was doing. So he is said to have said: "If God had given ferns a smell, they would smell like Fougère Royale."

Parquets Fougère Royale became the founder of the fragrance family Fougère perfumes, in which mainly lavender, bergamot and geranium play together and have become the DNA of men's perfumery. Hardly any men's fragrance was imitated more often until it disappeared from the market in the late 1960s. No wonder it was recently revived for younger noses.

However, Parquet did not live to see how "Fougère Royale" also revolutionized women's perfumes. It is doubtful whether "Chypre" by Coty, which came on the market in 1917, a year after his death, was influenced by this. Although both perfumes share some ingredients such as bergamot and oak moss, "Chypre" certainly also has other facets. Nevertheless, the fragrance philosophy of "Chypre" with its liveliness and power or, expressed differently, with its "rough shell and soft core" reveals much of the fresh spiciness of "Fougère Royale". Meanwhile, more than one hundred years after Parquet's death, the "Fougère" fragrance family is also well represented for women and is currently becoming more and more popular. These include fragrances such as "Acqua di Colonia Fougère" by O Boticário, "Fougère" by L'acqua di Fioria, or "E for Women Green Fougère" by Clive Christian.

▪ **The first fragrance of emancipation**
However, Parquet saw how Guerlain took over his idea of synthetic coumarin. So in 1889 the Guerlain perfume "Jicky", about eight years after the introduction of "Fougère Royale", was the first women's perfume with synthetic coumarin and vanillin. You can smell two intersecting fra-

grance families: "Amber-Oriental" and "Fougère". Originally, "Jicky" was created and brought to market as a men's perfume. But then emancipated French women discovered the perfume for themselves. After all, they were the first women in Europe to demand their rights in petitions, at assemblies and also on the barricades.

The Paris Commune of 1871, when women stood on the barricades and demanded equality, is considered a high point of the women's movement. In 1880, Hubertine Auclert, founder of the women's suffrage movement, organized a tax boycott with her fellow women: We leave the privilege of paying taxes to the men who vote on their amount in parliament and who claim the right to decide on state support. "In addition, Hubertine Auclert organized a census boycott with the motto:" If we don't count, why count us? "Finally, the women's movement experienced another increase with the radical suffragettes, who, at the beginning of the 20th century, demanded universal suffrage for women with passive resistance, disruptions of official events and hunger strikes.

With the increase of the women's movement, the traditionally feminine fragrance notes no longer fit the modern lifestyle of many women. They wanted to be different from their mothers. This was also reflected in the choice of perfumes, which, like "Jicky", were actually intended for the modern man. The perfume manufacturers quickly adapted to the new circumstances. For example, Jicky's perfumer Aimé Guerlain changed the story of his fragrance: originally dedicated to his nephew, he declared, in view of the changed circumstances, that he had created it in memory of his English girlfriend.

Guerlain owes a lot to Paul Parquet. Unintentionally, he gave the starting signal for further great perfumes with synthetic fragrance ingredients. The milestone for the so-called amber-oriental fragrance family was set by Guerlain in 1925 with "Shalimar". Here, the heart and the base shine with an oriental-fascinating, tempting mystery, to which synthetic vanilla (vanillin) also contributes.

Sure, other perfumers would have eventually come up with the idea of using the new synthetic aromas or brilliant wonders of the then chemical industry in their fragrance creations. Because it has been an old dream of perfumery to make flowers shine even more than in the light of the sun. In 1921 it was finally time: for the first time, aldehydes, the newly discovered light amplifiers, could be used generously in a fragrance. "Chanel No 5" was born. The most gifted perfume in the world has influenced an entire fragrance family with its synthetic brilliance.

What outsiders may not be so aware of: without the innovations of the chemical industry, there would be no more great perfume classics. Thanks to synthetic fragrance ingredients, it is today possible to recreate perfume masterpieces of past days without animal ingredients. They were previously common in many perfumes, such as civet (gland of the civet cat), ambergris (the excretion of the sperm whale—the possession and trade of ambergris was banned in the United States by the Endangered Species Act of 1973), musk (gland on the belly of the musk deer) and castoreum (secret of the beaver). They are fortunately mostly forbidden today, but unfortunately they are still partly used from breeding stocks. Animal odors such as ambergris have also become unaffordable.

I do not want to deepen the discussion about natural or synthetic scents here. There are good arguments for both sides. However, with natural aromas, the limits of growth are often reached. So the worldwide harvest of about 1000 tons of real vanilla can no longer cover the increasing demand. Here, the synthetic fragrance is the only solution.

8.7 Perfume & Self—Effects on the Ego or: The Most Beautiful of All Drugs?

Not only as a psychologist, one wonders why modern people perfume themselves at all. Only "to cover unpleasant odors" should hardly satisfy perfume lovers as an answer. What do you want to achieve with the use of your favorite perfume? There is a variety of contributions from neighboring disciplines such as sociology, aesthetics, medicine, philosophy, anthropology, ethnology, history and of course literature that try to answer these and similar questions. The following areas of psychology deal with this topic:

- Psychophysiology: It studies above all process sequences during smelling.
- General Psychology: It deals with the meaning of olfactory perception for the quality of life, that is, for the general well-being.
- Developmental Psychology: It is particularly interested in olfactory socialization and in innate and learned olfactory preferences.
- Social Psychology: It focuses on the effect of scents on the social environment.
- Psychology of aesthetic perception: It evaluates the factors of olfactory preferences.
- Clinical Psychology: It studies the effects of loss of smell and the use of aromas for therapeutic treatments.
- Personality Psychology: It provides insights into the fascination of perfumes and fragrance choice. Within this direction there are different schools and approaches, including the already mentioned self-discrepancy theory by Higgins (1987). According to this, people are more satisfied, the smaller they experience the distance from their current self to their ideal self. So you feel that you are getting closer to what you want. Ideally, the experience wish overlaps

with the current experience. For example, you want to become more active, fitter or more athletic and you also achieve this.

The "experience of being" and the "ideal experience" rarely overlap. The self-discrepancy theory therefore does not claim that the ideal state must be achieved in the "now". Rather, for personal happiness it is enough if experience wishes become tangible and one experiences them as approaching. If you transfer this theory to perfumery, you can in particular see favorite perfumes and their fragrance directions as transformation offers for the current experience and self. One tries to get closer to a desired ideal self or ideal experience more or less consciously (◯ Fig. 8.1).

The perfect choice of perfume is therefore both experience- and goal-oriented. It depends on the personal experience wish and how one wants to appear to others or in a situation.

In general, one can say: With perfumes one reduces the distance between how one experiences oneself at the moment and how one would like to experience oneself more now or in the near future. Simply put, the use of perfumes is always about self-optimization and -enhancement, about mood brightening, -change and -correction, combined with the experience- and goal-oriented desire to become more attractive to oneself and others. This can be shown in different ways, for example through the recognition one receives because one was the first to find and wear a perfume; because it smells good; because one feels comfort-

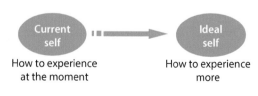

◯ **Fig. 8.1**   Current self—ideal self

able with the perfume; because it refreshes pleasant memories; because it brings happiness and success. The personal reasons are manifold.

This does not contradict the general view that a perfume should underline the personality of its wearer. Many fragrance experts even claim that they feel naked without their perfume. Even someone who has used the same perfume for all situations for a long time does this to complement his style, his aura or personality with a final, still missing touch. With this one stages oneself anew and comes closer to the desired image of oneself of the personality. This is confirmed by modern psychology.

Personality is always also a process. From the perspective of neuroperfumery, the choice of fragrance is still additionally influenced by the search for individual brain areas or regions for stimulation. One could even say: perfumes are a welcome over- or under-stimulation compensation for certain brain regions. But perfumes are—at least at the moment—not drugs that change consciousness or redefine reality. But they can trigger processes in the wearer and in the environment and are therefore powerful.

- **Why we "perfume" ourselves with a certain perfume**

The question of why we "perfume" ourselves with a particular perfume, why this fascinates us and we bought it, is much more difficult to answer than the general question of perfuming. The final influence of possible factors such as climate, season, smell on the skin, current mood, price, time of day, day of the week, bottle design, image of the brand, chance, influence of the best friend etc. is almost uncalculatable, and yet all these factors are closely interwoven. Often, the explanation for the specific choice of a perfume and its fascination also fails due to the language. It is almost impossible to express what a scent means

to one and how one experiences it. Or one may not admit the true reason for a fragrance choice, possibly one simply does not know which previous, unconscious olfactory experience has influenced the choice.

To this day, it depends on the eloquence of each individual fragrance lover, which is extremely brilliant with some, to describe his or her most beautiful of all drugs to others. Classification systems, the so-called fragrance families or fragrance directions, as I have already introduced them, are helpful, but often too general to really make individual fragrance experience and the why of wearing understandable. The history of perfumery has taken a long time to develop a language for describing scents. With taste, people were always a little further ahead. So already Aristotle in addition to the four basic taste directions sour, sweet, salty and bitter also called herb, sharp and hard, in the sense of harsh or unpleasant. This resulted in seven basic directions. Transferred to the world of beautiful scents, this number is of course not enough.

Carl von Linné (1707–1778), the great Swedish botanist and plant systematist, dealt with fragrance taxonomy in the 18th century. As head of a large garden, he found the time to work on his main work, the *Systema naturae* (System of Nature). The result for the classification of smells were seven classes: aromatic, floral, ambrosial (musk), leek-like (garlic), goat-like (goat), repulsive (rotten), nauseating (bad). This classification was only partially applicable in fine perfumery.

In 1895, the Dutch physiologist Hendrik Zwaardemaker (1857–1930) undertook one of the most famous experiments of the past 200 years, classifying invisible things. He developed 30 smell classes. Also worth mentioning is the smell prism developed by the German psychologist Hans Henning (1885–1946) for the three-dimensional representation of smells. The six cor-

8

ners of the prism represent the following six smell classes: rotten, fruity, resinous, smoky (burnt), spicy and floral. According to Henning's theory, each smell should occupy a specific position in the prism. But this did not impress the experts of later generations.

More successful in the 1960s and 1970s was the British biochemist John E. Amoore. He proposed the distinction between base smells such as etheric, camphor-like and musk-like as well as floral, minty and rotten. The discussion about the number of base smells did not end there. How much the perfume industry wanted to postulate a scientific theory of smell, comparable to the visual perception and the qualities of the basic colors, from which impressions can be derived systematically.

Contributions to the topic "base smells" came earlier from the most diverse areas. For example, the ethnologist Claude Lévi-Strauss (1908–2009) reported in his book *La Potière jalouse* (The jealous potter) published in 1985 about the South American Suya Indians, who divide people into four smell groups: strong or wild game-like, sharp, bland and rotten.

As early as 1927, the US chemists Charlton Crocker and Lloyd F. Henderson postulated four pure base smells: floral, sour, smoky and rancid. All other scents should, according to their theory, be derived as mixtures in different strengths (1–9). This resulted in the number 6432 for the scent of roses and the number 7683 for the scent of coffee.

William McCartney reports in *Olfaction and Odours* (McCartney 1968) that much earlier, around 1894, an attempt was made to create a smell alphabet based on chemical formulas. For classification and also for the creation of beautiful scents themselves, the restriction to a few base smells or formulas quickly reached its limits. That is why the perfumers themselves became active. Also, in the long history of perfume culture, an internal jargon had developed,

which was traditionally very similar to figurative expressions or comparisons.

English flavor researcher Roger Harper, to name just one, presented a total of 44 fragrance classes in 1968. These include, for example, fruity, aromatic, almond-like, minty, citrusy, sweet, Vanilla, soapy, metallic, animalistic and floral. They became an official language repertoire for perfumers in order to describe ingredients and fragrance creations uniformly. Nevertheless, one is dependent on the eloquence of consumers for the understanding of individual fragrance fascinations. For example, on the website fragrantica.de you will find a successful fragrance description by the blogger Aquaria for the perfume classic "Chypre" by Coty:

"I have long hesitated to write a review of this fragrance for fear of not doing justice to the fragrance. But what the heck, I'm not a perfume expert, I'm just going to write about how I feel this fragrance. First of all, I am surprised by the liveliness and power of this fragrance. Already with the first spray I smell cool, rough oak moss, which remains very prominent and powerful throughout the drydown. At first it forms the mossy background for aromatic green citrus notes. I smell a green, bitter Bergamotte in combination with bitter herbs. A very cool and green impression, sometimes on the verge of bitter, but beautifully fresh and full of energy. I also believe that I can smell green, supple labdanum. The fragrance is in this stage green, herbaceous, with a cool, rough surface. I see craggy rocks in front of me, out of which mosses and aromatic grasses grow, over which a cool, rough wind blows. So far, so masculine. But in the heart the fragrance also offers warmth. It becomes softer and more floral under the moss layer. Sweeter and warmer tones of jasmine and rose make themselves felt, based on an earthy and powdery iris. I also smell civet, not obtrusive, but noticeable; it gives the fragrance depth and animal warmth. And finally, a slightly powdery-spicy garden carnation seems to play along. For me, the fragrance smells like a fairy tale scene. I see an originally idyllic landscape with beautiful, bright flowers that were enchanted by a grumpy sorcerer and then completely overgrown with cool moss. And there the sleeping flowers patiently wait for a nose to come along, to smell them and thus to wake them up again. This magical process happened in the drydown; the rough surface and the soft core eventually resulted in a harmonious image. In short: an impressive chypre, not pleasing at first and second sniff, it requires some sniffing work, but once you have discovered the beauty inside this fragrance, you understand why it became a reference fragrance."

To better understand individual fragrance choices, one should still distinguish between "private fragrances", which one wears more for oneself and for one's own self, and "social fragrances", which one wears on a special occasion, from a fragrance psychological point of view. With the first group, the fragrance decision is often spontaneous and emotionally oriented towards what one needs for oneself and the current mood at the moment, what feels good, what one wants to experience and what gives one pleasure. With the second group, the decision is more cognitive in nature. One thinks more about how the fragrance is experienced by others, whether it fits the occasion and arrives accordingly. Of course, a fragrance can be both a private and a social fragrance. Also, a private perfume life can overlap with the anticipation of a social event. It's just not always easy to distinguish between these two types of fragrances.

Fragrance psychology is particularly interested in perfumes to which the wearer feels a deep connection, which trigger something in him, which touch him, which he needs for himself before he thinks about the acceptance of the fragrance by others. In their own perfume cabinet, private perfumes are usually in the minority. After their discovery, they are worn until they are consciously not used for a certain period of time. These are fragrances that one likes to re-purchase or receive as a gift. They have become quasi-part of oneself. Such perfumes are defended against everything and must never be changed. If they disappear from the shelf, their lovers equate it to a world collapse.

These fragrances serve only the own self. Their secret: With them one lives out one's ideal self. One enjoys the fragrance, even if one is not really aware of who one actually is or should be—artist, fashion designer, musician, film star, top athlete or entrepreneur. Sometimes one even has the impression that the fragrance leads one back into one's own past life, in which one could have been an Egyptian queen, princess, baron, adventurer or world explorer. Only then can one understand the passion with which perfume artists exchange fragrances on perfume blogs and pay tribute to certain fragrance preferences of celebrities and powerful people from history. In the way they experience the fragrance, they define their own ideal self-requirements. These are usually less oriented towards personalities from the past, but towards those from the present, such as stars from fashion, film, music and entertainment.

The ideal self as an instance of experience desires is also olfactory in the case of wanderlust. So we associate the smell of perfume with travel experiences, distant places and situations—regardless of whether we have already made or experienced them or whether we simply imagine them. This became particularly clear in times of travel restrictions such as Corona. What is fascinating is that perfume users often have similar or identical associations with a perfume. Here are five examples:

8

— For example, the smell of tropical sea is smelled by many untrained noses in "Marble Sea" (Atelier Oblique).

— For example, the smell of the beach of a city by the sea in the south is caught by many perfume users in "California Dream" (Louis Vuitton).

— For example, the smell of a Mediterranean sailing trip is smelled by perfume lovers in "Replica" (Martin Margiela).

— For example, the smell of sunscreen by the pool is smelled by sun worshipers in "Huile Prodigieuse" (Nuxe).

— For example, the smell of tropical forests, which one may only know from visits to domestic greenhouses, is reproduced by "A Chant for the Nymph" (Gucci).

Of course, the naming and mentioning of essential ingredients of a perfume support the goal of an imaginary long-distance journey. For example, "Bal d'Afrique" (Byredo) reveals where the journey is going. Nevertheless, in a blind test, the scent, which smells of cedar wood and other exotic ingredients, creates impressions that many perfume lovers associate with the Sahara and North African regions, especially when they want to travel or see them again. In other words, scents not only help the own self, but also the ideal self, especially in times when one has wishes and needs but cannot live them out, for example because the vacation is cancelled. Perfumes work, for example, against a need for wanderlust (or better said "far-woe," more figuratively: A longing for distant places) because the ideal self can at least travel on a scent journey in their thoughts.

For the sake of fairness, it should be said that the ideal self can not only be sent on a scent journey with the smell of perfume (for example with Eau de Parfums or Eau de Toilettes), but also, for example, with the smell of scented candles. All other types of room scents (such as fragrance dispensers, aromatherapy lamps, fragrance stones, diffusers or room sprays) can also act as mood makers. They not only aim at the olfactory improvement of the domestic aura, but also take you to distant worlds.

How do you like our shared scent journey? Isn't it highly exciting how controversially the topic "pheromones in humans" is discussed or how the first perfume of emancipation came about? The latter shows how the perfume industry has adapted to social movements and thus to the zeitgeist or how it must constantly adapt. In ▶ Chap. 15 I will give you further insights based on interesting examples, especially with regard to the development opportunities perfume creators have today.

Summary

In this chapter we discussed the second revolution of perfume and saw how it came to the first scent of emancipation. We also examined the ups and downs of perfume from different perspectives. So we discussed the important role of Charles Baudelaire for the development of modern perfume. But we also saw that not a few of the famous classical philosophers have expressed themselves disparagingly about the sense of smell and the effect of scents. I therefore introduced you to the "Godfather" of perfume— Friedrich Nietzsche—who has contributed a great deal to the image enhancement and fascination of the sense of smell and smelling. The romantic period also contributed a lot to the positive image of perfume. But we also got to know another side of perfume, which stigmatizes, excludes and spies on the olfactory. Pheromones in humans are a controversial topic in research and have caused an up and down of the perfume image with perfumes that did not keep their promises. Although the existence of pheromone receptors has now been discovered in humans, which can be activated by the fragrance Hedion (a synthetic fragrance with a floral character), one is not really sure whether the receptors are involved in sexual readiness. After reading this chapter, you can now weigh up good arguments for why you belong to the pheromone believers or unbelievers.

References

Berger S et al (2017) Exposure to hedione increases reciprocity in humans. Front Behav Neurosci 11:79

Canetti E (1992) Masse und Macht. Claassen, München

Chakkarath P, Weidemann D (2018) Kulturpsychologische Gegenwartsdiagnosen: Bestandsaufnahmen zu Wissenschaft und Gesellschaft. Transcript, Bielefeld

Classen C, Howes D, Synnott A (1994) Aroma: The cultural history of smell. Routledge, New York

Corbin A (1984) Pesthauch und Blütenduft. Eine Geschichte des Geruchs. Wagenbach, Berlin

De Villa M (2017) "Who are you, incomprehensible you spirit" Perfume in nineteenth and twentieth-century German literature. In: Ciani Forza (Eds) Perfume and literature, the persistence of the ephemeral. Linea edizioni, Padua pp. 179–198

Flammersberger T (2016) Baudelaire und die Moderne—Charles Baudelaire. Der Dichter der Modernität? Hausarbeit, Bayerische Julius-Maximilians-Universität Würzburg (Romanistik)

Gelstein S et al (2011) Human tears contain a chemosignal. Science 331(6014):226–230

Grammer K, Jütte A (1995) Der Krieg der Düfte—Bedeutung der Pheromone für die menschliche Reproduktion. Gynakol Geburtshilfliche Rundsch 37(1997):149–153

Higgins ET (1987) Self-discrepancy: a theory relating self and affect. Psychol Rev 94:319–340

Hurton A (1994) Erotik des Parfums. Fischer, Frankfurt

Largey GP, Watson DR (1972) The sociology of odours. Am J Sociol 77(6):1021–1034

Le Guérer A (1992) Die Macht der Gerüche. Eine Philosophie der Nase. Klett-Cotta, Stuttgart

Leistikow A (2019) Das Recht auf Schönheit—ein Plädoyer der Literatur. An Vio, Düsseldorf

McCartney W (1968) Olfaction and Odours: An Osphrésiological Essay. Springer, Berlin\Heidelberg

Miertsch B (2005) Parfum, donc souvenir?—Bedeutung und Einfluss von Geruch(ssinn) in 'Le Flacon' und 'Parfum Exotique' von C. Baudelaire (Fleurs du Mal). Studienarbeit, GRIN

Novalis (1978–1987) Werke, Tagebücher und Briefe Friedrich von Hardenberg, hrsg. von H. J. Mähl und Richard Samuel, München: Carl Hanser Bd. 1, p. 240

Oren C et al (2019) A scent of romance: human putative pheromone affects men's sexual cognition. Soc Cogn Affect Neurosci 14:719–726

Raab J (1998) Die soziale Konstruktion olfaktorischer Wahrnehmung. Eine Soziologie des Geruchs. Dissertation zur Erlangung des akademischen Grades des Doktors der Sozialwissenschaften an der Universität Konstanz

Reinarz J (2014) Past scents: historical perspectives on smell. University of Illinois Press, Urbana

Rimmel E (1988) Das Buch des Parfums. Die klassische Geschichte des Parfums und der Toilette. Ullstein, Berlin

Rindisbacher H (2015) What's this smell? Shifting worlds of olfactory perception. KulturPoetik 15(1):70–104. Vandenhoeck & Ruprecht

Savic I, Berglund H (2010) Androstenol—A steroid derived odor activates the hypothalamus in women. PLoS One 5(2):e8651

Shepherd G (2013) Neurogastronomy: how the brain creates flavor and why it matters. Columbia University Press, New York

Simmel G (1992) Soziologie. Untersuchungen über Formen der Vergesellschaftung. Suhrkamp, Frankfurt/M

Süskind P (1985) Das Parfum. Die Geschichte eines Mörders. Diogenes, Zürich

Tullett W (2019) The past stinks: a brief history of smells and social spaces: in the conversation newsletter (theconversation.com) published August 9, 2019. Boston

Williams M, Apicella C (2017) Synthetic copulin does not affect men's sexual behavior. Adapt Hum Behav Physiol 4:121–137. (2018)

Williams M, Jacobson A (2016) Effect of copulins on rating of female attractiveness, mate-guarding, and self-perceived sexual desirability. Evol Psychol 14(2):1–8

Wohlleben P (2019) Das geheime Band zwischen Mensch und Natur. Ludwig, München

Wyatt TD (2015) The search for human pheromones: the lost decades and the necessity of returning to first principles. Proc Biol Sci 282(1804):20142994

Ye Y, Zhuang Y et al (2019) Human chemosignals modulate emotional perception of biological motion in a sex-specific manner. Psychoneuroendocrinology 100:246–253

On the Way to the Future of Smelling

How Fragrances Increasingly Excite the Brain

Contents

9

Do you know how to help the brain today to be excited about a new perfume?

This question is quite controversial. Every year, about 2000 new perfumes come onto the market—certainly not all of them in the stationary perfume shop, but constant new launches of perfumes also compete for the favor of the user there and wait to be discovered as a new favorite scent.

Since the responsible persons of perfume brands are usually under enormous sales pressure, they need an ally, someone who helps to favor the purchase decision for their new perfume. This is traditionally the dealer. But he too must ensure that his perfume recommendation is crowned with success, and for this purpose a perfume is often recommended that is assumed to be easier to sell. Too specifically smelling perfumes are then shown to customers with reluctance, but exactly what many perfume lovers are looking for as a perfume for themselves.

Now there are new findings through the neuroperfumery and artificial scent intelligence how to make the choice of perfume more individual, customer-oriented and more accurate for a win-win for all parties, but also how fragrances and perfumes - like active fragrances or neuroscents - can fascinate even more in the future and thus offer more pleasure.

But let me first recap some of this at the beginning of this chapter and give you more information about what the current research on fragrance and its processing in the brain has found out. If you are already familiar with the findings of neuroperfumery presented in ▶ Chap. 1 and 4, you can read this chapter well. It gets really exciting again in ▶ Chap. 10, where it is about fragrance therapy.

9.1 Smell Search: The Scent has been on the Research Radar for a Long Time

The discovery of the olfactory nerves and thus the first research on smell is attributed to Claudius Galenus (129–199 AD), a Roman physician from Pergamon (in present-day Turkey) who also became the personal physician of Emperor Marcus Aurelius. The first assumptions about the functioning of smell perception can already be found in the work *De Rerum Natura* by the Roman poet Titus Lucretius Carus (97–55 BC). He assumed that the smallest particles can pass through slits of the sense organ with an adapted shape. Accordingly, pleasantly smelling substances should consist of smooth and round particles, while unpleasant, bitter-sharp odors should consist of compact and curved particles.

It was to take 2000 years before researchers discovered that not only in the nose, but on many body organs there are olfactory receptors (Hatt and Dee 2012). As already mentioned, recently biochemists from the American Monell Institute in Philadelphia, one of the strongholds of fragrance research, found that even the taste cells on the tongue have olfactory receptors (Malik et al. 2019). This discovery is due to the twelve-year-old son of the scientist Mehmet Ozdener, who pointed out that snakes extend their tongues to smell the air around them. Today one learns in biology lessons, in a nutshell, that each fragrance molecule when inhaled hits the elongated olfactory cells on the mucous membrane of the nose, is converted into an electrical signal and then in a split second rushes along the nerve paths mainly into the limbic system, our emotional center. There it is stored as memories in the long-term memory and coupled to certain emotions with memories.

Since olfactory perception is very complex and by no means fully researched even for professionals, scientists are not always in agreement on the subject of "smelling". Most of them assume that we can perceive at least 400,000 different fragrance notes. Some researchers even theoretically assume up to one billion. This is ensured by around 20 to 30 million olfactory cells. In any case, one can say that man can perceive more fragrance notes than colors, taste or acoustic impressions. The sense of smell outperforms any other sense in terms of perception.

In practice, however, no more than a few thousand fragrance notes can be distinguished and named correctly. Optimists speak of 5000 to 10,000 odors, with the error rate in correctly naming them at over 50%. It is unlikely that even remotely as many fragrance notes can be coupled to emotions and memories. Because much of what happens when smelling happens in the threshold area of perception and also unconsciously. Smelling is also not static, but runs like a moving image, almost like a short film. Fragrances are electronic patterns, and since they are in vibration, the areas of smell have to decode them psychodynamically.

The brain also makes a difference whether we like the scents or not (Rolls 2004). Depending on whether we only sniff or actually smell, other areas in the brain are also stimulated (Sobel et al. 1998; Kareken et al. 2004). The sense of smell is also different from person to person, which makes its research more difficult. It not only depends on psychophysiological and other biological differences and is strongly influenced by personal olfactory socialization, but is also shaped by culture, cultural specific knowledge, habitat and language, which have significant effects (Ferdenzi et al. 2017).

Smelling itself is a complicated but fascinating process. Many questions are still open, on which there is disagreement in research—for example, how specifically the conversion of a chemical stimulus (for example, the molecules of a smelled flower that the receptors in our nose perceive) into electrical smell signals takes place. Since 1996, there has been a dispute in research that is only partly understandable for insiders. So the physiologist and biophysicist Luca Turin developed a theory that the smell of substances is based on the frequencies of their molecules' vibration or on different vibration frequencies and thus becomes an electrical smell signal and does not work by means of receptor recognition of form (Turin 1996). However, the majority of fragrance researchers now hold the view that the brain smells by recognizing the form of the fragrance molecules. Nevertheless, some physicists still adhere to Turin's theory, because identical-looking molecules can smell very differently and vice versa molecules with very different structures can smell similarly.

When writing this book, research was just discussing the following view: odors are played directly to the piriform cortex (PC) via the olfactory bulb (bulbus olfactorius). The bulbus olfactorius represents the primary olfactory center and is considered a protuberance of the brain. In the piriform cortex, a section of the cerebral cortex belonging to the olfactory system, which belongs to the outer side of the human brain, the stimulation by fragrance molecules, which have now been converted from a chemical stimulus into electrical smell signals, is forwarded to other specialized brain regions. I have already described the manifold functions of the piriform cortex in various places. Here are some additional information.

The piriform cortex is a phylogenetically old structure that also occurs in reptiles, amphibians, and other mammals. The small area is located next to the temporal lobe, also called the temporal lobe, where in anatomy "temple" is used to refer to the region just before and directly above the ears. Current research attributes a great effect to the small area on other brain areas such as the hippocampus. This stores complex memories, but can be directly stimulated by the piriform cortex. So this part of the olfactory brain also affects our memory and, as already mentioned, even our perception. Apparently, the piriform cortex also processes what we see and creates an expectation of how a smell might smell. This affects the actual smell perception. As has been said, this was discovered through studies using functional magnetic resonance imaging (fMRI). Based on photos showing a person with a happy, neutral, or disgusted facial expression, the same smells were rated differently. However, the piriform cortex also seems to be an archive for long-term memories, which in turn affects the actual smell perception. But in order to function as an archive, the piriform cortex must first be informed by a signal from the orbitofrontal cortex, the maître des parfums, which coordinates the senses and is the seat of personality traits and thus a higher brain region, that an event is to be stored as a long-term memory. In other words, what we finally perceive as a concrete smell impression is influenced by visual perception, such as the networking with the other senses, but also by the meaning of the memory, such as our current experience and our personality, or rather modulated.

Is this the explanation for why, as recent studies show, we can smell even without the olfactory bulb, that is, parts of the olfactory brain, much more often than we think, a smell impression that is not at all due to the previous stimulation by certain fragrance molecules (e.g. a certain flower smelled)? Does our olfactory brain in cooperation with higher brain regions often create a smell impression for which the previous stimulation by certain fragrance molecules (e.g. a certain flower smelled) is not at all responsible?

Of course, one has known for a long time: the smell arises in the brain; but that the healthy brain can also produce its own smell—detached from an olfactory input—much more often than we are aware of, is a new finding. Apparently, it is enough for a smell impression if the brain is somehow stimulated or if it stimulates itself, for example during sleep. So far it was always thought that the smelling of smells that others do not smell at all, such as a clearly life-threatening fire smell, is a symptom of psychopathology and occurs in certain diseases such as schizophrenia. Also, as already mentioned, a smell hallucination is called a Phantosmia or Phantom smell, which occurs especially before epileptic seizures. Now the current smell research suggests that even the healthy brain builds up its own subjective world of smells and creates it anew, detached from external input. One could say: Even for the healthy brain, a certain degree of "smell hallucination" is quite normal because it can create a smell for itself even without a verifiable external stimulus.

Of course, I do not want to say that everyone always smells things differently, just because everyone has their own

9.2 · Nobel Prize-Finding: Do We Smell (Almost) Everything Twice?

205

9

characteristic and recognizable breath (Benchetrit 2000), which, in addition to the individual emotional-cognitive processing of the smell, also affects the sense of smell. If everyone really lived in their own world of smells, smell tests like the University of Pennsylvania Smell Identification Test (UPSIT) would not be possible for general use (Welge-Lüssen et al. 2002). In the UPSIT, 40 fragrance compounds are impregnated on a paper block in the form of microcapsules that need to be scraped off. In this way, the fragrance is released and is to be identified within four answer options. Even if 40 scents are evaluated in this multiple-choice test, one could argue that such smell tests can only identify gross olfactory dysfunctions, for example in the context of Parkinson's, and thus cannot really uncover the differences in fine smelling. However, in the practice of fragrance consulting, often differences in individual fragrance perception become clear. Then you are fascinated how differently one and the same fragrance is described by fragrance users and obviously their brain is stimulated in a very individual way. This brings us back to the question of how fragrances actually stimulate the brain.

9.2 Nobel Prize-Finding: Do We Smell (Almost) Everything Twice?

Recent studies now come to another surprising finding. From the piriform cortex, the olfactory signals are probably forwarded to three control circuits (Olsen et al. 2012). In each control circuit there are specific brain areas and networks that trigger neurobiological and psychological processes through the stimulation of certain fragrance signals. These are associated with specific feelings and memories as well as with our self-image, our wishes and needs as well as our values and even with other senses.

Each area and network is particularly receptive to certain fragrance notes. The individual brain areas, as mentioned, have favorite aromas that they particularly like to be stimulated with. If that is the current state of fragrance research, a very exciting question arises that cannot currently be answered by research:

How does the piriform cortex know which fragrance to give to which feedback loop, or how does it know which brain area wants to smell which fragrance first, even if there is a feedback loop that it always uses?

As mentioned, the piriform cortex appears to be much more decisive for smell perception than previously assumed. It is not just a relay station, but also seems to have a kind of olfactory consciousness. You can also just put it this way: The piriform cortex is probably our little man in the nose and our fragrance manager.

In the following, I will give an overview of the three feedback loops in the order 3, 2, 1, because the last one is the fastest. At the same time, I am offering a short summary with additional information on the brain regions already explained, which are important for smell.

In the **feedback loop 3** the orbitofrontal cortex (OFC) is in the center, responsible for the conscious recognition of fragrances and thus for the categorization of all smells fed to it. So it coordinates all olfactory impressions and brings them into a overall context, without which a perfume enjoyment would not be possible. I have already introduced the OFC as the Maître des Parfums in the brain. It also coordinates our senses and is therefore responsible for synesthetic and multisensory relationships. That is why we can, for example assign colors or music to olfactory impressions. It was also explained above that value relationships are created in the OFC, for example whether a perfume is worth its price. As the seat of the personality trait extraversion, the fragrance preferences of this area

9

are more than understandable. The fragrance psychology has known for a long time that fresh, citrusy notes like bergamot particularly appeal to extraverts or people with corresponding experience wishes. This can also be confirmed by brain research.

In the feedback loop 2, the so-called Papez circuit, the hippocampus (seahorse) is in the center. He is especially responsible for memory storage and thus for our long-term memory and learning. With the hippocampus we have an excellent sense of smell (Gottfried 2006). For example, we also remember smells from early childhood and associate them with events and feelings (Herz et al. 2004) into old age. Enemy No. 1 of the hippocampus is stress. Chronic stress can shrink this area through excessive cortisol secretion, accordingly one loses more and more the ability to smell, for example, the individual flowers correctly. Clinical studies have shown that this brain region can shrink by up to 26% of its original size (Sheline et al. 1999; Greenberg 2012). Understandably, the hippocampus has fragrance preferences that offer relaxation and stress prevention. The aromatherapy has long known a whole range of promising fragrance notes against stress. They are mainly from the flower and blossom area like osmanthus, cyclamen, roses, chamomile and lavender.

Feedback loop 3 and 2 thus point to specific types of recognizing smells. The first explains why smells can address us directly and cognitively. The second shows the close connection of fragrance, learning and memory or associations.

The **feedback loop 1**, the so-called almond-shaped amygdala loop, is particularly fascinating for psychologists—on the one hand because olfactory perception begins here initially unconscious and can remain unconscious, on the other hand because fragrances can modulate emotions through it. This explains the possible emotional coloring of a fragrance experience (Adolphs 2010). In the center is the already often quoted amygdala, responsible for emotional attachment and the ability to show emotions. Olfactory signals appropriately charged via the hypothalamus—the so-called chamber, which is responsible for pleasure and temperature perception, but also for sexual behavior—and via the hippocampus finally also in the orbitofrontal cortex.

The discovery of the feedback loops makes the following spectacular thesis more and more likely: We probably smell most of it twice. Namely, first unconsciously by sniffing and then by actually smelling or smelling again, at least part of the unconsciously perceived smells become conscious. We owe this thesis to the psychologist Daniel Kahneman, who—as already mentioned—received the Nobel Prize in 2002. He recognized two types of sensory perception:

- The fast, instinctive and emotional System 1 (autopilot), controlled by the amygdala, for example, is our oldest center for emotions. The actual task of the amygdala is to protect us as an early warning system, so to speak as a spy, from dangers in the environment. We know from perceptual psychology that the amygdala can scan something from the environment in 300 ms and alarm us if something is wrong. All of this happens for us at first in the threshold area of perception. We then know it as a subconscious feeling that something is wrong.

The speed of the amygdala in smelling is explained by the fact that it is connected directly and without detour with the olfactory cells on the mucous membrane of the nose via its feedback loop. Also with smells she—for us unconsciously—makes a preliminary decision what to other brain regions is passed on or is to be passed on and then becomes conscious. There must be some kind of secret pact between the piriform cortex and the amygdala. The latter will probably always be played the olfactory signals first. Of course, as already mentioned, there is a direct connection between the piriform cortex and the orbitofrontal cortex. But probably the instinctive and emotional system of the amygdala is reached faster, as the immediate proximity between the piriform cortex and the amygdala suggests.

We can assume that we smell much more than we are aware of, and that in the millisecond range. There is now evidence for this. So women, as mentioned, can very quickly smell differences in the immune system of men, which they do not have to be aware of and still have an impact on their behavior.

— The slower, more thoughtful, and more logical System 2 is primarily controlled by the orbitofrontal cortex. He consciously delays, recognizes, analyses, experiences and smells. In this way, he makes value decisions that lead to fragrance and purchase decisions. In research, however, one is not sure which system is ultimately responsible for the purchase decision. The only thing that is certain is that the amygdala decides on the emotional meaning and thus makes a preliminary purchase decision. However, she is only focused on her favorite scent. It is perceived centrally by her and always receives the emotional priority. What is presented after him has a much harder time impressing the amygdala. This reaction has to do with how our brain likes to work. It always prefers the fast, energy-saving way and therefore initially—to stay with the example—the previously favorite perfume. Brain scans confirm this. What is a favorite perfume/product/brand, called in the jargon First Choice Brand, is decided intuitively and immediately by the amygdala or (re-)recognized by it.

If the amygdala has chosen its favorite, it pushes the orbitofrontal cortex (OFC) to make a quick (purchase) decision. This usually happens unconsciously, is then felt semi-consciously in the threshold area and finally experienced consciously. This experience is expressed, for example, as fear of not getting something, missing something, losing something or not winning something—all before the OFC can make a conscious decision or show alternatives.

So the choice of the favorite perfume is immediately made in the intuitive system 1 (autopilot). Even if the amygdala makes the purchase decision, brain research has shown that it is ultimately the orbitofrontal cortex that decides whether one can afford something or wants to. The OFC, which coordinates all our senses and makes value decisions, often has the following problem already mentioned: The amygdala, which has already decided, pushes because fear also grows in it of missing something. So the question remains: Who really makes the decision?

9.3 Imaging Smell: The Starting Shot For Neuroperfumery

The real innovations in the field of olfactory research have arisen in recent years through studies with the new imaging methods in brain research, the so-called functional magnetic resonance imaging (fMRI). With it you can, as mentioned, almost literally watch the brain smell. That was the starting shot for the neuroperfumery. With the beginning of the 21st century, this field of research also became interesting for perfume practice, for example with studies on the topic "Where do we smell something in the brain?". Examples of such research activities are the Italian teams around Laura Romoli (Romoli et al. 2012) from the University of Trieste with a focus on consumer research as well as around Faezeh Vedaei (Vedaei et al. 2016) from the Thomas Jefferson University Hospital in Philadelphia with the specialism neuroimaging.

With fMRI it can be shown that individual fragrance compounds (aromas) have different cortical activation or that individual aromas stimulate certain cerebral regions and their networks first or even only these. This helps to understand how, for example, smell and emotion or olfactory experience, memory and self-perception are related. These are more complex processes than initially expected. So although a variety of brain regions are activated, it is far from clear what role they play in part. Also, existing assumptions about responsibilities in the brain are repeatedly relativized by new findings. For example, recent studies show that it is not only the hippocampus and the amygdala that modulate memories, but that they are also mapped in the hypothalamus (Hasan et al. 2019).

By now, one suspects what incredible complexity still awaits brain research. Alone, 30 different areas, some of which are spatially far apart, are involved in processing visual information supplied by the eye. So far, no higher-level center has been found that compares and coordinates the different visual information with each other. With smell, one is already a little further along, even if the research on smell is sometimes treated like a stepchild. Here one knows that there is a Maître des Parfums in the brain that coordinates the senses and is therefore responsible for synesthetic and multisensory connections.

The search for the seat of our consciousness and our self has turned out to be extremely difficult for research—not to mention the search for the soul. This is actually not surprising. Because what self, consciousness and soul are, where they have their seat and where they come from, has been one of the great questions of philosophy for millennia, but also one of the comparatively young psychology. Meanwhile, almost resigned tones come from brain research, which is also looking for the center of the self and consciousness. The common opinion can be summarized as follows: It is not to be assumed that consciousness and self, as the highest instances of the brain, control subordinate areas from a certain location. Rather, it is to be assumed that it is a matter of an interplay of different areas in the brain.

Originally, one had probably hoped in research that the classical division of brain areas, which goes back to the German neuroanatomist and psychiatrist Korbinian Brodmann (1868–1918), would serve as a blueprint for the research of connections. One hoped that one would find functional connections behind the 52 fields with which anatomy divided the cerebral cortex into. Also a further subdivision between adjacent fields, so that for example also the Brodmann areas 23a or 23b emerged, did not bring the desired insights.

By now, one discovers that spatially widely separated brain areas cooperate for more and more brain functions. In the past, the naming was made more difficult by the fact that there was no agreement on the association of different brain areas. What, for example, should belong to the limbic system or limbic association system? According to Brodmann, these are the areas 28 and 34. For a long time it was also not clear whether the amygdala should be attributed to the limbic system at all. With Brodmann it is in the area 25. Therefore, for a first overview, the brain is initially divided into large areas, so-called lobes, before they are further differentiated.

It would now exceed the scope of the book to introduce the names of brain regions commonly used in neuroimaging studies today with abbreviations such as vlPFC and to further discuss the details with the latest state of neuroimaging studies, which have increased almost explosively. I therefore refer the interested reader to so-called meta-analyses such as those by Buhle and colleagues (Buhle et al. 2014), in which the individual results of brain research are analyzed in detail for certain topics. In these articles, brain regions are much more finely differentiated and distinguished from each other than I can do here. For example, there are distinctions in dorsomedial, dorsolateral or ventrolateral prefrontal cortex (dmPFC, dlPFC, vlPFC), which play a role in emotion processing and partly in olfactory processing, in addition to the amygdala. One would also have to go into gender-specific differences and here into the lateralization of the brain or into the neuroanatomical inequality and functional task division and specialization of the cerebral hemispheres, which show differences in part in men and women. However, my interest in brain research with its many detailed findings only goes so far as they concern developments in the perfume industry and thus in neuroperfumery. To quickly locate brain regions and get a first impression of their functions, websites such as ▶ http://www.gehirn-atlas.de offer a good overview.

Back to the four major lobes of the cerebrum, which are differentiated as follows:

1. Occipital lobe with many individual areas that work together to process visual stimuli. This visual cortex stores images of what we have seen, e.g. a black cat coming from the left.
2. Parietal lobe. It adjoins the upper occipital lobe and is responsible, inter alia, for attention processes, sensory sensations and memories. The main task of the somatosensory cortex located therein is to store felt memories, e.g. those experienced during the last yoga exercise.

The two largest lobes, the temporal or temporal lobe and the frontal or frontal lobe, are mainly responsible for the connection between scent, personality and self-experience. They have individual regions that cooperate particularly closely here.

3. The temporal lobe is the second largest lobe of the cerebrum. The majority of it is located in the area above the ears and just in front of them, towards the nose. So it is no coincidence that the auditory cortex is located here, among other things, and that it is responsible for the processing of acoustic stimuli. For neuroparfumery, the temporal lobe is particularly interesting, not only because the olfactory tract ends at the uncus (hook), but also because in it, scent, emotion and memory are linked. The uncus is a small protrusion that points inward and encloses the amygdala directly above it. So the temporal lobe also houses the olfactory cortex with so-called cortical fields, of which the primary fields process and forward information of a certain quality, such as olfactory perceptions.

The olfactory cortex also includes the amygdala, which is attributed to the limbic system. The limbic system is actually not an anatomical, but a functional unit. Anatomy sees the hippocampus in the temporal lobe in addition to the amygdala. The limbic system also includes the hypothalamus, which is often attributed to the diencephalon in anatomy. There are reasons for this: The limbic system has been suspected for a long time to be a unit for emotion, drive, memory and learning. The amygdala and hippocampus with adjacent regions have always been in the center of research. Over time, the importance of the hypothalamus, responsible for pleasure and temperature sensation, but also for our sexual behavior, has been discovered more and more. This includes that, for example, it can also map memories.

The temporal lobe, the center for the processing and storage of scents, but also of acoustic stimuli, could also be referred to as the scent-sound maker and -storage, but one would have to add the word "emotional" in front of it and speak of the "emotional scent-sound maker and -storage". The temporal lobe has a far-reaching internal communication network and is also responsible for the emotional evaluation of memory content there. Furthermore, regions of the temporal lobe are responsible for facial recognition or reading in the eyes and relate this information to emotionally experienced content. Without this, one could not react adequately to experiences with others.

The temporal lobe is in turn closely interconnected with the insula, which I have already introduced in connection with the addiction center and the preference for chocolate notes. The insula is located in the brain under the temporal lobe and is hidden from view by it when viewed from the outside. The insula is increasingly being regarded as its own and thus fifth lobe of the brain and is, for example, responsible for an olfactory experience that we perceive as disgust in cooperation with regions of the temporal lobe. The main function of the insula is the perception and reaction to the interior of the body. For example, the insula is involved in the perception of an undefined, inner feeling of restlessness, which can further increase to stronger heartbeat, anxiety or pain. The insula thus seems to play a role in bodily awareness, for example in the emotional assessment of pain and in feelings of well-being and discomfort. Anxiety itself has its seat in the amygdala, which is in direct exchange with the insula controlling our emotional body feeling. This makes it much easier to understand the spontaneous strong emotional effect of odors when odor is suddenly associated with anxiety, disgust or physical discomfort. We can look forward to future research results on how exactly the amygdala and insula cooperate or complement each other not only in the evaluation of odors.

As I said, scents can also work unconsciously on others. In general, it can be said that they trigger unconscious, semi-conscious and conscious reactions in our physicality. It can even happen that an odor perceived unconsciously leads us to react to a person with physical disinterest, aversion or even disgust. Conversely, the same applies: For inexplicable reasons, we perceive a person as physically attractive and react accordingly positively.

These processes can also take place without the influence of a scent. Olfactory experiences from the past, which have been stored by the piriform cortex, the amygdala and adjacent brain regions such as the hippocampus, can—since facial recognition also runs through these regions—even trigger a physical reaction to a person from a certain distance, even if this person cannot be smelled. Scents and perfumes can therefore indirectly affect our physical feeling, our physical self, even if they cannot be smelled in this situation. Since humans have a very good sense of smell, just the short sight of a person with whom one was in contact in the past is enough to experience positive or negative physical reactions. The same applies to objects, which is why certain objects have an erotic attraction.

The fact that even unconscious or threshold-level olfactory stimuli can influence us has become a research focus of scent marketing. Researchers are particularly interested, as we have discussed, in the interplay with other senses (touch, taste, hearing, sight) (Krishna 2010), which can certainly also raise ethical questions. The professional literature contains many examples of how scents are used to complexly target consumers in combination with other sensory experiences, for example for certain target groups, in order to attract them olfactorily first.

A much-cited example is the sale of sneakers to children. The soles of these shoes have been made pleasantly fragrant edible. Since children usually have a better sense of smell than adults and are closer to the shoes when trying them on, it is expected that the children will be physically attracted to the smell experience. The sight of the happy children creates a positive image of the sneakers in the parents' minds. But the problem with scent marketing is always the dosage of the aromas and the reactions of different target groups to individual fragrance compounds. Quickly, a shoe store, for example, can smell like candy, which is not associated with quality by adults.

4. The frontal lobe is the seat of the Maître des Parfums, our perfume personality. Here it is decided whether a scent fits our self and what effect it has on our self-experience. This experience is located in a brain region that is responsible, among other things, for motor skills and speech production, but also for humor, consciousness and personality. The anterior part of the frontal lobe, the prefrontal cortex, is not only important for smelling. Briefly on anatomy: The prefrontal cortex can be divided into an orbital frontal (as we have already discussed), medial and lateral part; the lateral prefrontal cortex is divided into dorsolateral and ventrolateral areas. The prefrontal cortex is particularly large in humans and other primates compared to other animal species and contains special nerve cells. This is often given as an explanation why self-consciousness and control of emotions were possible in human evolution. In this region as well as in the anterior cingulate cortex (ACC) and other regions of the prefrontal cortex (dorsolateral prefrontal cortex), which all belong to the frontal lobe, there are particularly specialized neurons, the so-called spindle cells. They are also called Economo neurons after their discoverer and are associated with self-consciousness and social intelligence.

■ **On the difference between being aware of a scent and being conscious of a scent**

Consciousness is a complex concept. In neurophysiological research, it is distinguished from attention. Both processes are intertwined, however. Consciousness has the function of creating a continuous image of reality, e.g. the insight that a perfume always goes over well with others. Attention has the function of conferring relevance on the objects of thought, e.g. how a perfume is coming across now. This means that consciousness and attention also have a slightly offset temporality, with attention being more related to the moment. Of course, consciousness can also be momentary, but then it builds on previous insights, to which attention need not relate. Why am I pointing out the difference? According to current research, consciousness and attention are probably assigned to different brain regions, or rather: networks (Nani et al. 2019), which are not limited to the frontal lobe. In neuroperfumery, this means that a distinction must be made between scent attention and scent consciousness. We can look forward to future studies with interest. Neuroperfumery is probably going to have to expand its previous assumptions for both types of scent perception. In this context, one could expect that the amygdala and its network alone decide on scent attention and regions in the prefrontal cortex control scent consciousness. Since the amygdala smells first, scent attention would precede scent consciousness. This means that the amygdala, with its emotional network, could always influence our scent consciousness or color it emotionally. Now, for the feedback loop 3 mentioned above, a direct olfactory connection is also assumed between the piriform cortex and the prefrontal cortex, more precisely between the piriform cortex and the orbitofrontal cortex (Illig und Wilson 2014), which, by the way, also runs in the reverse direction. The orbitofrontal cortex, which is part of the reward system and creates multisensory links to other senses, could thus develop a scent consciousness that builds up without the direct emotional coloring by the amygdala and its network. This also means that there could be a scent consciousness, so to speak, out of the blue, without prior attention,. This is certainly somewhat hypothetical, because, as we have discussed, the amygdala always smells along and is also the fastest in olfactory processing. Nevertheless: The amygdala also decides what olfactory impression it passes on and thus whether it pays attention to a scent at all. So it is quite conceivable that only the orbitofrontal cortex is fed an olfactory information from the piriform cortex. This information fed purely to the orbitofrontal cortex could then, with regard to scent consciousness, have its own quality, for example a scent consciousness that is associated with reward and value and is integrated into other senses without any special attention being paid to it and without any emotional affinity to it. This would be a smelling controlled by the mind, where the value of a perfume or of ingredients is recognized, but which does not really touch one emotionally.

Hegel (1770–1831), as we have already discussed, demanded this kind of smelling in his philosophy, so that it would be compatible with the interests of art and spirit. It is a smelling led by the mind, that gets rid of its primary emotional function. The end result is a cognitive awareness of smell— also made possible by nature. In their training, aspiring perfumers are taught the cognitive smelling necessary for this. To what extent they learn to suppress the emotional reactions of the amygdala and its network to olfactory impressions is certainly a legitimate question. As with any socialization, something is left behind or distorted. Perfumers do not report by chance of an occupational disease. They still practice their profession with passion, but the spontaneous enthusiasm for certain olfactory impressions is lost.

Let me briefly go into different ways in which a perfume can be smelled or how to deal with olfactory impressions before we return to the prefrontal cortex.

■ **Different ways of smelling**

How do perfumers smell in comparison to the fragrance-loving layman, or: Why can't a perfume be created with the amygdala alone in perfumery?

The range of smelling extends from cognitive-recognizing-systematic-analytical to associative to holistic-perceiving-subtle-emotional. Perfumers are trained in their training to smell recognizing-systematic-analytical and thus classifying with the orbitofrontal Cortex, that is, cognitively, for example with citrus notes that he recognizes. Of course, the consumer can also smell associatively (reminds me of …), but in contrast to the professional, the fragrance-loving layman usually smells more holistically and experiences pleasant vs. unpleasant in a subtle emotional olfactory environment.

Some perfumers have kept the holistic sense of smell, but an analytically-cognitive, single-ingredient focused training in fragrance has its price. You are no longer quite as impressed with fragrances as a whole because you quickly identify fragrance building blocks, evaluate them in the overall composition and, for example, think about which other ingredients in which quality and quantity you could actually have used better for the composition. Of course, there are perfumes by other perfumers that you value very much and recognize as milestones in perfumery, but often you also distance yourself from the competition because you have your own perfume signature. You find, for example, certain ingredients to be inappropriately or wrongly selected, not valuable enough, too well-known, linear, flat, disharmonious, dominating or used with too little personality or the perfume overloaded, too simple or not composed. Especially when perfumers lose a fragrance briefing, i.e. when the customer, i.e. the brand manufacturer, decides on a different fragrance, the complaint is of course great. What really distinguishes perfumers is their style, the way they work, which fragrance building blocks they like to use in men's or women's notes, which fragrance families and areas of application they are good at, how they interpret a fragrance for a specific theme, which market and product knowledge they have and, most importantly, which ingredients they are allowed to use for a fragrance in order to stay within the price range. The most difficult thing for a good perfumer are unspecific customer briefings with international hopes. For example, the customer requests a young, dreamy fragrance that is well accepted in different countries and is supposed to beat certain benchmarks—i.e. fragrances that already exist in the market— in terms of acceptance. The perfumer has to interpret and analyze this for himself first. Does the theme play at the beach, in the meadow, or in the club? Accordingly, he must choose ingredients that are appreciated not only in the individual countries, but also in the target age group, which not only fit the concept, but also stand up to the benchmarks in terms of overall acceptance. You see: the orbitofrontal cortex as well as the entire prefrontal cortex are very busy here. If now additional information about bottle design, packaging color and advertising are added, which by the way all perfumers like to have as inspiration for their work, additional cortical-visual areas are required. What I want to say with this is: no perfume can be created on this level with the amygdala or the emotion center alone.

■ **Back to the prefrontal cortex**

A region of the prefrontal cortex in particular, the medial prefrontal Cortex (mPFC), is increasingly fascinating to clinical psychologists and psychiatrists, but also to fragrance psychologists. As we have already discussed, various mental illnesses can affect the sense

of smell. The medial prefrontal cortex is increasingly being identified as one of the main structures for psychiatric disorders, next to the Maître des Parfums. Of course, one cannot conclude from the proximity in space that there is an easier mutual interaction, but it is conceivable because, with many mental illnesses, the sense of smell is also partly very strongly affected.

The medial prefrontal cortex is associated with schizophrenia, autism, depression, obsessive-compulsive disorder, anxiety disorders such as phobias and post-traumatic stress or post-traumatic stress disorder (Marques et al. 2019). All of these disorders, as reported in part, also show up in smell or influence the sense of smell. There can be smell hallucinations, altered smell perception and smell disorders, but also— as with severe depressions—a complete loss of interest in olfactory stimuli. The medial prefrontal cortex also has various positive functions that it contributes to, such as emotion regulation, behavioral reinforcement, implicit associative learning and decision-making—all functions to which the sense of smell also contributes. We can therefore only look forward with interest to future research results that show which other secrets the Maître des Parfums in the orbitofrontal cortex and his neighbor, the medial prefrontal cortex, share.

9.4 The Third Fragrance Revolution: "Evening" of the Future of Perfumery

Studies with imaging methods offer the perfume industry a whole range of new opportunities. Thus, experiences of the aromatherapy in terms of the effect of certain fragrances on the brain can be given a physiological-visual basis. This leads to more detailed insights into how and where fragrance stimuli act on different areas of the brain. In turn, this can be used to derive which brain areas or networks are particularly receptive to stimulation by certain fragrance stimuli.

The latter is of interest above all to perfumers. They have so-called new active fragrances or neuroscents in mind for the future of perfumery, which are more than just smelling good. They then contain ingredients that are supposed to increase the fascination and effect for the experience, but also for the health, for example as Alzheimer's prophylaxis, as well as for the general mental well-being. The big keyword here is "mood & health modulation". Perfumes will be developed and fine-tuned more and more specifically in order to trigger quite specific psychophysiological chain reactions by activating certain brain centers and their networks. This is already partly possible today.

This development poses risks that the legislator must keep in mind. Accordingly, there are regulations on how perfumes may actually work. Nevertheless, there is room for manoeuvre, and not only the aromatherapy will benefit from it and reach a new level. This development will be observed in all areas of application of perfumery—from body care products to washing and cleaning agents—in addition to the fine fragrance. Just imagine a floor cleaner that not only smells good, cleans and contributes to well-being, but also brings targeted health benefits to all those living in the house, including four-legged friends.

Numerous studies have already been carried out in the field of fragrance effect in the 20th century (Sowndhararajan und Kim 2016). These were, as reported, often EEG (electroencephalography) studies. For this type of investigation, namely how and where fragrance stimuli act on different areas of the brain, voltage fluctuations are usually measured at 21 to 28 points on the head surface. So far, a large number of ingredients known from perfume creation and aromatherapy have been examined for their effect. These include, to name just a few: bergamot, jasmine, lavender, rose, ylang-ylang, sandalwood, eucalyptus and cinnamon. Already during the first EEG examinations it became clear what was then confirmed by fMRI studies: smelling is much more complex than is often assumed. That, as said, the two hemispheres of the brain (the right and the left) react differently to fragrance stimuli was to be expected. But it was surprising that individual brain areas of the respective hemispheres, such as the right amygdala or the right piriform cortex, apparently play a greater role in fragrance perception. It gets even more complicated when only sniffing or really smelling. Also the concentration and the duration of the fragrance stimulus play a role. There is still a lack of studies on the effect of individual ingredients on the brain, but also on aroma combinations as they are contained in perfumes.

Even though neuroperfumery is only at the beginning of research into the effect of scent on the brain, we are on the eve of a third revolution in perfumery. Perfumers will be given entirely new tools for their work. Currently, most of them create the first perfume formulations on a laptop before they are mixed in the laboratory. The computer then calls, among other things. Price and availability of ingredients. In the future, there will be AI (artificial intelligence) programs that allow perfume creations to be specifically optimized for certain effects, e.g. more cuddling effect. In the final stage, the perfumer only has to enter how a perfume is to be experienced or what it is to trigger as an experience, which prophylaxis or other effects are desired. Then the program creates the scent alone, based on the findings of neuroperfumery.

First approaches to such AI programs already exist. The IT company IBM, together with the fragrance manufacturer Symrise, has already introduced the artificial smell intelligence Philyra, named after the Greek goddess of scent, beauty and healing. With Philyra alone—that is, without human help—the first perfumes have already been created, as can be read in various newspapers. The Symrise perfumers are grateful for this. With the almost endless number of possible combinations of ingredients, the human nose is quickly overwhelmed in time. So some great perfumes were created by mistake. About 3500 starting materials are available for a perfume creation. Many scents have 60 or more ingredients. In addition, the dosage plays a role. In addition, one could still consider regional, national and global preferences for scents, areas of application (fine perfumery or functional perfumery, e.g. fragrance development for a moisture shampoo) and other factors such as target groups and benchmarks. The Symrise perfumers have calculated the number of possible combinations for only eight basic substances: There are 40,320. Smelling and especially mixing them would take a lot of time and a small fortune. In addition, the mixing room would be paralyzed.

Where artificial intelligence will lead perfume creation for perfumers is already foreseeable today. Of course, we will still need the noses, but rather to make sure that no (or just) mixing errors occur in the increasingly virtual world of perfume creation and that the new fragrance intelligence learns from humans for as long as it has learned. This can happen very quickly, as we know from other applications, for example in dynamic pricing with learning algorithms in online retail. In the next few years, it could even come to the decline of a superstar, namely the nose.

As mentioned above, neuroperfumery can already today provide first indications of how perfumes can be optimized for active ingredients using imaging methods. The fMRT can already very precisely measure the intensity of cortical reactions to certain external olfactory stimuli. In combination with other methods, it is then possible to see which aroma or which substance is particularly ideal in which strength to stim-

ulate a brain area such as the hypothalamus for a certain secretion of hormones. This can be visualized by changes in blood flow, more precisely by a different O_2 content in the blood, the so-called BOLD effect. This makes it possible to make activities visible in brain areas and to trace them back to metabolic processes. In other words, when corresponding areas are activated, there is an increase in metabolism, which results in the activated area reacting with an disproportionately increased blood flow or with a local oxygenation of the blood. This is the decisive parameter that influences the signal intensity in measurements. The oxygenated vs. deoxygenated blood of the subject itself becomes the contrast agent that shows the differences (Ogawa et al. 1990). From the comparison of two points in time, e.g. in the stimulated state on the one hand and in the resting state on the other hand, it can be seen how strong or weak an odorant is, which direct or indirect effect it has and how this can be confirmed with other methods. This offers the perfume industry exciting prospects, but leads to the following explosive question:

- **How strong can perfumes actually work?**

The legislator (e.g. in the EU) lists several ingredients that generally may not or may only be available on prescription (e.g. as nasal sprays) in "fragrance applications". These include in particular those with allergen or health risks or too strong physical influence, as is the case with hormones, for example. Nevertheless, the effects of individual substances and ingredients are of high interest to perfumers, because the legislator still accepts a grey area for perfumes. This is the case if they have a "slightly positive hormone-like or hormone-like effect".

But this topic is not yet settled. It is still being discussed what is meant by a "hormone-like effect". Scientists have different opinions about this. They are divided into two camps: pharmacy vs. perfume trade (perfumeries). The former has taken over many market shares in cosmetics, especially skincare, in recent years. Therefore, a lot is at stake for the perfume trade. One wants to avoid that a new innovative category of perfumes, namely active fragrances or neuroscents, can be sold exclusively in pharmacies by prescription. So it's about a lot of money, but also about image and competence.

Scientists who are close to the perfume trade—and this is still the majority—argue as follows: one must distinguish between endocrine active substances and endocrine disruptors. Endocrine active substances may have hormone-like or hormone-like effects, but the decisive difference is that only endocrine disruptors are also known to have harmful effects. An endocrine active substance may have a hormone-like effect, but this does not have to be associated with negative effects on human health. This is shown by many ingredients in food, such as soy products or beer. They have a weak endocrine effect, but no adverse effects and active fragrances/neuroscents could be developed accordingly.

We will discuss the topic of active perfume and the consequences at a later stage. First, however, an apolitical fragrance topic: the broad-spectrum effect of citrus aromas.

9.5 Multiple Brain Stimulation Using the Broad-Spectrum Effect of Citrus Aromas as an Example

As already discussed, the orbitofrontal cortex (OFC) and its network region are the seat of two personality dimensions: extraversion and conscientiousness. In addition, the OFC acts as Maître des Parfums. Its preferences for smells are also well confirmed by research. Although it should actually be neutral, since it is responsible for the recognition and categorization of all odors it receives, it is particularly well stimulated by fresh, clear-bright citrus notes—

e.g. lime, orange, grapefruit and mandarin. That's why you don't just feel it when you smell fresh bergamot notes, the "spirit in the frontal brain" is invigorated, in which the OFC is located.

It is also known from the psychology of scent that the desire to experience oneself actively, openly and dynamically is associated with perfumes from the fragrance direction "fresh-green-citrus". Bergamot is the classic key ingredient in this fragrance direction. Apart from chocolate and vanilla, hardly any other aroma is appreciated by men and women equally in smell as the inedible citrus fruit bergamot. The lively spiciness that bergamot gives to the top note of perfumes has attracted the famous, powerful and extraverted in droves since 1672, when the fragrance was first introduced from southern Italy to France and Germany. "Bergamot-addicted" were e.g. Louis XV., his mistress Madame Pompadour, Napoleon Bonaparte and also Richard Wagner. They used Bergamot scents (like "Aqua Admirabilis" or the Eau de Cologne by John Maria Farina) by the liter.

The fragrance was less sprayed than rather poured with pleasure on the body. The group of the powerful and extraverted has already set the standards for refreshing-invigorating Bergamot notes with their impatient noses, which demand immediate action, early on. They are still valid today. The current generation of Bergamot fans loves the mixture with additional freshness, for example by lime, grapefruit, mood-enhancing mandarin, but also kumquat, aqua notes and green plants. This promises a more complex and longer lasting "revival kick" and immediately puts new perfumes of this fragrance direction to the test of olfactory immortality. Whether new perfumes of this fragrance family are successful is regularly discussed by the merciless "Bergamot addicts" in blogs. Most agree that one of the still available "immortal classics" from this fragrance family is "Eau de Fleurs de Cédrat" by Guerlain (1920).

Neuroperfumery now provides the understanding of what makes citrus notes so special and why they can become addictive. They have a sophisticated broadband effect on the neocortex and stimulate fourfold (Romoli et al. 2012):

1. Citrus notes act in seconds, slightly delayed, first on the right, middle occipital gyrus (in the occipital lobe—cortex) and trigger the known, visually stored associations with the citrus fruit.
2. They then stimulate the left, postcentral gyrus (in the parietal lobe). It reacts to touch and movement. It is no coincidence that citrus notes are often experienced physically and actively. But this part of the brain is also responsible for recognizing shapes and their sizes. Not surprisingly, synesthetes—to a certain extent most of us—often experience citrus notes as flying triangles or flying saucers flying through the air at different speeds.
3. They act on the left middle frontal gyrus. This is responsible for color perception, among other things. That is why many see the citrus note family in bright yellow, orange, red, light green and white.
4. At the same time, the Gyrus frontalis superior is stimulated by citrus notes, which is the seat of the prefrontal and orbitofrontal cortex in the frontal lobe and now lets us experience the whole thing olfactorily. By the way, it could also be shown that stimulation of the Gyrus frontalis superior can lead to spontaneous laughter. Even slight stimulation leads to smiling (Fried et al. 1998). Perhaps this is the often described good mood effect of citrus notes.

The fourfold brain occupancy, if one can call it that, interestingly does not start with the olfactory impression, but starts—expressed in layman's terms—from back to front, with visual associations, form, touch, movement experience and color impressions, until the olfactory impression reaches the OFC. This is the course of finest multi-sensory entertainment in our head.

- **The claim on fragrances and their effect in the future**

This knowledge, how the olfactory experience process, here on the example of citrus notes develops, gives the future of smelling and thus the perfume some food for thought and possibilities. Olfactory stimuli or fragrance ingredients could be optimized so that they each trigger an optimal multisensory broadband effect on the neocortex. They would therefore be measured by whether they trigger a multiple, for example, as citrus notes a fourfold, effect and optimally stimulate brain regions involved in each case. Of course, the individual olfactory and taste memory of aromas plays a role. Nevertheless, a multiple claim of fragrance effect can be applied well to popular fragrance notes such as chocolate or vanilla. It would also show which specific fragrance note, for example, a certain chocolate note, has the best effect on different regions of the brain. In combination with self-descriptions of fragrance users, how they perceive individual notes, fragrance ingredients could then be "multi-psychosensorically" evaluated. In fact, research into the effect of fragrance ingredients has been going on behind the scenes of the fragrance industry for years. I will go into this in more detail in connection with the discussion of individual brain regions. Here just so much: The future of smelling and the perfume has already begun. One goal is the multiple brain activation.

9.6 Neuro-Fragrance Sales: How to Excite the Brain for a New Perfume

It is known from the practice of fragrance consulting that the first-smelling fragrance, if it is liked, has a greater chance of being bought. This is even true if the perfumes shown later are also liked. Of course, the first-shown perfume meets a "fresh nose". Nevertheless, recent findings from neuro-perfumery suggest: The amygdala does not know any (fragrance) ranking in terms of best perfume, second best, etc. (Barden 2013). If the first perfume is liked, at least the amygdala has already decided, even if the orbitofrontal cortex would like to smell further. In general, however, the amygdala sticks to the old favorite fragrance.

If you want to excite the amygdala for a new perfume, you can trick it at the fragrance presentation. For this purpose, you couple a new perfume with the previous favorite perfume by means of perfume layering, that is, by applying two fragrances one above the other. If a third perfume results from the combination that is liked, the probability is high that the amygdala will also adopt the new perfume as a favorite fragrance. The orbitofrontal cortex, which evaluates the value, likes it anyway. He gains a third, very individual perfume through the new perfume and receives three perfumes for the price of two. For consumers, this is an interesting opportunity to rediscover their own scent or to further personalize their favorite scents.

However, the perfume industry and trade also benefit from the findings of brain research, for example perfumeries. Because in addition to the large ones, countless medium and small perfume houses are also struggling in a tough competition for the favor of the end consumer. In Germany, as already mentioned, 2000 new perfumes come onto the market every year. Only 5% make it to the next year. A large part of the scents does not even cover the launch costs (Leistikow 2019).

The decision for a perfume is finally made in the store. The marketing responsible of the perfume industry therefore know the following situation only too well: When a new fragrance is launched, the advertising drum is beaten first—with success. But as soon as marketing and promotion measures are reduced, a certain disillusionment sets in. Consumers and trade do not really take the new perfume. Big brands are often at a loss because the consumer tests were promising. Those who have been active in the industry for a long time and successfully usually react to the mixed results more calmly and hope for the next fragrance launch. Sometimes, driven by the parent company, new measures are taken to achieve the desired success.

However, if the desired success failed to materialize at the first attempt, you can get tips from neuroperfumery for the second attempt. They are perfect for use in stationary perfumeries. However, you should not reveal them to the perfumer and the licensor who is responsible for the image of a brand.

Since those in industry and trade are often under enormous pressure to generate sales, they need an ally, someone who helps them to favor their perfume purchase decision. This is the customer's amygdala. It occupies a special position in the brain by making a first purchase decision. And when the amygdala wants something, it is difficult for the OFC to say no. However, when it comes to winning a new perfume, the amygdala presents a different problem: It reacts like a trader who explains, "I already have enough perfumes on the shelf and don't need any new ones". The amygdala is exclusively focused on favorite scents or its current favorite scent, that is, a perfume that it already knows and with which it feels comfortable. Higher brain regions are open to smelling something new, but the amygdala chooses the familiar. It is, so to speak, conservative and decides according to the motto "What the farmer does not know, he does not eat". It strives for security and familiar comfort and is satisfied with the once-found favorite scent. This attitude can be used to outsmart the amygdala in perfume advice.

If you want to enthuse someone for a new scent or be sure that a perfume gift arrives or boosts sales in advice by giving a sample, you should first ask the person about their current favorite perfume. Then you put it in front of the person in a visible way, without letting them smell it. For its amygdala, which scans optically in milliseconds, the sight is immediately positive because it is familiar to her. Now put the perfume to be introduced on the right next to the person's previous favorite perfume and say: "I would like to create a very personal perfume for you. We do this with a small, fragrant experiment." Then spray the lighter, fresher scent and then the heavier scent for your customer and yourself, each receives a scent strip—alternatively, cut coffee filter paper is suitable— with the two layered scents. The lighter molecules of the fresher scent evaporate more quickly and want to pass through the "heavier" scent molecules. This makes the mixture interesting immediately. If you spray the lighter on the heavier scent, you often smell that both perfumes separate quickly and do not really connect to a new scent.

For the first perfume layering, it is recommended to apply the favorite fragrance in a ratio of 2:1. So spray the favorite fragrance twice on the scent strip, then once the new perfume. The amygdala of the customer will immediately recognize that her favorite perfume is involved in the new fragrance creation. First, you should smell the scent strip yourself. Maybe the fragrance ratio also has to be adjusted, for example, 1:1 or 3:1. The amygdala of the other person can read very well from the eyes whether one is also enthusiastic about the new fragrance. Therefore, no ingredients should be named before the amygdala of the other has smelled. It would only distract them. Now hand over the scent strip to the other person with friendly, positive eye contact. If a third perfume arises from the combination that pleases, the amygdala of the opposite adopts the new perfume as a favorite fragrance with a high probability.

The emotionally special experience is that the favorite fragrance of the other is personalized by the perfume layering. He now has a very own perfume that does not exist a second time. The experience of layering perfumes can be offered to everyone as an event in a perfume shop or even as a home fragrance workshop, where everyone becomes a perfumer and new and older perfumes get a second chance. These are often wonderfully smelling perfumes, but which look less attractive due to the name, the brand, the bottle or the packaging. In such a case, a blind test is recommended, in which only the favorite fragrance is known and visible from the beginning, while the bottle of the other perfume is covered with a label.

If both scents belong to the same direction, you can start with the mixing experiments at will. For the best mixing ratio there are no rules, except that the amygdala must recognize its previous favorite fragrance in the perfume layering and that the new creation pleases you.

For women's notes, I have discussed the following fragrance directions above:
1. **Chypre-leathery**
2. **Fresh-green-citrus/Aqua- & Ozon notes**
3. **Gourmand-fruity**
4. **Floral-powdery**
5. **Floral-aldehyde**
6. **Floriental (Amber floral)**
7. **Amber-oriental**
8. **Woody-aromatic**

For men's notes, these are:
1. **Fougère**
2. **Fresh-green-citrus/Aqua- & Ozon notes**
3. **Gourmand-fruity**
4. **Leathery**
5. **Amber-oriental**
6. **Woody-spicy**

Here are my recommendations for which fragrance directions work best when layering perfume.

For women's perfumes:
- Amber-oriental fragrance notes with the lighter notes from the Fresh-green-citrus/Aqua- & Ozon notes spectrum.
- Gourmand-fruity with often lighter Floral notes.
- Chypre-leathery with often slightly lighter **Gourmand-fruity** notes.
- Woody-aromatic with lighter notes from the Fresh-green-citrus/Aqua- & Ozon fragrance family.
- **Floriental (Amber floral)**, i.e. warm floral notes, with mostly heavier notes from the **Chypre-leathery** fragrance direction.

For men's perfumes:
- Fougère can usually be combined well with all other fragrance directions.
- Fresh green citrus/Aqua- & Ozon notes with heavier notes from the Amber-oriental fragrance direction.
- Gourmand-fruity with **Leathery**.
- Woody-spicy with Fougère or **Fresh green citrus/Aqua- & Ozon notes**.

As a rule of thumb in neuroperfumery for the success and fascination of a fragrance creation: It must also touch its wearer inwardly. Then it acts on the amygdala and the emotional center pleasantly and on the regions of the anterior cortex, where consciousness, self and personality have their seat, attractive.

If you have read or skipped this chapter as offered, I will briefly summarize its contents for you …

Summary

For over 2000 years, research has been following the scent. Meanwhile, brain research or neuroperfumery knows about three olfactory feedback loops in the brain and two types of sensory perception that influence smelling, perfume and effect. We have also discussed how findings from neuroperfumery combined with AI are currently ushering in the future of perfumery and increasingly enabling a new category of perfumes: active fragrances or neuroscents. The big keyword here is "Mood & Health Modulation". This development certainly also entails risks that the legislator must keep in mind. The key question is therefore: How strong can perfumes actually work? However, neuroperfumery also increasingly leads to findings about what makes certain scents so special, for example citrus notes, and why one can become "addicted" to them. They have a sophisticated broadband effect on the neocortex and stimulate in many ways. This type of multiple brain stimulation provides ideas for the research and application of fragrance molecules and thus also for the future of smelling and perfumery.

Based on current findings from neuroperfumery, I have given tips for fragrance advice at the end of the chapter: how to inspire the brain for a new perfume or how to give it a little help.

References

Adolphs R (2010) Social cognition: Feeling voices to recognize emotions. Curr Biol 20:R1071–R1072

Barden P (2013) Decoded—the science behind why we buy. Wiley, Phil Barden

Benchetrit G (2000) Breathing pattern in humans: diversity and individuality. Respir Physiol 122(2–3):123–129

Buhle JT et al (2014) Cognitive reappraisal of emotion: a meta-analysis of human neuroimaging studies. In: Cerebral Cortex, Bd 24, S 2981–2990

Ferdenzi C et al (2017) Individual differences in verbal and non-verbal affective responses to smells: influence of odor label across cultures. Chem Senses 42(1):37–46

Fried I et al (1998) Electric current stimulates laughter. Nature 391:650

Gottfried JA (2006) Smell: central nervous processing. Adv Otorhinolaryngol 63:44–69

Greenberg M (2012) How to prevent stress from shrinking your brain. Published on August 12, (The mindful self-express) in Psychology Today. Ongoing blog

Hasan MT et al (2019) A fear memory engram and its plasticity in the hypothalamic oxytocin system. Neuron 04:029

Hatt H, Dee R (2012) Das kleine Buch vom Riechen und Schmecken. Albrecht Klaus, München

Herz RS, Eliassen J, Beland S, Souza T (2004) Neuroimaging evidence for the emotional potency of odor-evoked memory. Neuropsychologia 42:371–378

Illig KR, Wilson DA (2014) Olfactory cortex: comparative anatomy. In: Reference module in biomedical sciences. ScienceDirekt, Amsterdam

Kareken DA, Sabri M, Radnovich AJ, Claus E, Foresman B, Hector D, Hutchins GD (2004) Olfactory system activation from sniffing: effects in piriform and orbitofrontal cortex. Neuroimage 22(1):456–465

Krishna A (2010) Sensory marketing: research on the sensuality of products. Routledge, New York

Leistikow A (2019) Das Recht auf Schönheit—ein Plädoyer der Literatur. An Vio, Düsseldorf

Malik B et al (2019) Mammalian taste cells express functional olfactory receptors. Chem Senses 44(5):289–301

Marques, R.C. et al. (2019) Transcranial magnetic stimulation of the medial prefrontal cortex for psychiatric disorders: a systematic review. Braz J Psychiatry 41(5): 447–457

Nani A et al (2019) The neural correlates of consciousness and attention: two sister processes of the brain. Front Neurosci 13:1169

Ogawa S et al (1990) Brain magnetic resonance imaging with contrast dependent on blood oxygenation. Proc Natl Acad Sci USA 87:9868–9872

Olsen RK, Moses SN, Riggs L, Ryan JD (2012) The hippocampus supports multiple cognitive processes through relational binding and comparison. Front Hum Neurosci 6:146

Rolls ET (2004) The functions of the orbitofrontal cortex. Brain Cogn 55(1):11–29

Romoli L et al (2012) fMRI study of smell: perceptual, cognitive and semantic components of cortical elaboration of 3 familiar aromas—Lecture. German Research School for Simulation Sciences, Jülich

Sheline YL, Sanghavi M, Mintun MA et al (1999) Depression duration but not age predicts hippocampal volume loss in medically healthy women with recurrent major depression. J Neurosci 19:5034–5043

Sobel N, Prabhakaran V, Desmond JE, Glover GH, Goode RL, Sullivan EV, Gabrieli JD (1998) Sniffing and smelling: separate subsystems in the human olfactory cortex. Nature 392(6673):282–286

Sowndhararajan K, Kim S (2016) Influence of fragrances on human. Psychophysiological activity: with special reference to human electroencephalographic response. Sci Pharm 84(4):724–752. (School of Natural Resources and Environmental Sciences, Kangwon National University, Chuncheon 24341, Korea)

Turin L (1996) A spectroscopic mechanism forprimary olfactory reception. Chem Senses 21(6):773–791

Vedaei F et al (2016) The human olfactory system: cortical brain mapping using fMRI. In Neuroradiology (in Press e16250). In: Iranian Journal of Radiology 14(2):e16250

Welge-Lüssen A et al (2002) Grundlagen, Methoden und Indikationen der objektiven Olfaktometrie. Laryngo-Rhino-Otologie 81:661–667

Scent Therapy: Scents for More Joie De vivre

Working with Primal Perfumes in Scent-Assisted Therapy—a Look into the Future at What Else Might be Possible for Increasing Well-being

Contents

© The Author(s), under exclusive license to Springer-Verlag GmbH, DE, part of Springer Nature 2023
J. Mensing, *Beautiful SCENT*,
https://doi.org/10.1007/978-3-662-67259-4_10

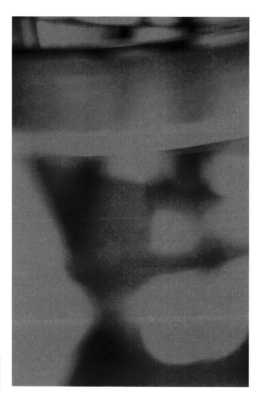

10

Trailer

Are you interested in new approaches in scent therapy? Do you want to become your own aromatherapist? Recent findings from baby scent research provide very interesting hints and instructions for adults on how to work beneficially with scent.

Wouldn't it be great to be your own perfumer, who can create perfumes for your own enjoyment without any great effort? Not just any, but primal perfumes, which we smell as humans in our development first and which have a special effect on us? And to find out which specific scents have a beneficial effect on certain needs and which brain regions are involved? These are just some of the incredibly exciting topics in this chapter—one of the highlights of our journey through the world of scent.

10.1 Aromatherapeutic Application: Two Exercises For Introduction

Before we discuss the latest findings from baby scent research and experience the therapeutic power of primal perfumes for ourselves, I would like to introduce you to the topic of "scent-supported self-therapy" with two exercises and give you more background information on the effects of scents and aromatherapy.

There are now several studies that confirm, for example, that citrus notes—as well as aqua and fresh fougère notes—increase concentration. Some even claim that citrus room scenting - ideally applied in essential oil quality - can increase IQ in students. However, it can certainly be said that these fragrance directions have a stimulating effect on the prefrontal cortex as well as the orbitofrontal cortex and their networks, which are also discussed in connection with personality dimensions such as conscientiousness and extraversion. Conscientiousness is one of the five main dimensions of the popular Big Five personality model. This includes the factors that are particularly important for professional success, such as self-control, responsibility, accuracy, ambition and endurance, as well as the ability to plan and organize oneself. Extraversion provides the necessary dynamics and activity. All these factors together support self-confidence and personal success. This can also be confirmed by studies. For example, studies show that pronounced conscientiousness and extraversion often result in higher salaries and the achievement of leadership positions.

But even with the greatest of wills and sense of responsibility, obstacles can weaken motivation, inner strength, inspiration and ambition. Often one can also not let go of something that is constantly on one's mind. This affects self-confidence and creativity. In such a situation one needs something that builds one up again, makes one take a deep breath, invigorates and brings one back on one's own path. For this I present two fragrance-therapeutic exercises that do not take longer than ten minutes. They appear simple, they are, but one should not underestimate their effect when repeating them several times. The first exercise is based on a multi-sensory journey, a "scented flight"; the other exercise, "Scented Power Posing" (in German something like "duftende Kraftaufstellung"), promises that one will find inner strength again through psychohormonal processes triggered in the body and feel like a winner.

10.1.1 "Scented Flight"— Letting Go, Inspiration, and Creativity

This fragrance therapy is primarily intended for people who responsibly drive things forward, but who may currently be lacking inspiration and creativity due to too much stress. It works particularly well as a multi-sensory installation of fragrance, color and form and lets the senses merge for a greater effect, so that one can experience the unity of the senses in their effect on oneself. To experience this unity, one only needs some space on a table. This puts one in good Bauhaus tradition, because already Kandinsky (*Concerning the Spiritual in Art*, 1911) allowed synesthetic experiments to merge color, form, texture, sound and even dance for optimal inspiration.

This is how it works: First, one chooses a fragrance from the fragrance direction **FRESH-GREEN-CITRUS** that is pleasant and invigorating and that one can associate with fruits from this direction. Ideas for the

choice can be found on fragrantica.de and in a perfume shop with expertise. This is especially the case when one is advised by a "maître des parfums" or "expert en parfum". Both are additional training courses that the trade offers its employees. If possible, one should be shown certified organic and vegan natural notes in essential oil quality from this fragrance direction. They are particularly ideal for aromatherapeutic applications.

There are usually **FRESH-GREEN-CITRUS** notes in a lighter concentration as Eau de Toilette (EDT). Furthermore, this fragrance direction is also available as an even lighter Cologne and as a body spray (body mist). For this fragrance therapy, the top note of a sparkling-sparkling Cologne is particularly suitable, which develops within the first ten minutes after spraying. Many fragrances from the **FRESH-GREEN-CITRUS** direction in higher concentration, for example as Extrême, can become somewhat spicy over time.

The quickest and most effective way to achieve new inspiration is to fuse the fragrance experience with other sensory impressions, so that the fragrance is supported by color, form, music, and ideally also tactile and gustatory. Furthermore, one knows from creativity research that the combination of sensory stimuli is beneficial in the imagination. For example, by visualization, yellow, light blue, light green, and turquoise paper triangles lying in front of one begin to fly like leaves in the wind. It is then found that the flying triangles have a **FRESH-GREEN-CITRUSY** smell. Those who practice this will become part of the movement and simply fly in their imagination. Being able to fly like Peter Pan—even just in imagination—is very inspiring and also stress- and even anxiety-reducing. Here are tips for a first "scent flight" at home or at work:

10

- The **FRESH-GREEN-CITRUSY** fragrance / essential oil should be stored in a dark place and slightly cooled. If the air in the office is a bit stuffy or if summer temperatures prevail outside, put the bottle in an ice bucket or cool it in the refrigerator beforehand.
- The place where you smell and start the "scent flight" should be decorated in turquoise and light blue with a little yellow (◘ Fig. 10.1).

In color psychology and mythology, these tones are also the colors of water, space, carelessness, rebirth, and the future. Ideally, one has a few objects, accessories, or sculptures in these colors to get in the mood for their flight. This color world can also be well supported by pictures, such as pictures of water in which the sunlight sparkles.

For the first flights of scent, videos that were filmed from the perspective of a bird in flight are recommended, especially if they convey a sense of space and show southern waters. You can find such videos on YouTube, for example the video "Fly Away to a Tropical Island!" (▶ Fly Away to a Tropical Island!). Background sea noise supports the effect. Of course, you have the best flight with a 3-D glasses.

- After everything is installed and you have sat comfortably for two to three minutes, you spray or drip the scent on your wrist or better on a scent strip so that the scent is not influenced by the smell of arm and watch bands or hand and body creams. First you sniff the scent, then you smell it. When smelling, you should pay attention to relaxed

breathing, just like in yoga, abdominal breathing. For smelling, you take about ten minutes. You start with the scent flight and let yourself be inspired by a video for about a minute at the beginning. Then you become active yourself and integrate the fruits you smell in the image into your thoughts. For example, a lemon you smell floats in blue and turquoise water, as you can see in the video. In your imagination, you can also see how gentle waves play with the fruit, how it drifts towards a southern beach and how the light beautifully reflects on the waves and the fruit.

At first it is not quite easy to connect two sensory impressions—one that you see and one that you smell—in one film. But that already works well after at the latest two exercises. And the brain thanks for the refreshing relaxation with new inspiration and creativity. I recommend a scent flight as a break when you feel mentally exhausted. Afterwards, the prefrontal cortex deserves a gustatory pleasure as further stimulation. How about a lemon sorbet?

10.1.2 "Scented Power Posing"— Recharge For Mind and Body

An inner body posture that radiates confidence, energy and sovereignty. This is an advantage of increased testosterone levels. It causes a body posture in which one feels more energetic and successful. On the other hand, psychologist Amy Cuddy and colleagues from Harvard Business School found that certain power-posing exercises (power postures) can increase testosterone levels in women and men in a few minutes and be reflected in their appearance (Cuddy et al. 2013). To also stimulate the orbitofrontal cortex for more extraversion, these exercises were further developed by me as

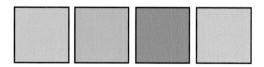

◘ **Fig. 10.1**   Color examples

"Scented Power Posing". To get the most out of this exercise, you only need four things:

- a room where you can withdraw for a few minutes to do the exercise standing, preferably in front of a large mirror,
- a scent / essential oil from the fragrance direction **FRESH-GREEN-CITRUS** that delights as a "pick-me-up" or motivator and is applied to a scent strip or an unscented handkerchief. This fragrance can be more sparkling and citrusy than the the scent used in "Scented Flight",
- an iPod etc. with headphones, on which you can load and play a certain music that I will introduce later, and
- the will to do the exercise with passion and seriousness.

This is how "Scented Power Posing" works:

Stand in front of a mirror in a proud winner's pose as wide as possible: raised chin, arms still stretched up, fingers spread like Mick Jagger at the concert to the victory sign (V).

At the beginning, snap your fingers from time to time in the rhythm of the music that I will introduce in a moment. Stand with your arms stretched and your fingers spread to the V-sign all the time. Feel like a sprinter who has just run over the finish line as the first and is now being celebrated. Tell yourself a few times that you have won and will win again (you—preferably address yourself by first name or nickname—"have won and will win again"). Enjoy this feeling. By the way: Upstretched arms are a quite natural, innate winner's pose. Even blind people who have never seen them react like this when they win. Make yourself as wide as possible, the whole room belongs to you. Feel strong, sovereign and like a winner.

After standing for about 30 to 60 s, turn on the music. I recommend the song "I Will Survive" by Gloria Gayner, which you can also sing along to. If you triumphantly enter the stadium in your mind,

a triumphal chorus like the finale of Beethoven's Fantasia for Piano, Chorus and Orchestra (▶ Choral Finale from Beethoven's Choral Fantasy) is also appropriate. The chorus sings for about four minutes, the last three minutes are particularly triumphantly building. No matter what music you choose—support the lyrics by repeatedly vigorously nodding with your arms still outstretched.

The whole song by Gloria Gayner lasts about five minutes. At half the song, as well as with the chorus, lower your arms and apply your "pick-me-up" fragrance, e.g. the "Eau Sauvage" spray by Dior, to a fragrance strip. For the next two to three minutes, you should first sniff it, then smell it and listen to the rest of the music. If the saxophone starts in the song (or the drums in the chorus singing), give yourself the promise "I can do it" while continuing to smell, but always stretch one hand up in a V-sign and fidget vigorously a few times in rhythm with the music. After the music has faded, smell a little longer in silence and think about what you have already achieved in life. Finally, say out loud to yourself: "I'm proud of you", which you can also repeat to yourself a few times.

10.2 Fragrance Enjoyment: Olfactory Cuddling "On Demand"

Light is energy in the form of electromagnetic radiation that we experience as color depending on the wavelength. The Hypothalamus is amazingly sensitive to the different wavelengths of the main colors blue and red. This has mainly to do with body temperature. From color therapy we know: If the hypothalamus experiences the body as too warm, it prefers short wavelengths, which we experience as cool colors in the blue spectrum. If necessary, sweating is then triggered. If it is too cold, the hypothalamus is open to being stimulated by longer wavelengths, which we experience as warm tones in the red spectrum. If that is not enough, the muscles are made to tremble to generate real heat.

The example of color shows—as we also see it for smell—that the hypothalamus and its network use the senses strategically for themselves, and not only on a purely physiological level, but especially from our point of view with psychological intention. However, as the supreme control center of hormone secretion, it is mainly focused on balance and survival, but also has its own agenda.

Let me go into this in more detail in relation to the connection between scent and reward, because certain olfactory stimuli associated with "edible" have a special, also olfactory therapeutic effect on the brain. In this chapter I will show you approaches to self-therapy with scents for more joie de vivre.

The hypothalamus is, as already mentioned, part of the dopaminergic system and thus of the reward system. In this network it decides on the dopamine release and thus on the emergence of the emotion of joy, that is, the feeling of reward and pleasure. These feelings arise because the hypothalamus is fond of sweet things. Just think of the craving for sweets during a diet or the late-night search for ice cream or chocolate. Here the hypothalamus—even if not alone—is at work. It is part of the reward system, also called the mesolimbic system, which consists of a neuronal network of various brain areas and connections. More correctly, one should speak of a multitude of reward centers that are linked together in a large circuit.

Olfactory stimuli such as the scent of a delicious piece of Sacher-Torte or chocolate cake can act as reward stimuli on the brain. This then makes the limbic system, to which the hypothalamus belongs, and the insula react. This in turn stimulates struc-tures located in the prefrontal cortex. The limbic system and the insula generate a real craving. This is, not only when it comes to chocolate, more than understandable.

Imaging methods showed that in most people the insula is very gladly stimulated and activated by the scent of chocolate as well as by sweet aromas, olfactory and gustatory, (Han et al. 2019). The resulting emotional craving is detected as a conscious desire in the cerebral cortex. It then gives instructions to still the desire. If you get closer to the goal or give in to the craving, the midbrain comes into action. Here it is especially the nucleus accumbens in which a feeling of happiness arises. In combination with other brain areas such as the amygdala, which process the excitement as pleasurable sensations, dopamine is then released.

The reward system was already discovered in 1954. They implanted rats with an electrode in the brain to stimulate it, which emitted weak electrical currents at the push of a button. These could be triggered by the animals themselves and were apparently experienced as very enjoyable—even more enjoyable than food, because left to their own devices, rats regularly stimulated their brains every five seconds, leaving food behind.

Studies have now shown that, of all the rewards, food is considered the primary and most universal one and can therefore have a strong effect on the reward center. It apparently does not matter whether the stimulus is presented gustatory, visual or olfactory (Frasnelli et al. 2015). This is a very important statement for a scent-supported therapy with food scents, such as milk/milk mousse notes, which we want to use in our therapeutic work.

In this context, Katharina Schoen (2018) from the Technical University of Dresden carried out imaging studies to check whether scents from the food category or food-associated scents activate regions of the human brain that can be assigned to the reward center, compared to flower scents. This was confirmed. Since the reward system in the brain is quite branching, one also expects a wider activation of food scents than of flower scents. There are also indications for this.

One finding for self-therapy is therefore: With scents from the "Gourmand-fruity" fragrance direction, one can activate one's own reward system.

In addition, studies have shown that the scents perceived as edible intensify the effect of the reward scent on the reward system when hunger sets in. Finally, it could be shown that the reward system has real scent affinities that activate it particularly. First, it could be shown that a sweetness perceived in scents activates regions of the brain that belong to the reward feedback loop more strongly (Stice et al. 2013). Then it was observed that fruit scents cause stronger activation in the reward centers (Frasnelli et al. 2015). The fragrance preferences of the reward system can thus be described as fruity-sweet-edible, which typically characterizes the Gourmand-fruity fragrance direction. It is therefore no coincidence that the hypothalamus, which is responsible for the sense of taste and sexual behavior, prefers it to be a little sweeter and fruitier in taste and smell. Sweet and red are innate taste and color preferences. They trigger the vital sucking behavior in infants. The combination of sweet-fruity-red not only signals to people that something is edible.

As an adult, one often distances oneself from too sweet or sweet-fruity scents. Nevertheless, the so-called gourmand-fruity scents are the fastest growing fragrance category in recent years. They may not always smell sweet, but they smell like dessert. It is the smell of preferred, edible aromas such as caramel, nougat, cream and raspberry, as well as above all the scent of the world's most popular flavors chocolate and vanilla. Understandably, gourmand notes in particular, such as in the chocolate direction, interest and stimulate the hypothalamus and its "reward network". So chocolate has always been said to have an aphrodisiac effect in literature. Because in the hypothalamus, sexual hormones such as testosterone and estrogen nerve cells are activated, which are responsible for libido and arousal. This is all additionally reinforced by chocolate. It contains tryptophan, which is converted into serotonin. This explains the special sense of pleasure when eating chocolate.

Chocolate as a gustatory experience is certainly superior to chocolate smells in the final effect as a stimulant, because these are primarily learned. Nevertheless: The close connection between the hippocampus, our long-term memory, and the hypothalamus has built up a strong classical conditioning à la Pavlov over time. So a previously innate or automatic reaction to a taste experience and its effect can be triggered again by a scent. This can generally be assumed for most gourmand notes. They are able—just like music and other sensory pleasures—to trigger the happiness hormone dopamine in the hypothalamus and its network.

The gourmand fragrance direction apparently has a greater effect on the brain than previously assumed. This is also confirmed by experience from practice. For all those who are on a diet and love gourmand notes, they have an additional benefit. They can keep the hypothalamus in check during eating attacks or craving for sweets. So losing weight with gourmand notes is easier, also because the hypothalamus and its network are given the impression that another edible pleasure is still to come. Practice has also shown that gourmand notes have a pain-relieving effect and, depending on the type of their creation, bring the hypothalamus and its reward system into a Piña-Colada-holiday mood. Milk notes are also pain-relieving, as we know from baby research and as I will show below.

■ Olfactory cuddling "on demand" with a sense of happiness

But when the hypothalamus is stimulated, more happens. It also secretes the hormones vasopressin and oxytocin. Vasopressin plays a role in the body's water regulation. It activates water reabsorption and, for example, reduces nocturnal urination; it constricts blood vessels and has a blood pressure-increasing effect; it regulates body temperature, affects sexual arousal and motivation; it is involved in memory and learning performance and, together with oxytocin, controls emotions that then have an effect on social behavior. Only recently it could be shown that oxytocin in particular significantly influences female care behavior.

The hormone oxytocin supports close emotional bonds between people, for example between mother and infant during breastfeeding. As a bonding hormone, it makes one's own partner appear more attractive. It also promotes monogamous behavior and mutual trust (Hurlemann et al. 2010). No wonder that in recent years, perfumes and nose sprays enriched with oxytocin or vasopressin have appeared in some countries. They are banned in the EU because hormone-enriched perfumes are not allowed to be sold in perfume shops. Nevertheless, they could be obtained from foreign mail-order pharmacies. An effect was promised as early as 15 minutes after spraying. Insiders are also aware of odorless, pure oxytocin sprays that are secretly sprayed shortly before business meetings or the first date in order to achieve more trust.

Vasopressin and oxytocin also have their downside. Too much vasopressin retains too much water, the sodium content in the body is increased, which can lead to lethargy, for example. Oxytocin apparently amplifies envy (why is not yet known), but it also has something impressive and can therefore be advantageous for a business transaction or a first rendezvous when the person opposite is associated with "successful" or also with "desired by others". Numerous studies have now confirmed the effect of vasopressin and oxytocin as nose sprays. For this reason, oxytocin is also used successfully, for example, as aroma therapy in autism (Gordon et al. 2013).

However, several fMRI studies show that perfumes do not need to be enriched with the hormone in order to have an oxytocin effect. Because oxytocin is also secreted when there is no physical contact. Apparently the hypothalamus is stimulated by the smell of warm milk skin, especially in combination with the scent of certain culinary notes such as chocolate. Insiders confirm that the perfume industry is already working on aromamodulation in this direction using fMRI and has achieved good results. The aim is to find out which molecules cause the milk, skin and chocolate odors and in what mixture they stimulate the hypothalamus particularly for the secretion of oxytocin.

But one goes one step further and uses a so-called precursor technology in scents (Whitehouse 2019). It was developed by Givaudan, one of the world's largest perfume manufacturers, employing over a hundred perfumers worldwide. A so-called fragrance precursor, an odorless molecule, is implemented into the perfume. It releases fragrance molecules at certain triggers such as oxygen, light, water, but also body heat. This allows perfumes to be timed, adapted to certain situations and of course made longer lasting. What is particularly interesting is that this technology not only works with perfumes, but also with other possible applications such as body care products, detergents and cleaning agents. In other words—to stay with the topic of oxytocin: The industry is well on its way to delivering us "cuddling 'on demand'", and that with a sense of happiness!

10.3 Fragrant Anti-Stress: How and Where Scent Works Against Stress in the Brain

It is not by chance that it was possible to show, using imaging methods (fMRI), how yoga, meditation and mindfulness training particularly benefited the hippocampus and adjacent brain regions. This was known before. That is why there are already various fragrance prescriptions in aromatherapy to help the hippocampus and its network cope with stress. Because, as we have said, the biggest enemy of the hippocampus is stress. According to clinical studies, it shrinks the hippocampus by up to 26% through excessive cortisol release. This leads to greater forgetfulness and thus to the fact that we can increasingly rely on our knowledge when we have problems. This also affects other brain areas such as the amygdala and the OFC, resulting in negative self-evaluation and sleep disorders. Since deep sleep is essential for stress management, a self-destructive cycle begins.

Clinical studies in recent years have confirmed the special effect of meditation on the management of stress, inner restlessness and even fears—and that even in combination with scents. In this way, processes are not only stopped, but the brain is literally rebuilt. This is then referred to as neuroprogramming or positive reprogramming of the brain (Bernhardt 2017). The connection of psychotherapy and meditation as mindfulness-based stress reduction has proven to be particularly effective for this purpose. This program goes back to the molecular biologist John Kabat-Zinn (Kabat-Zinn 2006), who already worked with so-called scent landscapes in meditation in the late 1970s. The success of his own and other further developed methods in the treatment of stress and anxiety, even panic attacks, depression and sleep disorders, has

been proven several times (Seppala 2013; Stahl and Millstine 2013).

Aromatherapy can also report several successes. Various fragrance notes are used against stress and inner unrest. Some, as already mentioned, such as Osmanthus, Cistrose, Rosenabsolue, Chamomile and Lavender do not come from the flower and blossom area by chance. Because they address the hippocampus and its network specifically. However, the hippocampus with its complex functions does not limit its preference to a certain fragrance direction. In addition to floral notes, it is also activated by edible fragrances. This is not surprising, as it not only receives information about stress, but also about physical satiety, and is also part of the reward system.

The hippocampus is, of course, also responsible for storing olfactory memories. This raises the question of how it learns and categorizes scents, especially when it comes to, for example, a flower scent that is still unknown to him. As I said, studies show that the anterior piriform Cortex helps the hippocampus smell and evaluate. The piriform cortex offers the hippocampus and other brain areas involved in fragrance storage much more than just a sensory forwarding of a fragrance impression. First, it provides emotional, visual information before smells are perceived (Schulze et al. 2017). The piriform cortex therefore suggests to the stress-prone hippocampus, for example, that something that looks like a small, white flower also smells like that and can be smelled and experienced relaxed. In other words, the piriform cortex generates instructions on how a olfactory stimulus is to be experienced and processed. In this way, the entire olfactory system and also the hippocampus benefit from the information of the visual system.

Through the cooperation of the piriform cortex with the amygdala, which evaluates stimuli emotionally, it is possible to pre-select olfactory stimuli according to their potential emotional effect—for example, those that promise the hippocampus relaxation and relaxation. Simply put, this makes it possible to search for a scent that, for example, promises the hippocampus a relaxed, relaxed feeling. Interestingly, nature supports the search for scents when a feeling of stress arises, as long as the mood does not turn into the emotion of anger. Hoenen et al. (2017) were able to show in a study that an increase in cortisol is associated with better odor identification performance during emotional and physiological stress reactions, while increased anger is associated with poorer odor identification.

10.3.1 Plant Peptides with Anti-Stress Effect

The neuroperfumery and -biology are constantly expanding the knowledge of stress management, with the latter being able to extract and potentiate plants better thanks to new methods. New ingredients, especially plant peptides with special effects, are perennial favorites in relevant trade journals. A still young topic are hyperforin and flavonoids, such as biapigenin and rutin as powerful plant substances with anti-stress and antidepressant effects, which could be incorporated as bioactive ingredients in fragrance products in theory.

Linalool, which is found, inter alia, in lavender, has been known for longer. The *Journal of Agricultural and Food Chemistry* reported that linalool keeps blood values within the normal range during stress and at the same time reduces the activity of genes that are excessively active in stressful situations (Nakamura et al. 2009).

In classical perfumery, and aromatherapy bioactive substances directly and indirectly come into fragrance formulations via their plants. Their effect is sometimes re-

duced by manufacturing processes and dosage forms, for example in alcoholic solutions typically used for perfumes. Actually, perfumes should provide the main olfactory enjoyment when sprayed, but they could be much more than just smelling good and also be designed differently, for example as deliciously smelling bioactive alcohol-free active perfumes or alcohol-free organic plant peptide fragrances.

There could actually be perfume applications for external and internal use. The Romans already used rose water in two ways: as a fragrant perfume for external use and as a mouthwash for fresh breath. Occasionally you also read about the idea of a deodorant pill that is supposed to change body odor towards a favorite perfume. You can further fantasize and think of an inner stress blocker scent.

Cosmetics are already further ahead in this respect. They already provide today's beauty supplements for beauty, attractiveness and well-being. The greatest prospects for success are likely to be in the perfume sector with wellness sets consisting of one product each for internal and external use, for example a perfume and a luxury wellness nose spray that smells the same. But here too, the legislator (e.g in the EU) has his say: nose sprays may only be marketed as drugs that are introduced into the nose. There is a distinction made between locally effective and systemically effective nose sprays. Most of the time, the term "nose spray" refers to a locally effective, for example decongestant, spray. If the active ingredients are absorbed through the nasal mucosa, it is a systemically effective spray that acts on the whole body.

Now comes the critical point: according e.g to EU legislation, perfumes can only be applied externally. This excludes their use as nasal sprays. Perfumes are sprayed, applied or poured as a "splash" and rubbed or smeared on the skin as a "solid perfume". But they can also be applied with a stopper or other applications such as roll-on or pipette, which are known from aromatherapy. When using a pipette, you can decide for yourself whether the liquid comes before, under or into the nose.

Manufacturers of spray pumps are already providing ever finer methods of perfume application. So the perfume pump manufacturer Rexam brought a spray pump specifically for natural ingredients to the market. One could now combine pumps with pipettes for a new category of bioactive active perfumes or essential oils, which, for example, support meditation, to "spray pipettes". The user would then decide for himself how bioactive active perfumes are to be used.

For the full immediate effect, there is another innovation. The so-called "scent burst technology" and its further developments make it possible that the parts important for the effect of a head note of a fragrance are already highlighted at the beginning of the application. This gives a special kick, but also immediate help à la emergency drops, as they are known from Bach flower therapy.

10.3.2 **The Rush On Plant Peptides**

A lot is expected of plant peptides in order to support individual brain areas such as the hippocampus. Peptides are body-own substances composed of amino acids, which in the human organism can trigger specific reactions and influence a variety of body functions. Meanwhile, a veritable run has broken out in the research of plant peptides. Some have already been confirmed in their surprisingly good anti-aging and anti-inflammatory effect in skin care as further developed active ingredients—such as Rubixyl©, a biomimetic peptide inspired by seagrass from the fragrance manufacturer Givaudan, which consists mainly of hexapeptide 48-HCl.

A few years ago, it was now discovered that there is a very high density of so-called

Delta-opioid receptors (DOR) in the skin, but also in the olfactory bulb (bulbus olfactorius) and in the amygdala located in front of the hippocampus, which can be stimulated by related plant peptides. If these receptors are stimulated, an emotional reaction is shown, which the pharmaceutical industry already uses as a mood enhancer in depressive states. The patients also felt comfort and well-being and quickly achieved an emotional improvement, even with mild pain (Broom et al. 2002; Navratilova et al. 2011). A comprehensive description of the current peptide research would exceed the scope of this book. In addition, it is not easy to assess to what extent the individual discoveries are scientifically supported. For example, an international team of scientists has found a peptide in an African medicinal plant from the coffee family, Kalata B7, which is similar to the human neuropeptide hormone oxytocin (Koehbach et al. 2013), as I discussed above.

Another study reports on a discovery among a tribe in Central America. In the popular fragrance ingredient cinnamon there is not only cinnamaldehyde, the main component of the strongly aromatic cinnamon oil, but also epicatechin, a vitamin-like plant substance that also occurs in cocoa. It is used by the Kuna tribe in Panama against dementia, but does not really smell. Allegedly, studies show that the two molecules have the potential to directly address Alzheimer's, stop the disease like a biological shield, and even protect against its outbreak (George et al. 2013). In the meantime, however, the hoped-for success of these molecules in Alzheimer's has again become quiet.

Whatever the results are evaluated— the fact is that the basic research in the fragrance and cosmetics industry is currently building whole "peptide libraries" for their analysis in order not to miss the future of active fragrance cosmetics.

Accordingly, research on the effects of fragrances—in particular with regard to stress—has developed and become interdisciplinary in recent years. Special tests of cognition and emotion psychology are used in neurobiological fragrance research as well as in fragrance and cosmetics product development. With the aforementioned peptide library, the imaging methods of brain research in combination with psychological tests such as the Trier Social Stress Test (TSST, Kirschbaum et al. 1993) can now be analyzed, for example, what different substances cause in stress and whether fragrance and cosmetics products with corresponding ingredients offer a response to stress and its psychophysiological processes such as increased sweating.

> As one of the first companies, the Hamburg company Beiersdorf used the TSST to develop a new Nivea deodorant.

10.3.3 Olfactory Tools For Assessing Well-being

The further development of the so-called NIH-Toolbox is also promising. This is a test battery that determines cognitive, sensory, motor and emotional functions. A team at Monell Center, one of the world's leading research institutes for fragrance and taste, has enriched the NIH-Toolbox with another dimension. It can now also be used for odor identification (Dalton et al. 2013). With the olfactory analysis tool, which can be used in less than five minutes with "Scratch and Sniff" or fragrance cards that release the smell when rubbed, a complete multi-dimensional test is now available that can be used for a variety of questions. NIH evaluates, for example, experienced stress such as the feeling of being overwhelmed and unpleasant feelings such as sadness and anxiety.

It is known from olfactory research that early stages of Alzheimer's and Parkinson's are accompanied by impaired sense of smell. Therefore, scents are used to diagnose health and well-being. Since the ability to identify odors is very different from individual to individual and the sense of smell is further severely restricted under stress and strain, NIH can be used in the future to carry out so-called cross-validations. In other words, one can determine in a certain rank or with a certain range what is an unobjectionable impairment of the sense of smell or where an aroma therapy or another therapy is already advisable. Validated clinical smell tests are used together with other tests as an indication for aroma therapies, but also to exclude other impairments such as sinusitis or anosmia, that is, a loss of smell or a high degree of impairment.

As I said, about 3 to 5% of the German and certainly also the world population suffer from anosmia. It is caused, for example, by head injuries in an accident and can even lead to a 15- to 20-% impairment of the sense of smell in the case of minor injuries. The NIH-Toolbox, which can already be used to determine the sense of smell in three-year-olds, could also be offered as a permanent service in the perfume retail trade—because who would not like to know how their own sense of smell is?

Alternatives to the NIH Smell Tool are the UPSIT (University of Pennsylvania Smell Identification Test), the CCCRC-Test (Connecticut Chemosensory Clinical Research Centers) and Sniffin' Sticks. These are four approximately 14 cm long felt-tip pens filled with 4 ml of a specific fragrance. They are presented for about three seconds under both nostrils. NIH in combination with Sniffin' Sticks allows for a very fine and comprehensive diagnosis, which can also include the results of 16-step threshold tests.

To determine the threshold, repeated ascending and descending concentrations of the same fragrance are presented. Usually, work is done with phenylethyl alcohol smelling of roses. Since impairments of the sense of smell affect quality of life in the form of taste and smell enjoyment, but also the use of fragrance and social contacts, for example by over-perfuming, and can indicate possible diseases at an early stage, it would be of interest to at least have an annual smell check at their perfumery store for certain groups of the population. It would therefore pay off for trade to invest in this service.

❯ fMRT devices and active fragrance advisors in stationary perfumeries

However, the future of smell diagnosis for wellness and health prevention will be able to do much more. As a technology, it already exists. Research and medicine work with it. So it has developed into an important procedure in the imaging routine diagnostics. For the perfume retail trade, it is interesting for several reasons. Thus, harmless and easy-to-use fMRT devices for imaging procedures are becoming more and more portable and cheaper in terms of purchase price. This could enable stationary trade in cooperation with a specialist to offer a completely new fragrance and smell advice: the active fragrance advice.

❯ The future of perfumery and fragrance-supported therapy requires new training—to become an active fragrance advisor, with which you can become a certified fragrance therapist with an additional qualification.

The cooperation with doctors is already common practice in stationary perfumeries today, for example when a plastic surgeon informs perfume customers at care evenings about the possibilities of injections. A win–win for both, as medicine and perfumery complement each other very well for the

pre- and post-treatment of the skin or for the prolongation of the results of cosmetic surgery. Of course, you don't need fMRT devices for this additional advice.

However, its use would be very interesting for another area of stationary perfumery and would attract customers. Certainly, the use of fMRI machines would have to be adapted for a wide public (Wang et al. 2013). But there is no question that this harmless method without ionizing radiation will greatly enrich the treatment and diagnostic offerings in spas, wellness or fragrance cabins. Because fMRI can not only be used for the diagnosis of olfactory experience and memory, that is, for wellness and health prevention, but also in a very spectacular and fascinating way as an imaging and thus scientific perfume advice. It gives customers insights into which specific scents and fragrance directions trigger the best olfactory-induced emotions. That should interest a younger high-tech audience in particular. The method also visualizes the human brain with great, detailed images that can be taken home as a printout or spread via social media. It answers questions about which perfumes emotionally optimally stimulate the brain and are best for one's own enjoyment and well-being. For example, you can see which processes are triggered by the favorite perfumes in the emotional center and whether the mixing of scents (perfume layering) can increase their psychological effect for oneself.

The imaging perfume advice could then specifically suggest perfumes for specific effects, for example to experience oneself more relaxed with an individually tailored meditation scent or to discover the best olfactory mood enhancers for more joie de vivre and cheerfulness. Trade could thus offer a new fragrance category with its own shelf: the neuro-perfumes.

I am sure that not only perfume connoisseurs, but also new target groups will queue up for the one who will offer the first imaging perfume advice. Especially the stationary perfumery, which is always striving to win new customers, must offer them something special to keep them. I therefore suggest to the perfume associations to offer an recognized additional training for perfume specialists who want to specialize in fMRI and olfactory tools for the assessment of well-being as an active fragrance or neuroscent advisor and then as a scent or aroma therapist with the corresponding qualification.

With the training to become a active fragrance or neuroscent advisor one could also qualify for the diagnosis of smell within wellness and prophylaxis or on mood & health modulation. These additional training courses could be integrated into the training to become a perfume and cosmetics expert. The German perfume retail group "First in Beauty" already offers, in cooperation with the perfume association, as mentioned above, training to become a "maître des parfums" and even a "maître de cosmétique". An additional training to become a "maître des sciences de parfums" and possibly even a "maître des sciences de cosmétique" could make the perfume trade even more fascinating.

The first cheaper and portable devices for olfactory fMRI examinations have been in the experimental stage for a few years (Sezille et al. 2013). They would have to be further developed for the perfume industry. So far, MRI scanners have been unaffordable for many clinics. However, purchasing groups such as the "Beauty Alliance", in addition to Douglas one of the largest perfume providers in Germany, could develop their own fMRI devices for imaging perfume advice, wellness and prophylaxis for their often more than 100 perfume members. After all, a whole-body scanner would not necessarily be necessary for consultation in cosmetic or perfume cabins.

In the field of wellness and prophylaxis, cooperation could take place with doctors, but also with clinical psychologists. In par-

ticular, on the topic of "stress and prevention", cooperation e.g. with fitness studios, meditation, yoga, Pilates or Ayurveda schools or their trainers could significantly enrich the perfume industry. This would leave the success of the neuroperfumery in practice, but also the future of stationary perfumery with a completely new offer of experience, wellness and health services, nothing in the way.

10.4 Self-therapy: Healing the Soul and Smelling Better with Primal Perfumes

Each human being had their first olfactory experience as a fetus, namely when they smelled the amniotic fluid of their own mother. This was followed immediately after birth by the skin-milk smell of the mother while breastfeeding. So from the very beginning of our existence, we have come into contact with two very positive olfactory worlds. Since these two primal smells are complex in composition, one should actually speak of primal perfumes. However, there are variations in the good-smelling primal perfumes, especially because the odor of amniotic fluid smells different for everyone due to the mother's diet, for example, but also because the fetus excretes small amounts of urine into the amniotic fluid. Each of the two olfactory worlds is experienced by the fetus or newborn only unconsciously. However, they are usually linked to positive psychological experiences and effects in the case of uncomplicated pregnancy, uncomplicated birth, and breastfeeding. The primal perfumes also have their own positive, stimulating effect on the young brain, specifically on delta-opioid receptors, especially in the amygdala.

The research already knows a lot about the positive psychological effect of the second primal perfume, the skin-milk smell while breastfeeding, but so far only little about the effect of the first primal perfume, the amniotic fluid smell. In newborns, the smell of amniotic fluid is already stored and they can remember it well. It triggers deep feelings of well-being, almost happiness. This was found out when newborns were allowed to smell the amniotic fluid again. Apparently, the smell of one's own amniotic fluid as well as the skin-milk smell are ideal for stimulating, for example, delta-opioid receptors in the amygdala and the olfactory bulb. This leads to a greater sense of comfort and well-being in babies and adults, up to and including mood enhancement.

The recognition of the first primal perfume is understandable for olfactory research. Unborn babies already have a fully functional olfactory brain with amygdala and hippocampus at the end of pregnancy, which gives humans an phenomenal sense of smell. It can therefore be claimed that as an adult, one carries the positively experienced smell of their first primal perfume deep within them. This also explains why vanilla, as a component of the second primal perfume, that is, the skin-milk smell, is still highly valued as an olfactory and gustatory experience in later life and is perceived as appetizing and, above all, as warm and comfortable. By the way, vanilla, like almost all edible aromas, is more of an olfactory experience. As already mentioned, biochemists from the Monell Chemical Senses Center in Philadelphia have found that even the taste cells on the tongue have olfactory receptors, which probably intensifies the olfactory impression even further.

It can further be claimed: Fragrances that are reminiscent of the individually experienced smell of amniotic fluid and the skin-milk smell of the mother are associated with positive psychological effects in the form of a kind of primal well-being and happiness for most people—provided that the person has not experienced any special prenatal, birth or postnatal stress. This can make primal scents into primal active

scents and be of great psychological and psychosomatic benefit. It can therefore be assumed that these primal active scents not only have a positive and calming effect on one's own amygdala, but also give a deep sense of a wonderfully different world of life, since the first primal smell was mainly smelled in the weightlessness of the warm and protective amniotic fluid, so to speak in a state that only few experience through meditation.

It can therefore be expected that a deeper and more direct meditation experience can be achieved more easily with the support of a scent with notes of the first and second primal scents. I am thinking here, for example, of a Metta-Bhavana meditation supported by a primal active scent. The success of the scent-supported meditation will of course depend on how well the recreation of the individual's amniotic fluid smell or the skin-milk smell of the mother succeeds.

10.4.1 The Power of Primal Scents and Their Creation

Perfumers can already look back on successes in recreating the skin-milk smell. Some perfumes available on the market from the milk or milk mousse fragrance category, for example "Fragrance Condensed Milk" by Demeter or "London Sweet Milk" by Jo Malone, create a certain relaxing sense of wellbeing and can remind us of our earliest childhood experiences. No wonder, because most of these new fragrances smell warm, creamy and edible. They are loaded with sweet Vanilla, lactose and slightly musky notes as they occur in mother's milk. However, the milk impression is only an illusion, because milk is not used for the creations themselves.

Recreating the smell of amniotic fluid is more demanding. This scent is more individual in composition because, as already mentioned, it is more strongly influenced by the mother's diet and also by the fetus's urine.

In the following, I will introduce the primal base scent of amniotic fluid in its facets and give a guide on how to get closer to it—even if you are not a perfumer and do not have the appropriate ingredients—with the technique of perfume layering.

Ideally, a amniotic fluid smell set with the essential ingredients for self-mixing would be great. But I don't think that's something that's available to consumers yet. So when creating ours, we'll be relying on high-quality brand perfumes of individual scent profiles. Of course, alcohol-free organic perfumes that are similar in quality to essential oils would be the better choice. But in most perfume shops, it's very difficult to find perfumes of this quality in the scent profiles we need for our experiment. So in order to become the perfumer of our own primal-effect perfumes, we have to make compromises. There's something else to consider as well.

In order to recreate the smell of amniotic fluid, everyone will have to rely on their own nose, that is, their Amygdala, their Hippocampus, and their Hypothalamus, to become a successful private perfumer. There will be many attempts with error. And the actual base smell of amniotic fluid will probably smell different than its recreation, just because our memories and our taste and smell preferences have changed over the years. That's quite normal.

From a self-therapeutic perspective, what is essential in recreation is an olfactory journey to oneself, in which one becomes active for oneself and takes oneself as the goal. Apparently it is a journey into one's own past, but not exclusively. Because current experiences and desired experiences accompany one on the journey to the deepest self. So it's about getting on the trail of scents and perfume creations that can do more than just smell good. They speak to

you very deeply, and you associate them with a sense of well-being and even deep happiness that you even experience currently. They are therefore excellently suited for your own scent therapy. Ideally, the search for them and their discovery is even the therapy itself.

Most of the time, one experiences perfumes that others have created. But now one is challenged oneself. This self-directed scent journey not only has a great emotional benefit, it also stimulates creativity and artistic expression. So it may be that after initial difficulties, one discovers the talent within oneself to create primal perfumes or, better said, primal-effect perfumes for oneself and others.

But the activity of the private perfumer for his own olfactory brain has another advantage: the industry is working on intelligent wearable designer jewelry with a regulatable timer that emits small amounts of fragrance at the user's choice. A first step in this direction is, for example, the prototype of the Mini-Aroma-Shooter, which you can hang around your neck and fill with any perfume. So far, the technology of wearable E-Scents is still in its infancy. But it could offer mobile independence for self-therapy with fragrance as well as for all aromatherapy types. Ideally, you would not only have access to pre-made fragrance cartridges, but also to fragrances specifically created for the amygdala. These could then be applied discreetly for personal mood change, stress prevention, inner restlessness and insecurity, but also for increasing the feeling of happiness.

- **Amniotic fluid: first smell in thermally comfortable weightlessness**

When researchers discovered a few years ago that unborn babies can already smell their mothers' amniotic fluid, the interest was initially great, but without drawing any conclusions for the perfume and fragrance therapy from it. Amniotic fluid is, as already mentioned, the first smell that humans perceive, it is the primal perfume. But how does it smell for a fetus? Benoist Schaal, scientific director of the European Centre for Taste Research in Dijon (Centre Européen des Sciences du Goût), and his team were among the first to investigate this question. They confirmed that amniotic fluid has a basic smell, but also that it offers a range of smells that depend on the mother's diet.

Of course, the French research group recommends a varied diet to every mother so that the unborn child can experience a whole range of taste and fragrance notes right from the start and thus get used to them early on. Schaal explains the connections as follows: "What the mother eats is transmitted to the fetus via the placenta, and the fetus definitely inhales amniotic fluid. The amniotic fluid comes into contact with the olfactory receptors and is constantly renewed, so it also provides new stimuli and stimulates the nose. This information reaches the fetus's brain, where it is imprinted until after breast milk, namely preferences and diversity in food choice" (Benoit Schaal in the ARD broadcast "Breastfeeding is Dufte" from 26.04.2009).

The first smell is for most unborn in a paradise-like state. Because as early as the fourth week of pregnancy, the fetus is completely surrounded by protective amniotic fluid. The amniotic fluid allows the fetus weightless movements. With increasing brain development, he smells accordingly in a weightless state.

At the same time, the amniotic fluid protects the unborn from shocks and provides a constant comfortable ambient temperature. This increases his sense of security and well-being or his basic trust—provided that later no complications arise at birth.

Approximately from the 14th week of pregnancy, a almost perfect balance is created. The baby begins to drink the amniotic fluid, which is constantly being replenished. A residue of what is drunk then returns in the form of fetal urine back into the amniotic fluid. Of course, one does not know what the unborn smells as amniotic fluid exactly, since the smell depends on the mother's diet. Nevertheless, there is a first primordial smell, which the fetus probably smells again and again.

According to Benoist Schaal, the first perfume perceived smells quite complex of something honey, buttermilk and caramel, but also of urine and slightly rancid fat. The latter two factors give the amniotic fluid smell from a perfumer's perspective something slightly animal. Perfume connoisseurs will take notice of this keyword. Because since the beginning of perfume history, animal notes have been used to make scents irresistible human. The accompanying small, subtle stink made them really likable and attractive for people. In modern perfumery, many animal components have been removed from scents for various reasons such as animal welfare or allergy risk. Nevertheless, there is a magical group of ingredients that could be used specifically for the creation of a primordial perfume.

Since in general, the fetus with its increasingly developing amygdala experiences the environment in the amniotic fluid as very pleasant and sheltered, this should also apply to the first primordial smell. Because from the 5th to the 7th week of pregnancy, the development of the amygdala responsible for emotional feelings and smell and the hypothalamus begins, which inter alia controls the body and sense of pleasure. From the 13th week, the development of the hippocampus, which forms our sense of smell with the amygdala and the hypothalamus, begins. From this point in time, the primordial smell and other smells in the amniotic fluid could be stored. It is now assumed that babies can remember the smell of the amniotic fluid from the 28th week of pregnancy at the latest. Because from then on, the olfactory nerves are also functional.

■ **The therapeutic power of primordial perfumes**

The experience of the primal perfume takes place initially unconsciously and remains so because the brain develops, so to speak, from bottom to top. The spinal cord, the brain stem and certain limbic structures such as the amygdala function as the first. From the 7th month of pregnancy, the small brain is biologically able to remember original smells. So two to four day old newborns immediately recognize their own amniotic fluid and, as Benoist Schaal describes, turn their heads to the soaked towels, start sucking and put their fists in their mouths.

The smell of amniotic fluid has a surprisingly positive psychological effect. The first original perfume is even more efficient than the mother's milk smell to calm newborns (Varendi et al. 1998). The effects even occur at fragrance concentrations that are below the threshold of adults, i.e. which are not perceived by adults. So babies can smell finer than adults.

Of course, these are very primary processes that take place much less explicitly. Nevertheless, the positive psychological effect of the first primal perfume for fragrance therapy, possibly even in its effect on adults, is a sensation—all the more so because we can assume that the first primal perfume is stored deep in our olfactory brain for life, even if it is overlaid by many subsequent smells. So there is at least one smell, probably rather a chain of smells or fragrances or—to put it another way—real primal active perfumes, which are deeply associated with pleasant feelings of warm, protected weightlessness and trust on a psychological-biological level and have a corresponding effect.

A fragrance therapy with primal active perfumes alone or especially in combination with certain cognitive therapies should therefore have clinically verifiable effects for certain therapeutic goals such as more self-confidence and inner satisfaction and be ideal. At the same time, fragrance therapy with primal active perfumes provides an explanation for its effect, because with therapies it is often not quite clear why they work or why they do not work in individual cases.

However, fragrance therapy with primal active perfumes also had to lead to entirely new opportunities for psychological self-discovery, especially in the context of certain meditations, and this for two main reasons:

On the one hand, there are deep-seated feelings from prenatal times that are also the goal of, for example, the Metta-Bhavana-Meditation: to create a warm, relaxed, almost weightless atmosphere in which one feels welcome and taken care of. The typical wishes with which the meditation begins in view of oneself are biologically anchored wishes, as the fetus would formulate them: "May I be safe. May I be happy. May I be healthy. May I be loved. May I live with ease."

On the other hand, there is an original primal scent or an original primal perfume for everyone that allows access to biologically anchored wishes from prenatal times because it is associated with the corresponding feelings. Such an original primal perfume of the amniotic fluid would be ideal and highly effective from a scientific point of view for a "Scented-Loving-Kindness" meditation, even more effective psychosomatically than the already very emotional skin-milk smell of the mother.

The discovery of highly effective primal perfumes as a possible therapeutic approach also for adults comes at the right time for scent- and aromatherapy. The view that scent is particularly suitable for the initial focus of a meditation is gaining ground more and more in mindfulness research—

especially if it is complex and one can concentrate on the different facets of a smell.

■ **Recreation of the original primal smell as an primal active perfume**

The taste and smell of the amniotic fluid are a very complex mixture that nature offers to the young brain as an primal perfume. As mentioned before, for perfumers it smells of honey, buttermilk and caramel, but also of urine and slightly rancid fat. Amniotic fluid, scientifically called amniotic fluid, is usually a clear or slightly milky liquid. It contains, inter alia, sugar, proteins, potassium, sodium and trace elements and—as mentioned before—some urine.

Possibly fetuses, who have a much finer nose than adults, unconsciously perceive the basic smell of the amniotic fluid as a mixture of three large fragrance components: as a very attractive, sweet, soothing, slightly milky, nourishing and familiar-protective, physically-animal smell, which is additionally defined by the nutrition of the mother.

You can try to get as close as possible to your individual amniotic fluid smell with a self-experiment. Of course, no one has a conscious memory of their amniotic fluid. But the main components are known and might lead the nose intuitively and unconsciously in the right direction. For the creation of your own original perfume, you have to let your nose, i.e. your amygdala, your Hypothalamus and hippocampus, simply drift without any considerations.

— First, you learn to know the three big fragrance components or fragrance directions "sweet", "milky" and "physical-animalistic" of the amniotic fluid in different intensities and interpreted by different perfumes. Ideally, as said, alcohol-free organic perfumes would be used which are oriented on the quality of essential oils. But so far, they are only very difficult to get in the necessary fragrance directions in most perfume shops. So we have to make compromises and work with brand perfumes which, however, offer very fascinating possibilities in terms of smell.

— Secondly, you choose the perfume from each fragrance direction that suits you best. This has to be done purely emotionally, i.e. whether it triggers a deep sense of well-being or an inner touch. You must not let yourself be influenced by aspects such as bottle, haptics, color, brand, image, name, value or fragrance descriptions. Otherwise, you would smell with the orbitofrontal cortex and thus with higher brain regions, which make it more difficult to have an emotional access to the perfume. Maybe you smell the different perfumes blind at first and sort them by pleasant and less pleasant. If you have found the most pleasant perfume for the respective fragrance direction, you rate it on a scale. It is important to take your time and not to smell more than three to four perfumes in a row. The orbitofrontal cortex could still smell, but the amygdala quickly finds too much fragrance to be strenuous. The purpose of the exercise is to get in touch with the amygdala, which actually only responds to favorite scents when choosing a fragrance, second- or third-best fragrances do not interest the oldest region of our olfactory and emotional center—let alone whether a perfume fits your personality or could also please others.

— Third, the selected perfumes are brought together into a final scent for the creation of the personal primal perfume by perfume layering according to a special key. The three fragrance directions can be mixed surprisingly well. For all genders, it is only each individual person who can judge how well the first creation has succeeded—whether the primal active perfume triggers subtle associations and feelings of warmth, security, well-being, security and enjoyment, combined with a light, floating sense of happiness. This takes patience, because even many perfumers often revise their creations more than 30 times, often only changing the ratio of the individual ingredients to each other. "Cool Water" for example, a classic with only 16 essential fragrance components, was revised more than 20 times by the great perfumer Pierre Bourdon until it was right for him. Therefore, it is recommended that you have enough fragrance samples of each perfume on hand for layering your own primal active perfume. In any case, you need numerous fragrance strips, which you can also cut from unscented coffee filters. In contrast to professional perfumers, finished brand perfumes are used here instead of individual fragrance components. Since these are dissolved in alcohol, you always have to wait a little until it has evaporated.

In the following, I will give a practical guide to creating your own primal active perfume in four steps.

1. *Find the right sweetness for your own primal active perfume*

   To find the right sweetness for your own primal active perfume, it is best to visit a perfume shop. For this, first smell the perfumes suggested below in the top note on a perfume strip, which develops within the first ten minutes. For this, hold the perfume strip first at a distance of half an arm's length, then a quarter arm's length and finally bring it up to an inch to the nose. This is how you can best determine which sweetness level is most pleasant to you. But before smelling it, you should just sniff it first. If you have found the scent from this direction that appeals to you the most or at least emotionally, evaluate the perfume on a five-point scale that ranges from 1 for "rather not" to 5 for "enthusiastic". Here are the details for the evaluation.

   - If the top note of a perfume smells intense enough on the perfume strip at a distance of half an arm's length or if you do not want to bring the scent any closer to your nose because you do not like it or it is too intense, give the perfume a 1.
   - If the scent is enough for you at a distance of a quarter arm's length or if it is not unpleasant to you but you do not want to bring it any closer to you, give it a 2.
   - If you like the perfume or can't get enough of it, give it a 3 for "I like it", a 4 for "I love the scent" and a 5 for "I am emotionally enthusiastic". When doing this, you should not skip a scent direction or decide on a perfume for the final creation in each scent direction, as this would not reflect the complexity of the fruit water scent.

In perfume, sweetness is most enjoyable in gourmand notes. They remind us of edible delicacies like marzipan, caramel, honey, chocolate, coconut, candied fruit, and other sweets. Here are some perfume suggestions on the theme of "sweetness" for evaluation:

- "heliotrope" by Etro for women and men. A delicious and indulgent gourmand sweetness, reminiscent of almond, marzipan, and orange blossom, with extra warmth, softness, and appetizing qualities imparted by vanilla.
- "Virgin Island Water" by Creed for women and men. A light, aquatic gourmand note with a fresh, sweet interpretation of coconut, like one might imagine a tropical fruit-infused water.
- "Miel & Vanille" (Honey & Vanilla) by L'Occitane en Provence for women. A gourmand note that celebrates honey, caramel, and vanilla with a touch of cinnamon.

– "New York Nights" by Bond No. 9 for women and men. A caramel smoothie for those who like it a little less sweet.

2. *Finding the right "cream" for your own primal active perfume*

Perfumes with "milk" and "milk mousse" scents are suitable for the right "cream". Here are my perfume suggestions in different facets, which are also rated on a five-point scale.

– "Matin Calin" by Comptoir Sud Pacifique for women with a milky-warm, sweet-vanilla headnote.

– "DKNY Stories" by Donna Karan for women. Lightly spicy-fruity in scent with vanilla-infused, milky-powdery chords.

– "Condensed Milk" by Demeter Fragrance for women and men. A milky-vanilla scent with a hint of white musk as a bodily undertone.

– "London Sweet Milk" by Jo Malone for women and men. A milky-caramel scent with hints of tea with milk.

3. *Finding the right physically-animalistic scent for your own primal active perfume*

In modern perfumery, many animalic fragrance materials of animal origin, such as civet or musk deer, are no longer used. Although civet is still bred on farms, animals such as musk deer are protected in the wild. For ethical and environmental reasons, these fragrance materials are almost exclusively produced synthetically or naturally simulated notes are used that convey the impression of animalic smelling fragrance materials.

In classical perfumery, mainly musk, ambergris, civet oil and beaver musk were the most desired animalic fragrance materials. For example, civet is a very special substance that was obtained from the scent glands of the civet cat and smells like cat urine. Strongly diluted, however, it is perceived by humans as very pleasant and attractive. The same applies to musk, which originally came from the musk deer's musk gland.

There is still a natural, animalic substance, ambergris, which is highly sought after by the perfume industry and correspondingly expensive - however, in some countries (USA and Australia) it is also illegal to collect, store or sell it. Ambergris is simply excreted by the sperm whale without human intervention. The rare oil obtained from whale secretions was already appreciated in ancient cultures. The gray amber (ambergris) was used as a remedy and spice. Today, ambergris is also increasingly interpreted naturally identical. In small quantities in perfume, it smells sensual, emotional and relaxing as well as balsamic-sweet, reminiscent of human skin. Suddenly "1001 Nights" is in the air. You feel comfortable in your skin and have to smell it again and again, which is due to the slightly aphrodisiac effect of the ambergris smell.

The smell of ambergris is also often combined with oud to give animalic scents more depth and roundness. The rare fragrance material obtained from the agar tree was first mentioned in the Indian Vedas, one of the oldest surviving texts in the world. It is the secret of many exquisite Amber-oriental perfumes and one of the trend fragrance materials of recent years. Oud smells woody-sweet, with a slightly burnt-caramelized note that can have a light honey tobacco undertone depending on the quality.

Here are the perfume suggestions for the physically-animalistic theme. The selected fragrance for the final creation of one's own primal active perfume will also be rated on a five-point scale.

– "Amber Musc" by Narciso Rodriguez for women. A balsamic-amber creation from the "Amber-Oriental" fragrance family, in which Musk, Oud, Patschuli and the smell of ambergris with Orange Blossom in the top note set the tone.

- "Ottoman Amber" by Merchant of Venice for men. It is also balsamic-amber with a vanillized Sandalwood note.
- "Peau de Soie"" by Starck for women and men. A subtle, powdery-woody creation that smells like summer skin, but is also created with something dark-green-vegetable, like a vegetable chord.
- "Théros" by Ys-Uzac for women and men. A tender Oud note that smells like summer skin, to which the smell of Amber and Salt give a familiar and pleasant physicality.
- "Perfect Oud" by Mizensir for women and men. A mild, dark woodiness is combined with something animal leather as well as with something spicy and even flowery in the first impression. Overall very harmonious and balanced, as is the signature of Alberto Morillas, the perfumer behind the creation.

4. *Perfume Layering for your own primal active Scent*

So far, a total of three perfumes have been selected, one from each fragrance group. Each perfume was rated on a scale of 1 to 5. For creating your own primal active Scent, I suggest the following further procedure:

From experience, perfume layering works best if you first spray the fresher or lighter perfume on the scent strip and then the more intense one over it. The fresher perfume usually has more quickly evaporating molecules that want to pass through the more intense perfume. This often results in a surprisingly interesting effect.

According to the rating of the perfume on the scale, it is sprayed as follows:

- Rating with "1": barely and only very briefly spray on the scent strip
- Rating with "2": only spray once briefly
- Rating with "3": spray once
- Rating with "4": spray twice
- Rating with "5": spray three times

Here is an example:

You like "Virgin Island Water" by Creed best from the Sweet fragrance group. It was rated "4" on the scale. From the "Cream" group, "Condensed Milk" by Demeter was selected and rated "2". Of the physically-animal scents, you like "Amber Musc" by Narciso Rodriguez best. It was rated "3". Now apply the three perfumes to a scent strip as follows:

The lightest of the three perfumes, "Virgin Island Water", is applied twice to the scent strip first. Then "Condensed Milk" is sprayed over it once, followed by "Amber Musc", also once. The scent strip is now slightly wet with the three perfumes. Professional perfumers therefore proceed differently when they are looking for new compositions. They generally dip individual scent strips into bottles with the respective fragrance bases, after removing the spray pumps. They then hold the three scent strips at different distances in front of their nose at the same time. This allows for a finer combination of scents.

Even laymen can proceed in this way with the perfumes. However, one should wait one to two minutes before smelling, so that the alcohol evaporates somewhat. However, for perfumers, smelling different fragrance bases is easier, as their smell is usually less concentrated than a finished mixed Eau de Parfum. Therefore, it is no problem for trained noses to smell even five scent strips at the same time.

But what the layman shares with professional perfumers is the luck of the diligent. It is not uncommon for chance to play into your hands, and a combination that you have not even thought of and did not plan turns out to be a hit.

When creating the primal active perfume, the layman only counts on his own sense of smell. He alone decides how close he has come to his target with his creation. In any case, your own active perfume should evoke first subtle associations and sensations of warmth, security, well-being, security and ideally also of enjoyment in combination with a light, detached feeling of happiness. Every small step in this direction makes your own l primal active perfume valuable for your own fragrance therapy.

- **The therapeutic smell of mother's milk and breast milk**

In trying to get closer to the basic smell of one's own amniotic fluid, one also comes into contact with the first smells as a newborn. Both overlap to some extent, with the smell of breastfeeding, of course, the milk-skin smell and thus the fragrance directions "milk" or "milk mousse" in the foreground. It can be assumed that this mixture of smells must be much more present in the olfactory memory when breastfeeding and initially overlays the memory of the smell of amniotic fluid—especially if this smell is also encountered as a child, for example when breastfeeding younger siblings. But this is also the case if the mother herself was breastfed. Then the mixture of smells can also be perceived consciously, refreshed and stored. As a rule, one then likes perfumes from the milk direction. This means that if the phases of life as an infant, toddler or adult were experienced positively when breastfeeding, this world of smells is also a promising approach to a scent therapy. It can be equated with the olfactory world experienced by the fetus in terms of its psychological benefit.

Studies have shown that experiences made while breastfeeding have a very intense effect on the amygdala and adjacent brain regions of the infant, but also on those of the mother. Certainly, the first smell experienced positively by the infant after birth must have a very special meaning and attraction. It was experienced as powerful, since newborns are equipped with a much finer nose than adults. It is also essential that infants have an innate, biological affinity for this world of smells and tastes. When breastfeeding, the taste experience is not in the foreground for the infant, but the sense of smell is 80%.

Nature also helps with the olfactory impression. Between 10 and 40 scent glands are located on the nipples of humans. The Centre Européen des Sciences du Goût even found that different psychophysical effects depend on the number of these scent glands. Babies of mothers with many scent glands developed better, gained weight faster and were more active. So a lot of mother's scent helps a lot, and this is not only true for the amount of food intake. It is now proven for babies that the mixture of smells of mother's breast and mother's milk has an enormous psychotherapeutic effect.

A number of studies even conclude that the second primal smell not only has a psychologically relaxing effect, but is even physiologically pain-relieving. It conveys more comfort and relaxation than the smell of vanilla alone, which most adults value as the second most pleasant smell after chocolate (Porter and Winberg 1999; Badiee 2013; Wei and Tsao 2016; Neshat et al. 2016). It is therefore only too understandable that this primal perfume, the smell of the mother's breast with breast milk, is stored very prominently in babies' memories. It was found, for example, that even after weaning, two-year-olds could still remember it well.

Even adults keep the smell of the mother's breast with breast milk centrally in their memory. Of course, it is overlaid or suppressed by other fragrance preferences because fragrance and taste preferences change with age. For example, during fragrance socialization, the smell of sweat on the skin is stigmatized more easily, while babies perceive it more clearly and positively. Nevertheless: from a perfumer's point of view, the mixture of smells of mother's breast and breast milk, to which the smell of baby care is added, is directly and indirectly omnipresent. Otherwise, the universal preference of adults and the associated associations of well-being cannot be explained by white musk notes. They smell like part of the mixture of smells. White musk notes are used in perfumery as a general term for sweet, creamy, powdery skin notes with a light fruity freshness. Perfume examples are "Candy L'Eau" by Prada, "Eau d'été Flower" by Kenzo or "Eau de Musc" by Narciso Rodriguez. Even more: entire fragrance categories rely on this fragrance ingredient and its associations, especially in the heart and base notes. These include the floral -oriental fragrance category with perfume classics such as "Ombre Rose" by Jean Charles Brosseau, "Le Bain" by Joop! or "Coco Mademoiselle" by Chanel.

■ **Bonding smells**

To increase the scent impression of sensual wellbeing, in addition to white musk roses, jasmine and lychee are especially used and fused into a gentle, emotional and slightly fruity fragrance experience. In the aftertaste of these creations, an more or less conscious feeling of personal comfort is established, which expresses itself in emotional self-closeness and interpersonal closeness. Therefore, fragrance psychologists also like to call these perfumes binding scents.

The attractiveness of slightly milky skin scents is naturally promoted because the bonding hormone oxytocin is released during breastfeeding by both the mother and the child. In addition, the oral phase also lasts until the age of one year in infants. Through the mouth and the associated food intake, the libidinal development and the feeling of pleasure also begin. This is an-

other reason why babies find smells from skin and mother's milk irresistible.

But mothers also get their money's worth. The smell of babies triggers the release of endorphins, body's own messenger substances, in mothers, which are responsible for feelings of happiness and wellbeing. Therefore, it is no wonder that skin fragrances create a sense of wellbeing worldwide. Studies confirm this. For example, the British fragrance manufacturer QUEST (today the company Givaudan) found out through EEG studies that, above all, sweet, musky scents, as they occur in mother's milk, create a noticeably relaxing sense of wellbeing even in adults.

Based on these findings, the following tip can be given for the creation of one's own primal active perfume: To increase associations and sensations of security and wellbeing as well as of enjoyment in combination with a slightly detached feeling of happiness, one could experiment with a higher milk or milk mousse content. In this way, one might spray the perfume chosen from this fragrance direction and rated 4 or 5 two or three times to make it more present in the creation of one's own primal active perfume. But maybe you will also come to the conclusion that this perfume does not offer the desired effect and you would rather use another scent. But since this fragrance direction is of central importance for one's own primal active perfume, I will mention ten more perfumes:

– "Lait Concentré" by Chabaud Maison de Parfum for women
– "Tome 1 La Pureté" by Zadig & Voltaire for women and men
– "Plus Plus Feminine" by Diesel for women
– "Dent de Lait" by Serge Lutens for women and men
– "Good Girl Gone Bad Extreme" by Kilian for women
– "Jeux de Peau" by Serge Lutens for women and men

- "Dulce de Leche" by Demeter Fragrance for women and men
- "Mirror Collection—Dis Moi, Miroir" by Thierry Mugler for women
- "Sultan Gâteau d'Or" by M. Micallef for women
- "Something Sweet" by Lise Watier for women

If your own primal active perfume gives you a slight case of goose bumps or a nice feeling of well-being when you smell it, you have developed a great therapeutic tool for yourself. So by creating your own primal active perfume, you have invested in yourself, both in terms of time and money. This is not always easy. Because in general you feel responsible for everything, just not for your own well-being. You believe you don't have time for yourself and put your own interests last. Often, when you take time for yourself and spend money on yourself, you feel guilty. If you do it anyway, it often shows a certain urgency. Perhaps you have increasingly felt the need to do more for yourself and escape from everyday life because you were no longer able to cope with all the responsibility, pressure and stress, and you didn't even receive recognition for what you had done. Not to mention the consideration of your own feelings and needs by others.

With your own primal active perfume, you can achieve more inner strength, self-confidence, well-being and joie de vivre again. Within an accompanying therapy, it olfactorically influences the mental and physical condition in a positive way directly and at the same time mostly unconsciously via the amygdala, the hippocampus and the hypothalamus. This fragrance therapy with numerous elements of a fragrance-supported self-coaching is particularly effective when it focuses on specific emotional wishes and goals mentioned below. Some of them are deliberately not too big, unattainable or too specific. Perhaps

some of these goals were even the reason why you created your own primal active perfume.

In my experience, for example, in the context of exercises with "fragrance-supported loving looks/self-awareness", as I will show later, it can be beneficial to use primal active perfume. Of course, a fragrance-supported self-therapy cannot address all the topics listed below at the same time. But if you focus on one topic in your work, it will also positively influence other areas. These include:

- To feel more comfortable within yourself.
- To develop more composure and confidence in yourself.
- To feel more inner strength within yourself.
- To experience more cheerfulness and enjoyment.
- To gain more self-confidence.
- To build up more inner lightness.
- To experience more satisfaction.
- To trust yourself.
- To forgive yourself.
- To find yourself more likeable and attractive.
- To show yourself to others cheerfully and with understanding.
- To become more winning for yourself and others.

When creating your own primal active perfume, you may not have been aware of the curiosity involved, but you were following an inner need to reconnect with a time in your life that you hoped would have a positive effect on the present. So when you created your own primal active perfume, you may have already noticed how good certain elements smelled to you. Probably a lot of fantasies and feelings went through your head that you experienced or could have experienced as a baby or toddler. You may also have gotten a vague sense that there was an easier time in your life—a time with much more security, trust and well-being than you feel as an adult.

If your own primal active perfume evokes positive feelings in you, this is a possible indication that you felt good as a fetus, as a baby and perhaps even as a toddler. Probably you were cared for and loved during this time. This feeling may have been lost later on the way to adulthood. But only each individual can judge this for themselves.

In general, one can say: If you react positively to your own primal active perfume, there was probably a pleasant time at some point—even if you currently do not feel comfortable in your own skin and have made negative experiences that still burden you. This good time is probably not or only partially conscious to you, maybe you only know about it through stories. But there is no reason to believe that this good time could not come back. Even if you are convinced that there were no or too few positive experiences in the early years of life, this is only because these experiences are difficult to access in the subconscious. But they are just waiting to give you back strength, self-confidence and satisfaction. With the help of your own primal active perfume, you will probably find the easiest access to these unconsciously or less conscious positive feelings and experiences. The amygdala as the deepest center for emotion has stored them together with the hippocampus and hypothalamus. Therefore, your own primal active perfume is a very personal happiness medium that connects you with the good times, so to speak, and makes them come alive again.

10.4.2 The Power of Scent-Supported Loving Looks For Self-Therapy

Psychotherapy knows methods of treatment that, from my own experience, I know can be profitably enriched with a scent therapy or scent-supported techniques. I would like to show this using the example of affirmation in combination with primal active perfumes. Under affirmation one understands a self-affirming sentence that one repeats to oneself over and over again in order to program thoughts, so to speak (e.g. "It will be easier for you from now on, more and more every day").

The efficacy of affirmations is controversial in therapeutic practice, but the reciprocal relationship between thinking, feeling and acting is generally recognized in psychotherapy. However, those who opt for a therapy based on affirmations should not be skeptical about the effect of their auto-suggestive self-dialogues. Dr. Marcel Wilhelm from the Department of Clinical Psychology and Psychotherapy at Philipps University Marburg therefore asks his therapy participants right at the beginning of the first exercises to expect something positive: Then you can see it and it is easier to notice that it has become better. Otherwise you quickly overlook the positive because you are focused on the negative.

Another finding from therapy and emotion research: In experiments it was found that people who address themselves with "you" or, even better, with their own familiar (first) name, are more relaxed in emotional, stressful situations (Moser et al. 2017). Furthermore, studies using EEG and MRI showed that self-talk in the third person is more effective for processing and re-evaluating emotional content. This could be due to the fact that the use of one's own name—because on the one hand familiar, on the other hand something self-distant—reduces amygdala activity and thus the brain as a whole needs less energy for the emotional processing of things. The connections are still somewhat unclear because many brain regions are involved in the evaluation and re-evaluation of emotional content. However, it can be postulated for self-talk to influence one's own mental and physical well-being that the therapeutic effect is greater if one addresses oneself with "you"—even better with the familiar name or nickname—instead of with "I".

Affirmations are used in particular to increase self-confidence, e.g. to overcome self-doubt, but also to improve mood, so to speak as mood modulation and for an increased sense of well-being, and here above all as a "tool" for self-therapy. They are, so to speak, a self-suggestion that derives from the word "self-influence". Under Autosuggestion a process is understood by which a person builds up, indeed trains, automatisms to believe in something until they run almost unconsciously. In this context, self-affirmations are a self-induced influence on the psyche, in which mental visualizations (imagination) often increase the effect. The success is also more likely the more often they are repeated.

There are phases of life in which one has doubts about oneself and is unsure whether one should articulate one's own wishes and needs at all in relation to others, but also whether one can achieve one's own goals at all. There are also phases in which one has a so-called slump, the mood is depressed and one does not feel comfortable in one's own skin. In all these cases one needs someone who speaks well to one, especially when one feels alone; who encourages one, builds one up again, but also makes one smile, even laugh again. This is a essential goal of "scent-supported loving looks" (Petzhold 1995). As scent therapy, "scent-supported loving looks", as I now present them, are especially intended for all those who only have themselves at the moment. It is therefore also not a therapy with therapists in the proper sense. It is a therapeutic self-coaching with the support of elements of autosuggestion and affirmation, based on a scent therapy.

Autosuggestion and affirmation, that is, self-affirming sentences that one repeats to oneself over and over again and brings to one's consciousness, are partly underestimated in their therapeutic effect in psychology. Of course, part of the reason for this is that there is evidence that autosuggestion with affirmations can have a negative effect on people with low self-confidence. After all, self-affirming statements make the situation more or less consciously even more desperate, because one does not know or hardly any positive experience in this context. But I think that for self-therapy with primal active perfumes, in which the scents are already experienced as positive during the creation itself and afterwards, this effect of deterioration of the condition in case of low self-confidence hardly ever shows itself. That autosuggestion with affirmations is a winning method for people with moderate to high self-confidence to increase satisfaction and trust in oneself will only be mentioned here for the sake of completeness. It is no coincidence that athletes in particular use this form of inner dialogue very successfully, for example to block out negative thoughts and to tune in positively to a competition.

10.4.3 Scent-supported Self-coaching: Exercise Example "The Power of Loving Looks"

This exercise example serves to brighten the mood or to positively modulate the mood and has the goal of freeing oneself from self-doubt, strengthening oneself and rediscovering life-affirming feelings through the power of loving looks.

For this exercise you need a mirror in front of which you can sit well and see your face, especially your eyes, centrally. Look for a room where you can be undisturbed, which has a warm light and little outside noise and where you can talk to yourself undisturbed. If you can, repeat the exercise on the following days for 20 minutes each. It takes some time to really internalize an affirmation, but you can expect the first positive results for yourself after three to four days. But the first day of exercise is often the hardest. The positive experience of your own primal active perfume helps, but doubts as well as old destructive thinking habits can sometimes flare up again before they are replaced by new, self-affirming thoughts.

On the day of the exercise, take your primal active perfume or spray the perfume strip with it and sit comfortably in front of the mirror so that you can see your face well. You can sit in an armchair, in a meditation position on the floor or on a cushion.

Start the exercise as follows:

Breathe in and out for about three minutes and feel your stomach rise and fall. Feel how you can feel your breath in your nose. Experience how it feels to breathe in and out. Keep your eyes closed.

When you have calmed down a little, start sniffing your primal active perfume with your eyes closed and then smelling it. You should find your creation pleasant and familiar, and the scent should relax you further. But now you can also discover another nuance in your primal active perfume: it also smells like a happy baby and cuddling. While you smell, see a baby in your imagination who is doing really well. It wants to play with you, kicks enthusiastically, smiles at you and stretches its arms out to you. Alternatively, if it is difficult for you to visualize a happy baby, you can also think of a baby animal, for example a young dog or a kitten. The baby animal looks at you with trusting eyes, gently and affectionately. Obviously it wants to cuddle with you.

The goal of both fantasies is to trigger a loving, warm, non-critical look in yourself. This is sometimes not so easy because you quickly focus on things in the mirror that you do not like so much about yourself. Therefore, take the time you need to gain a loving, warm look at yourself through baby fantasies, with which you can then look at yourself in the mirror. Smell your primal active perfume again and again to intensify the fantasies. When you think you have this look, look into your eyes in the mirror with the kindness as if you were looking at yourself as a baby, and then address yourself with the examples below.

The most important rules for autosuggestion and affirmation are:

— Take yourself and the exercise seriously and focus only on yourself.
— Speak clearly and distinctly, and in the present tense, because this is what the brain responds to most for a positive inner dialogue.

A general rule for all autosuggestive self-dialogues is also:

— Do not use the subjunctive mood like "would", "were" or "should".
— Also avoid negations, because they are difficult for the brain to understand and have correspondingly less effect.
— Speak to yourself as you would like others to speak to you: Pay yourself the respect, appreciation and recognition you deserve.
— Try to project and express self-confidence in your inner dialogue as much as possible, especially during the exercises, and suppress all self-doubt and hesitation.
— Do not carry out the inner dialogue doggedly, but smile inwardly because you are now taking a break and you know that good and positive things are coming to you.

Here are examples of affirmations that you should personalize for yourself:

— "(First name/nickname), you, I take care of you."
— "(First name/nickname), you can trust me."
— "(First name/nickname), you can do it!"

If the affirmations you give yourself evoke a familiar and positive feeling when you smell and look in the mirror, you are on the right track with your self-therapy.

At the end of an exercise, let your baby or toddler smile in their thoughts and let them shout with pure well-being and joy for life and wave their arms, or let your animal baby show trust, playfulness, and sounds of joy. Maybe you can already smile or be happy and make corresponding sounds yourself—and thus experience spontaneous, carefree, and happy feelings and thoughts as well as life-affirming and mood-modulating. This means that the power of fragrance-supported loving looks begins to work for you—and thus for others.

Let me go back to the question at the end of this section why loving looks that one gives to others are so valuable for oneself as well as for others. I showed above that the amygdala reads in the eyes. The loving look relaxes them. This can go so far that there is actually a premature stop of disagreements, mistrust, aggression and anger. Certainly not always, but it is difficult to escape the loving look of a person. For sure it helps with many unnecessary disagreements between people. What you can experience for yourself through the fragrance-supported exercises as an increasingly positive and relaxing self-esteem, shows itself to others, namely that you radiate more self-satisfaction, security and inner joy. As a result, you automatically gain more sympathy, also because you begin to react more sovereignly in situations, but also smile more. Studies show: Laughter is contagious, because a happy partner increases the probability of being happy oneself by almost 10%. With good-humored people, the probability that one's own mood will benefit is even more than 30% (Fowler and Christakis 2009).

I spoke above about the right length of eye contact. The length and who looks at whom is the actual secret of the power of loving looks. But it also depends on who begins to give the look, or who it comes from. The social psychology has plenty of findings about this. From a well-documented theory (Stereotype Content Model) we know that people, among other things, are particularly successful when they meet someone who believes they have a higher status. They may not compete with this person, but they also radiate more competence and human warmth at the same time. It is also known from personality psychology that people who are, for example, extraverted-active or give this impression, are experienced and evaluated as positive much faster by their social contacts after only a few seconds than others. With passive behavior, on the other hand, the observer is more likely to assess a person who radiates warmth and competence as lower, even if this person radiates warmth and competence. Accordingly, a loving look is evaluated in the context. In combination with passive behavior, it can quickly be interpreted as nice, but not competent. The other person feels: The person needs me. In combination with active behavior, self-confident and competent appearance, however, the look is even more attractive. The other person feels: The person appreciates me. Especially by the radiation of inner strength and sovereignty while being active, a loving look becomes an unexpected gift that one does not even expect. This can lead to the fact that the other person not only feels appreciated, but also honored. In fact, social psychological research shows: The best way to influence others for leadership claims is to show activity with a radiation of inner strength and sovereignty, especially in combination with competence and human warmth (trustworthiness and friendliness) (Cuddy et al. 2013).

» *Actually, I had already considered shorten the chapter, but I just wanted to share the many new findings about fragrance and self-therapy with you.*

Summary

As an introduction to the topic of "scent-supported therapy", we discussed two exercises on how to use the fragrance direction "fresh-green-citrus" for one's own scent therapy. One exercise is based on a multisensory inspiration for creativity enhancement and is called "Scented Flight", the other is based on a physical and mental exercise for new inner strength and is called "Scented Power Posing". Subsequently, we followed up on current findings from baby scent research for adults, which have given new and valuable impulses to self-scented therapy. In this context, we discussed self-therapy with primal perfumes or with primal active perfumes and gave instructions on how to create and use them for oneself. Primal active perfumes are the first scents that humans smell in their development. Clinical research confirms their great positive psychological effect on infants, which even leads to pain relief. We postulate that this olfactory experience is still present in adults even unconsciously and can be activated for scent therapy. For the future of perfumery, which would like to offer even more wellness, plant peptides are expected to be of particular importance in supporting individual brain areas, e.g. against stress and its effects.

10

References

Badiee Z (2013) The calming effect of maternal breast milk odor on premature infants. Pediatr Neonatol 54(5):322–325

Bernhardt K (2017) Panikattacken und andere Angststörungen loswerden: Wie die Hirnforschung hilft, Angst und Panik für immer zu besiegen. Ariston, München

Broom DC et al (2002) Behavioral effects of delta-opioid receptor agonists: potential antidepressants? Jpn J Pharmacol 90(1):1–6

Cuddy A, Kohut M, Neffinger J (2013) Connect, then lead. Harv Bus Rev 91(7/8):54–61

Dalton P, Doty RL, Murphy C, Frank R et al (2013) Olfactory assessment using the NIH toolbox. Neurology 12:80

Fowler JH, Christakis NA (2009) The dynamic spread of happiness in a large social network. Br Med J 337(768):a2338

Frasnelli J et al (2015) Food-related odors and the reward circuit: functional MRI. Chemosens Percept 8:192–200

George R, Lew J, Graves D (2013) Interaction of cinnamaldehyde and epicatechin with tau: implications of beneficial effects in modulating Alzheimer's disease pathogenesis. J Alzheimers Dis 36(1):21

Gordon I et al (2013) Oxytocin enhances brain function in children with autism. PNAS 110(52):20953–20958

Han P et al (2019) Sensitivity to sweetness correlates to elevated reward brain responses to sweet and high-fat food odors in young healthy volunteers. NeuroImage 208:116413

Hoenen M, Wolf OT, Pause BM (2017) The impact of stress on odor perception. Perception 46(3–4):366–376

Hurlemann R et al (2010) Oxytocin enhances amygdala-dependent, socially reinforced learning and emotional empathy in humans. J Neurosci 30(14):4999–5007

Kabat-Zinn J (2006) Coming to our senses: healing ourselves and the world through mindfulness. Paperback—deckle edge. Hachette Books, New York

Kirschbaum C, Pirke KM, Hellhammer DH (1993) The 'trier social stress test'—a tool for investigating psychobiological stress responses in a laboratory setting. Neuropsychobiology 28(1–2):76–81

Koehbach J et al (2013) Oxytocic plant cyclotides as templates for peptide G protein-coupled receptor ligand design. Proc Natl Acad Sci U S A. ▶ https://doi.org/10.1073/pnas

Moser J et al (2017) Third-person self-talk facilitates emotion regulation without engaging cognitive control: Converging evidence from ERP and fMRI. Nat Sci Rep 7:4519

Nakamura A et al (2009) Stress repression in restrained rats by (R)-(−)-linalool inhalation and gene expression profiling of their whole blood cells. J Agric Food Chem 57(12):5480–5485

Navratilova E, Hruby VJ, Porreca F (2011) Delta opioid receptor function. In: The opiate receptors. The receptors. Springer, Totowa, S 307–339

Neshat H et al (2016) Effects of breast milk and vanilla odors on premature neonate's heart rate and blood oxygen saturation during and after venipuncture. Pediatr Neonatol 57(3):225–231

Petzhold HG (1995) Psychotherapie & Babyforschung. Band 2: Die Kraft liebevoller Blicke. Säuglingsbeobachtungen revolutionieren die Psychotherapie.

Junfermann, Paderborn, Innovative Psychotherapie und Humanwissenschaften 56

Porter R, Winberg J (1999) Unique salience of maternal breast odors for newborn infants. Neurosci Biobehav Rev 23(3):439–449

Schoen K (2018) Gegenüberstellung von Essensdüften und Blumendüften im Hinblick auf ihre Verarbeitung im mesolimbischen System—eine fMRT-Studie. Dissertationsschrift der Medizinischen Fakultät Carl Gustav Carus der Technischen Universität Dresden

Schulze P et al (2017) Preprocessing of emotional visual information in the human piriform cortex. Sci Rep 7:9191

Seppala ME (2013) 20 scientific reasons to start meditating today. Published on September 11, (Feeling it) in Psychology Today

Sezille C, Messaoudi B, Bertrand A, Joussain P, Thévenet M, Bensafi M (2013) A portable experimental apparatus for human olfactory fMRI experiments. J Neurosci Methods 218(1):29–38

Stahl B, Millstine W (2013) Calming the rush of panic: a mindfulness-based stress reduction guide to freeing yourself from panic attacks & living a vital life. New Harbinger Publications, Oakland

Stice E, Burger KS, Yokum S (2013) Relative ability of fat and sugar tastes to activate reward, gustatory, and somatosensory regions. Am J Clin Nutr 98:1377–1384

Varendi H et al (1998) Soothing effect of amniotic fluid smell in newborn infants. Early Hum Dev 51(1):47–55

Wang J, Sun X, Yang Q (2013) Methods for olfactory fMRI studies: implication of respiration. Hum Brain Mapp. ▶ https://doi.org/10.1002/hbm.22425

Wei Y, Tsao P (2016) Effects of breast milk and vanilla odors on premature neonates' heart rate and blood oxygen saturation during and after Venipuncture. Pediatr Neonatol 57(6):548

Whitehouse L (2019) Givaudan launches new sustainable fragrance precursors: In cosmetics design-Europe 07.02.2019

Sales Psychology of Fragrance Consulting

Relaxation in the Unconscious for More Beautiful Smelling. How to Increase the Sense of Well-being and Satisfaction of Clients and Consultants, but also in General in Human Contact

Contents

© The Author(s), under exclusive license to Springer-Verlag GmbH, DE, part of Springer Nature 2023
J. Mensing, *Beautiful SCENT*,
https://doi.org/10.1007/978-3-662-67259-4_11

11

Trailer

This chapter is about how to win people over, and that's of course especially important in times of stress.

That raises a central question: Why are some people in sales and consulting so much more successful than others and win customers over more easily?

In this chapter I will give you a lot of information from neuroperfumery and psychology about how to improve interpersonal contact and also how to win people over. You will also find a formula for sales and consulting that has proven itself in the practice of perfumery.

With this knowledge you will be well prepared for ▶ Chap. 12 , in which I will introduce you to the experience of perfume, how to make the experience of perfume even more fascinating.

11.1 The Epicenter of Success and Interpersonal Contact

The Amygdala (two almond-shaped protrusions in the brain), is one of the most fascinating areas in our emotional center, the limbic system. It not only plays a special role in the emotional perception of scent, but also decides on our emotional world and social skills, how successful we are in dealing with other people, how pronounced our intuitive emotional intelligence is and how we can deal with it. New results from neurobiology and attachment research are bringing to light more and more connections that decide on satisfaction, success and the course of interpersonal communication.

Why is it easy for some people to accept a warm embrace when they meet someone for the first time, while others retreat like they are frozen? When we meet someone for the first time—why do some people smile at us and come up to us actively, while others turn away and retreat? Why are some people in sales and consulting so much more successful than others and win customers over more easily? Researchers like Stephen Porges from the University of Illinois in Chicago have pursued such questions and came to the conclusion independently of each other that the reason is to be found in the amygdala and its neighboring nervous systems.

For psychologists, the consulting and sales situation in stationary retail is particularly exciting. Here you can see very quickly how findings from social psychological and now also from neurobiological research prove themselves in practice. For this purpose, the stationary perfumery is a particularly interesting terrain, since emotional goods are offered to a large extent within a direct contact. Accordingly, modern brain research is in the center of modern sales psychology with many exciting new findings about the amygdala. Because the amygdala has a great influence on the first contact as well as on the well-being of customer and consultant. How important feeling good is in stationary perfumery and expected by customers, I have discussed in ▶ Sect. 4.1 and the following sections.

The current findings about the amygdala relativize some aspects of classical sales psychology in their importance. For example, typical questions that are not as important as assumed. These include, for example, with the greeting of the customer, so-called open questions supported by an open body posture, such as "Can I help you?" Or "What are you looking for?", Even if you really want to know how to help the customer. Because it is the amygdala that decides to a large extent about the success of the fragrance advice before the first word is spoken. And she relies less on questions, answers or even body language. Because she smells and feels very fine for us unconsciously, but she is also a very accurate observer, indeed an excellent spy, who does not miss any inconsistencies, especially in the eyes and facial expression. For one's own unconscious, she can read within milliseconds in the eyes of others. That is why the amygdala is especially in new situations—especially with new social contacts and the assessment and development of the first impression—almost in constant alert or on the lookout. Nature has originally created it as an olfactory early warning system. Since we humans have lost the sense of smell at a distance, our early warning system is initially visually controlled. This applies in particular to the vigilance when contact with unknown or less known and the increased attention of the amygdala to the further development of a situation.

This also applies to the sales situation, especially when contact with new customers—but not only there. Even if the customer is a known regular customer and you think you know each other well, the amygdala can jump in at any time already at small inconsistencies perceived by her. This usually happens unconsciously at first, until a gut feeling comes up. It signals that something could be wrong. Until then, customer and consultant believed that the situation actually corresponded to the expectation.

An example of something that the amygdala really doesn't like and that causes it to warn higher brain areas and our nervous system is, for example, a positive statement without eye contact to a perfume. The same applies if the eye contact was too short and only lasted one second or less, or if it was uninterrupted for more than 20 s.

Social psychological studies at the University of Freiburg show that the vague feeling of unease generated by the amygdala in cooperation with the nervous system is often justified. The researchers first examined the effect of too much eye contact and found that it has a negative effect on the judgment of particularly skeptical customers. This type of customer stiffens up when there is too much eye contact. The scientists also found that if a consultant was lying, he insisted on too much eye contact and thus revealed himself.

So can too much eye contact be harmful in a sales situation? The research clearly confirms this, not least because the amygdala of skeptical customers is particularly on the alert. Therefore, the modern sales psychology based on neuroscientific findings recommends that when promoting products, especially with this type of customer and with new customers, eye contact should be used more sparingly. Specifically, this means: If you as a consultant are not sure how the customer—especially if he appears skeptical—generally feels about the topic, it is better to build up the eye contact slowly. It is better to reduce the eye contact to two to three seconds at the beginning and look at the customer's mouth once again. Only after a positive feedback should the direct eye contact of seven to ten seconds be sought.

Can too short eye contact also be harmful in consultation? How should the eye contact be so that it is winning, or what are the worst mistakes? You can calm down your counterpart's amygdala simply by the length of the eye contact being right. This can be practiced, although there are cultural differences for the right duration. In Central Europe, a direct eye contact of seven to ten seconds is perceived as positive. If there are several people involved in the conversation, an alternating eye contact of three to five seconds is experienced as pleasant. Everyone feels equally important.

There are countries and cultures in Asia, Arabia and the Caribbean where a direct eye contact of seven to ten seconds is already considered too long. Here a period of three to five seconds is recommended. Eye contact is also used less frequently, as it is not experienced as polite or positive in all situations.

Perceptual psychology shows that there are also differences in Europe and the USA. French people prefer an eye contact length of about seven seconds, Americans of just under ten, only in Germany may it last up to twelve seconds. What lies beyond that is perceived as increasingly dominant. Anyone who maintains eye contact in our culture for longer than twelve seconds is already communicating through their eyes who the boss is or who is in the right. In fact, a number of social science studies show that leaders in meetings and during recruitment communicate their status through longer eye contact.

Most people experience longer eye contact as stressful. This can quickly lead to an increase in cortisol levels, i.e. to increased secretion of stress hormones. This causes you to lose sovereignty, strength and energy, which are necessary for a successful negotiation.

If the eye contact is too short, this can quickly be interpreted as a lack of interest in the other person or in the consultation. This can also communicate a status, in the sense of: one does not need it; one is not responsible; one wants to get rid of someone.

Regardless of how long a gaze should ideally be, the following is important in our cultural circle: Never talk about a product without making eye contact. Otherwise, the customer's amygdala will perceive it as inconsistent. Studies show that salespeople who avoid eye contact while explaining product benefits send two contradictory messages:

1. A verbal message explaining the benefits.
2. A nonverbal message that neutralizes the benefits.

11.2 First Friend or Foe—also in Fragrance Advice

For the first impression, the amygdala automatically scans the gaze and facial expression first and then the person in their environment. During this scan, she only knows rough categories, but they influence the whole subsequent contact. How often have we already heard: The first impression is decisive, and we should therefore pay attention to voice, body language, appearance and clothing in order to achieve a positive result. Surely nobody will disagree with that. But brain research has now found out what exactly happens during the first impression.

Traditionally it was assumed that the first impression is formed within three to five seconds and then becomes conscious to us as an impression. In fact, the Amygdala scans for us unconsciously in cooperation with the "Fusiform Face Area" (FFA)—a brain winding that resembles a spindle that is necessary for the recognition of faces—within milliseconds. It does not have to be conscious to us what the amygdala registers on the other. What does the amygdala focus on first? Immediately—especially at the beginning of a fragrance consultation on the side of the consultant as well as the customer—she makes three emotional classifications in the following order:

- Friend vs. enemy,
- attractive vs. unattractive,
- indifferent.

The emotional, unconscious reactions of the amygdala could be described as follows:

- **Friend:** potential friend—no danger—build or maintain contact—has my interests in mind.
- **Attractive:** Win person for myself—put in a good word.
- **Indifferent:** Do not know—be careful—but could also be an enrichment.
- **Unattractive:** Keep person at a distance—get rid of quickly.
- **Enemy:** Potential aggressor—alertness—does not have my interests in mind.

Even if both sides feel good after the first contact, the amygdala remains active as a conversation partner, so to speak, as an interpersonal sensor for emotional input. Because experience has shown that feelings can change quickly, and friendship can quickly turn into mistrust and then hostility. But often it is far less dramatic, but of just as negative effect on the consultation situation. The amygdala switches from attractive to unattractive in the classification, and thus to "get rid of quickly".

In the history of mankind, it was vital to have one's own, well-functioning spy. In particular, when two strangers met in lonely wilderness, neither could know what the other was up to. Therefore, one had to be careful. The amygdala was therefore mainly created for vital physical decisions. These include, for example, flight or the feeling

of danger. In comparison to our ancestors, however, we modern humans are much more often in social contact with others. Most of the time it is not about physically vital things, but about a purely psychological state of mind.

We make between 80 and 100 decisions every day, involving the amygdala to a greater or lesser extent. Even though these decisions are not really of interest to her, she still automatically goes into "alert mode". After all, you never know how things could develop. This strains the amygdala, which we also feel emotionally. This means: We under-challenge her in her ability because it is rarely about life-threatening decisions. At the same time, we over-challenge her by the multitude of decisions. Because most problems have solved themselves within three days. But this can not be predicted in individual cases. Therefore, the amygdala is today primarily a constantly overused watchdog for psychological security and emotional well-being. The amygdala is actually biologically only oriented towards three questions together with the autonomic nervous system:

- Do we feel safe?
- Do we feel danger?
- Is there a threat in the room? (Porges 2003).

- **Eye scanning**
In determining safety, danger and threat, the amygdala proceeds very cleverly. She secretly and initially not consciously perceived by oneself and others extremely quickly reads in faces, in particular in the eyes. For her, all kinds of direct and indirect eye contact as well as the "inner eye" such as the size of the pupils are interesting, because the eyes betray the state and upcoming reactions in advance.

This ability to perceive can be trained. Poker players will confirm this gladly. It would exceed the scope of the book to discuss all types of eye contact and visual features in detail, so here is only a list of the most important ones:

- Often it starts with changes in the direction of gaze in conversation, which can be seen well through the white in the eye, the so-called sclera. The sclera is three times as large in humans as in many other primates.
- Duration and dominance of eye contact.
- Frequency of eye movements. This includes an unreturned, averted or empty gaze, or how someone looks.
- Blinking, rapid blinking, slow blinking, twitching, which indicates exhaustion, stress and fatigue, winking and twitching with the eyebrows.
- Whether you can hold the gaze or avoid it, whether you take quick glances at yourself or fast glances at the other.
- Situational gaze, e.g. during greeting; expectant, supportive direct gaze during a presentation or speech; socially expected gaze, e.g. as a sign of empathy.

The amygdala remembers everything: wide open eyes, narrowed to small slits or swollen eyelids. The most amazing thing is that when scanning the eyes, it mainly gains information about the inside of the eye. She is actually unimpressed by the eye color, as long as it is not atypical or something else has been learned.

The amygdala is rather focused on biologically laid-out warning patterns that are further reinforced by specific life experiences. Clinical psychology and medicine have to agree with her in many respects. For example, the amygdala registers the following aspects:

- **Changes in pupil size:** Dilated pupils can be a sign of current or future difficulties, for example due to drug use. On the other hand, dilated pupils can also be perceived as attractive because they correspond to a beauty ideal.
- **Larger or protruding eyes:** Over the course of life, one learns that these features can, for example, indicate high blood pressure, heart palpitations, irri-

tability, and other symptoms. Above all, irritable people keep the amygdala particularly well in view.

- **Red eyes:** They occur with fatigue, pain, or cold. Here the amygdala behaves cautiously because it is also about protecting its own immune system.
- **Atypical colors or discoloration of the iris:** Physicians know that a potential heart attack or stroke can already be hinted at early on in the eyes. They know, for example, the importance of the white-yellow ring around the iris, which indicates disorders in lipid metabolism with elevated cholesterol and triglyceride values and increases the risk of heart attack. Of course, the amygdala does not know this so explicitly. Nevertheless, it is initially suspicious of anything atypical.
- **The light in the eye:** The amygdala is particularly interested in two types of light. On the one hand, there is a bright, almost shrill light, a so-called fright light. It arises when the nervous system is activated during increased attention, fear or anxiety as well as physical aggression. For this purpose, it dilates the pupils over two muscles. This allows more light to enter the eye. Nature has probably regulated this in order to be able to perceive more of the environment, but perhaps also to indicate emotions. Because more light in the eyes looks like a warning signal to other amygdalas. But there is also a warm eye light that conveys security to the amygdala and from which sympathy, empathy and love can be read. This look plays an important role in therapy, in interpersonal relationships and especially in sales advice in connection with emotional products. I have already spoken about this type of look above when we pointed out the great power of loving looks in the context of a fragrance therapy. We will discuss this type of look in more detail below in its importance.

In summary, it can be said that the amygdala reacts most strongly when it experiences light of fear, aggression or anxiety as well as dominance in the gaze through changed pupil size. In order to avoid danger, it is correspondingly attentive to the eye area and first focused on the inside in the eye of the other person (Pessoa 2013).

Eye scanning with the resulting knowledge is not a discovery of modern times, it was reported about it already in antiquity. "As soon as you see into pure, shining eyes, know that such eyes show the honesty of the owner," wrote the Athenian philosopher Polemon around 300 BC.

Also pure, shining eyes and their color lead to false conclusions. So documents from the 9th century report that despite ransom, there were terribly violent raids by the Vikings in the Rhineland, which one could actually take for angels because of their blue eyes and their blond hair. Only their smile didn't fit when they suddenly pulled out their short sword hidden under their clothes. For many it was then too late despite the amygdala's warning.

■ **The amygdala relaxes with a real smile**

With the ability to eye scanning, the amygdala also has a sense of time. It determines what is too long or too short. That also applies to the assessment of a smile. In order to be convincing for the amygdala, the smile should last at least six seconds at the beginning of a contact, three seconds for the build-up of the smile and another three seconds to express it. For the amygdala it is important that the smile is shown in eye wrinkles. New studies confirm: Those who smooth out eye wrinkles, for example, by injecting collagen or hyaluronic acid, make it more difficult for the other person's amygdala to recognize whether it is a real smile or not, up to the wrong assessment. The body language also plays a role when smiling. The best smile is accompanied by a slight tilt of the head and an inviting gesture.

It is possible to consciously see with the amygdala, as mentioned before. Some people get very far, for example, mind readers and mentalists. Especially when the other person is already a little tired, one can read quite well in eyes and face. This is used by negotiators in talks until late at night. The former German Foreign Minister Hans-Dietrich Genscher was known for achieving the best negotiating results after midnight. He had stamina and was a very good observer.

■ **How the amygdala analyzes and defends itself**

The messages from the amygdala to the cortex are much faster due to their central position in the brain than from the cortex to the amygdala. By reacting very quickly, before rational recognition or evaluation can take place, the brain learns to react more quickly to dangers and inconsistencies. As far as I know, there have been no studies so far to find out who has a more active amygdala: the customer when entering the store or the salesperson when approaching the customer. I assume that this is probably the case with the customer. Because the amygdala of the salesperson is more familiar with its environment and therefore less alert than that of the customer. On the other hand, if a customer enters a store to complain, it is of course the other way around.

During a consultation or sale, the customer's amygdala rarely reports strong emotions such as anger or conflict, but rather misunderstood questions and remarks as well as small inconsistencies that are due to unintentional glances, feelings and reactions of the salesperson and can prevent or make a business transaction more difficult.

How does the amygdala recognize inconsistencies in finer sensations that are latent in the room? The answer is surprising because it reminds us of the systematic approach of a computer program. When the amygdala reads in the eyes and anticipates the next reaction, it compares this with three data: what someone is doing or saying right now, what he did before and how he behaved in the past. Then the amygdala decides together with the autonomic nervous system: Does this make sense, is this consistent? This means: The amygdala and the autonomic nervous system live and analyze in four times:

1. next moment (near future),
2. what just happened (recent past),
3. what has already happened (past experience),
4. what is happening right now (present).

If the information from the eyes is not enough for her to assess the present and near future, the amygdala also scans the face at the same time (Pessoa 2013).

All of this happens within milliseconds. The amygdala becomes hyperactive when it detects ambiguity in the overall context. We feel this—if at all—as an initially vague, light inner unrest. Something is wrong between the feeling of feeling good and safe at the moment, and the reactions that indicate the next moment. Experiences from the recent and more distant past can intensify this. This triggers a neurobiologically controlled system. Even if we are not aware of it cognitively, our body has already started a sequence of processes that can lead to adaptive defensive behavior. Stephen Porges (2003), professor of psychiatry, has developed the Polyvagal Theory to explain these processes).

In terms of the sales situation, there are four typical defensive reactions on the customer's side that can occur quite suddenly and that everyone knows from themselves because they are deeply anchored in our biology:

1. The first reaction is *caution*. It starts with gaining time, waiting, holding back or withdrawing first, which is communicated, for example, by "I need to think

about it again", often in combination with spontaneous loss of interest in buying.

2. The second reaction is *fight*, e.g. as latent or open arrogance, complaint, interrupting or comprehensive critical questioning.

3. The third reaction is *flight*. It usually starts with impatience, no time left or leaving the store prematurely. But it also shows when the customer apparently turns to another person or another product without reason.

4. The fourth reaction is *rigidity*. The customer is, for example, hardly or not accessible, appears passionless and uncommunicative or simply stubborn.

People differ in their adaptive defense behavior or in how well the amygdala and the autonomic nervous system work for them. Some can withstand the stage of caution for longer, others react immediately with fight, flight, or even sudden paralysis. The reason is certainly experience that one makes as an adult. But many developmental psychologists and psychoanalysts also attribute these reactions primarily to early childhood experiences. Even infants need social contact strategies and must develop them when neurobiological inconsistencies are reported. These include feeling good and trusting while breastfeeding while experiencing stress, for example because the parents are under time pressure. But only if the infant feels generally safe is he able to learn contact strategies when things do not go as planned and something unpleasant is in the air.

Only the feeling of security lets the amygdala and the nervous system mature strengthened so that one does not fall into unnecessary overreactions and can stay calm. So the feeling of principle security is everything decisive—not only how one processes a situation biologically-nerval or how the body assesses it, but whether one can support contact-initiating, pro-social behavior in difficult situations again. A nerve closely associated with the amygdala, the so-called myelinated (fast-conducting) vagus, helps with this. It promotes calm by inhibiting the influence of the sympathetic nervous system on the heart. This acts as a stopper and prevents unnecessary overreactions in a social situation. These include, for example, an increase in heart rate, which manifests itself in stronger heart contractions and shallow breathing.

11.3 The Formula: How to Win the Other Person's Amygdala

In sales consulting, a central question arises: How can I win or win back the amygdala of my counterpart?

In sales trainings, the importance of competence, knowledge, approach, type of questions, will, self-motivation, self-presentation, smile, posture, gestures, voice and appearance is pointed out. This is all true, but it is only half the truth. With all the well-intentioned advice, an essential finding remains unaddressed. It was discovered in the therapy research in the 1990s and is the secret of great therapists. Nevertheless, it can be learned by everyone, which the current emotion and neurobiological research confirms. It is the power of loving, unexpected glances at the right moment (Petzhold 1995). But something else is added, which potentiates the whole thing in combination: one must surprise the amygdala with an unexpected reward, that is with a chance of winning.

In addition to the adaptive defensive behavior, the amygdala shows another neurobiological primal reaction: it is afraid of losing something or not getting it. So the amygdala immediately sets off the alarm if a situation threatens to result in a loss. You could also say it the other way around: if there is any way for people to do it, they want to win and definitely not lose. So you have to offer the amygdala a gain, and it

has to come from the moment and be unexpected.

I suggest the following self-test:

Imagine that you are looking for an anti-aging care product in your perfume shop and indicate in the consultation that you do not want to spend more than 50 EUR. Which information do you find more appealing now?

A1. The consultant tells you that you could save 20 EUR, because there is a corresponding care product that offers the following …

A2. The consultant tells you that the estimated 50 EUR would be disadvantageous for you. If you spent 20 EUR more, you would receive a care product with the following additional effects and thus something to gain: …

Which statement speaks better to the amygdala? Both statements can address the amygdala of customers positively, but only if they are experienced as a gain. What is decisive is whether one feels emotionally comfortable when saving or bargain hunting. Especially the perfume shop, which offers products that are linked to emotions, mood wishes and identity, has pitfalls for savers. If there is any way for people to do it, they want to win and definitely not lose or get less than they want.

Saving works for the amygdala if it is experienced as a gain. However, if saving results in the dull feeling of loss or the impression of getting less than one would like, the amygdala reports. Therefore, anyone who can afford it financially will find statement A2 more attractive, in which a gain is guaranteed.

With A1, the feeling of gain depends on the trust in the competence of the expert. In both cases, therefore, the conveyed feeling of gain is essential. Ideally, there is a gain guarantee for the amygdala in every situation, i.e. something that enthuses it.

The formula to win or win back the amygdala of the other is therefore:

> Real smile plus the power of loving looks plus unexpected prospect of gain.

But this formula can also be increased. To feel comfortable, you need to have trust and security in the right hands. In addition, you must have the feeling that you are being dealt with on a very personal level and that the best way to enjoy and almost feel lucky is to arise during the entire situation. In the counseling situation, the conveyed feeling of trust and security also goes hand in hand with competence and honesty. The customer must get the impression that his interests are in the foreground. Of course, this also applies to other areas of life.

» *Can you name your personal "why"?*

How can you convey honesty and competence to the amygdala at first contact, without overwhelming the other with expertise? This is done through the principle of the "golden circle". It was discovered when we looked into what inspired leadership had in common. The difference lies in the three-step "why—how—what".

The amygdala in dialogue with higher brain regions is first interested in why a person does something, and only then in the "how" and only later in the "what", namely what someone offers. People who first communicate the "why" create trust with their self-claim. They are assumed to be competent, having a unselfish interest and therefore potentially more honest. At the beginning of an anti-aging consultation, you could say the following why-sentence:

» *"I believe that every person is beautiful in their own way. Therefore, he deserves a care that expresses this."*

Applied to perfume advice, you could say:

» *"Perfumes smell different on every skin. That's why it's important to me to find scents that really fascinate and work for you"*

At the beginning of my seminars and lectures, I therefore ask the participants to write down some of their why-convictions right away. The shorter, the better. They can also be half-sentences that are incorporated into the conversation shortly after the welcome.

If you ask me about my personal "why" and give me 30 s, I usually say the following two sentences with a certain passion:

» *"As a clinical psychologist, the secret of the perfume industry with its ability to transform through perfumes has always fascinated me. I want to show people that they can experience themselves anew with perfumes that are made for them. They can be more attractive to themselves and others and find more joy and happiness. They deserve it."*

This will certainly not be understood by everyone immediately, but that is not the point. What is important is that you hear and feel that I have a mission and that from now on my counterpart is in the center: Because he or she deserves it.

The extended formula to win the other's amygdala is now called:

> Real smile plus power of loving looks plus "why" plus unexpected chance of winning.

Depending on how you extend the "why", you can already indicate a win in the opening of the conversation.

Something else is especially important for the opening of the conversation. From the psychology of emotions we know that people who find themselves more sympathetic begin to unconsciously synchronize their movement patterns. If one leans forward in a conversation, the other does too. Translated into the sales situation, this means: To build sympathy with the customer as a consultant, it helps to adapt to his movement dynamics. As a rule of thumb, if the customer is moving quickly towards me, I will approach him with almost the same speed or indicate with a quick gesture that I will be there for him soon. The same applies to a slow approach. If the customer makes the famous right turn and sneaks along a shelf, I will first hold back until he (eye) contact.

The consultation could then begin based on the extended formula:

— The customer comes in.
— The consultant shows a real smile, approaches the customer at almost the same speed as this one and greets him with the power of loving looks.
— The customer names his wish (e.g. finding a new perfume for himself). The consultant lets his "why" flow into the conversation.
— To offer an unexpected chance of winning, the consultant could, for example, point to a scent test and say: "There is a connection between color and perfume preferences. Here you see different colored shapes. Tell me which one you are most attracted to at the moment. Then I will present four matching perfumes to you and tell you what color psychology has to say about it. Then I will name the valuable ingredients of the individual perfume creations."
(I will introduce you to a corresponding scent test later in the book.)

If the customer shows interest in a fragrance after smelling it first on a fragrance strip and then on his skin, the consultant can increase the feeling of gain and anticipation of the perfume by giving the customer the bottle with the original cap in his hand. This often increases the value of what is offered, because in recent years there has been a trend towards higher quality closures that offer a valuable tactile experience. But the art is not simply to give

the perfume in the hand, but at the same time to address the personal gain of enjoyment, attractiveness and pleasure with positive lightness or to share it with the customer. Therefore, the ingredients should only be mentioned at this point. So you present luck and pleasure to the touch, which can also be experienced sensually-erotically. How far you can go with it, of course, depends on the customer. In doubt, restraint is the best advice so that the customer can decide for himself in peace.

In addition to olfactory and tactile attraction, you can always refer to an additional benefit of the perfume for the purchase decision. This way you win more rationally deciding brain areas such as the orbitofrontal cortex. Additional benefits can be manifold for a perfume and are influenced by the general fragrance direction.

So a perfume can be additionally ideal in the following situations:

— at a job interview or team meeting, because it – typically for chypre notes – exudes self-confidence and a sense of responsibility,
— at a party, because it – typically for fresh-fruity gourmand notes – conveys joie de vivre,
— during a flirt, because it – typically for slightly buoyant floral notes – leaves things open,
— when wanting to seduce, because especially amber-oriental notes blend with the body odor,
— when wanting to cuddle, because the so-called "florientals" invite one to dream,
— when trying to evoke trust, because many woody and natural notes radiate inner harmony,
— when visiting the gym, because green-fresh notes, especially lightly cooled, bring new energy,
— when wanting to lose weight, because gourmand notes trick the brain into thinking that there is something to

snack on and one can better withstand cravings,
— when in not such a good mood, because especially gourmand-fruity notes convey more joie de vivre and fun to the brain,
— when more concentration and creativity is needed, because citrus freshness refreshes the brain,
— on Monday, because aqua and green plant notes give a natural kickstart,
— on the weekend, because powdery floral notes and especially white flowers let one relax so wonderfully.

The extended formula with six tips to win over the amygdala of others for oneself now reads:

❯ Real smile plus synchronization of movement plus loving look plus "why" plus unexpected prospect of gain plus additional benefit.

Of course there are other techniques, but at this point I would like to mention another effective seventh tip that I have already mentioned: scarcity.

■ **Scarcity**

If one feels that one can still get something despite scarcity, not only the amygdala, but also the reward system, the nucleus accumbens, is activated with anticipation. The latter is located above the amygdala. It contains dopamine receptors of type D2, the stimulation of which triggers an anticipation with a feeling of happiness. If, in the consultation, one creates an artificial scarcity in order to achieve an emotional climax at the end of the consultation, as when watching an exciting film, this certainly has an unpleasant aftertaste. From a sales psychological point of view, however, this is a very efficient tool.

In a sales situation this could look like this: A customer is presented with some wonderful perfumes, one of which he falls in love with and would like to buy. It doesn't matter if he already knows the price or not. Since the consultation is carried out with a tester with a cap, the salesperson can arrange it so that the customer does not see and therefore does not know how many perfumes of this new favorite fragrance are still in stock. He is confirmed in his choice and at the same time made aware of a scarcity, for example with the following words:

❯ "This particular perfume really smells wonderful on your skin. We don't always have this perfume in stock, but I can order it for you. But let me first take a look to see if I can't get one more from our perfume treasure chest."

When the salesperson comes back from the warehouse/storeroom and happily hands the customer the last remaining perfume, one can imagine his anticipation and relief.

❯❯ *Was this chapter useful to you? I think that even if one is not active in consultation, the current findings on the right length and the right use of eye contact are particularly helpful. After all, one usually wants to win over one's counterpart.*

Summary

In this chapter we have mainly dealt with the question of why some people are so much more successful in counseling and sales, especially why they can win customers more easily. In this context, we have seen that knowledge of the amygdala is essential, especially for all those who work in a perfume shop and want to offer a service or product successfully. Not only processes that decide the fragrance and purchase decision run via the amygdala, but also unconscious experience and thus interpersonal interaction is controlled from there.

This led us to the question of how not only the customer's amygdala, but also his own can be positively influenced as an advisor in order to increase the feeling of well-being for both in perfume advice. We were able to show that the right eye contact is essential, also because the amygdala initially reads faces unconsciously for us. We also discussed how the amygdala reacts to loss or scarcity and the prospect of gain and what consequences this can have in fragrance advice. We also saw the positive effects of

an "why" introduced at the right place in the fragrance advice. This leads to a formula for fragrance advice that has proven itself in practice.

References

Pessoa L (2013) The cognitive-emotional brain, from interactions to integration. The MIT Press, Cambridge, MA

Petzhold HG (1995) Psychotherapie & Babyforschung. Band 2: die Kraft liebevoller Blicke. Säuglingsbeobachtungen revolutionieren die Psychotherapie. Innovative Psychotherapie und Humanwissenschaften 56. Junfermann, Paderborn

Porges SW (2003) The polyvagal theory: phylogenetic contributions to social behavior. Physiol Behav 79(3):503–513

Welcome to the Experience Perfumery

Smelling Unexpectedly or: How to Make Perfumes in Practice-even more Fascinating for Yourself and Others

Contents

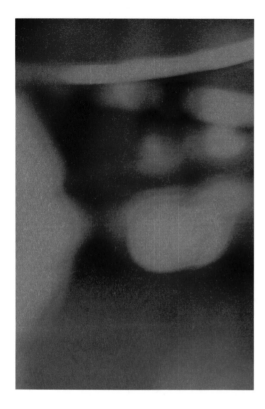

12

Trailer

As a psychology student, I had the opportunity to gain my first practical experience in perfume advice in a perfume shop in my university town. As a newcomer, I was partly surprised how perfumes were advised. Of course, there was already good "storytelling" at that time, for example, what inspired the creation of individual perfumes, but perfumes were also unintentionally much fascination taken away by the sales advice.

First, I was certainly polite as a beginner in the perfume shop and trusted the experience of my colleagues, but I soon realized that the dramaturgy of the perfume advice was often not right. The customer was at the beginning of the perfume advice cheerfully excited, but then experienced no further increase in experience over time, and there were cases in which the customer became impatient and left the store less happily. This made me think about how one could bring about an increase in experience for the customer and

also for oneself as an advisor, and I discovered that surprise or unexpected smelling plays a very important role in customer satisfaction and perfume advice.

This was the first step towards the experience perfume shop with neuropsychological perfume tests that are based on the connection between perfume, color and self-experience and recommend to the customer within seconds partly unexpected perfumes to smell. So that you can check these connections for yourself, there is a neuropsychological perfume test waiting for you in this chapter. I am convinced that you will have fun!

12.1 Myths in Perfume Sales: How to Become a Perfume Psychologist

My career as a perfume psychologist began about 35 years ago. As a psychology student, I had the following key experience: I wanted to buy a perfume gift for a girlfriend in a perfume shop in my university town. The saleswoman started the dialogue with a classic W-question: "How can I help you?" ("Wie kann ich Ihnen helfen").W-questions are still part of sales psychology training courses today. They are certainly justified with customers who are under great time pressure or know what they want. But no matter how a classic W-question may sound, it does not arouse great expectations in the customer, let alone anticipation or curiosity about the shopping experience. Even worse: Well-posed W-questions are the killer of uniqueness, based on the advisor, the business and also on the customer. He has already heard a thousand different W-questions and already suspects how it will continue after this first contact.

So I explained at the time that I was looking for a women's fragrance as a gift. Then something completely unexpected happened to me: The saleswoman asked me about my friend's hair color. This was particularly surprising to me because I had learned in the physiology course as part of my studies that the sense of smell is closely related to the emotional center, the limbic system, and that the experience of fragrance is also influenced by psychobiological factors. These include, for example, the personality of the perfume user, his or her own scent and the nature of his or her skin, his or her olfactory socialization or how scents and their ingredients are learned and associated with him or her, as well as climate and sociocultural factors. I was baffled and asked how hair color and fragrance choice could be related. This is how it was explained to me: "Women with blond hair love light, floral fragrances, women with black hair amber-oriental, richer notes, women with brown hair heavy-flowery and women with red hair chypre perfumes, so fine-herbal fragrance notes."

I asked if this method would also work if women dyed their hair. To answer the question, the store owner was called, which was the beginning of a long friendship. I repeated the connections I had learned at the university between fragrance and emotion and was invited to convince myself of the success of a fragrance sales typology based on hair color. This happened over several days. The result: There was no significant relationship between perfume preference and hair color—with one exception: Red-haired women actually showed a greater affinity for perfumes from the chypre fragrance family. More precisely: Not quite young red-haired women with light skin and freckles. When asked about their fragrance preference, it became clear that hair color only played an indirect role. Much more decisive was the fragrance in addition to the better adhesion of the perfume on the skin.

When I presented the results of my small study to the sales staff of the perfume shop, a small world collapsed for them—even though they had actually always experienced that there was no connection between

fragrance preference and hair color. They even asked the owner of the perfume shop to continue advising customers on fragrance choices on this basis. After all, they had no other clues except for the fragrance direction used before. So I quickly developed a new perfume advice method that was based on the relationship between color and fragrance preferences. It was known as the "Farbrosettentest" (Color Rosette Test), which then also found resonance in the perfume industry (Mensing and Beck (1988) The psychology of fragrance selection. In: Van Toller S, Dodd GH (Eds) Perfumery: the psychology and biology of fragrance. Chapman & Hall, London). The fact that this happened, I owe to the owner (Christa Beck) of the then perfume shop in Freiburg/Germany. Because she asked me to present the connections between fragrance, emotion and personality to the then managing director of the Parisian fragrance house Guerlain.

- **First deep psychological analysis of fragrance choice**

As a clinical psychologist, I was commissioned by the fragrance house Guerlain to investigate the psychological basis of fragrance preferences. The cooperation with a psychologist was not only new for Guerlain, but also for the entire perfume industry. Guerlain wanted to understand the users of the perfume "Shalimar" better at that time. Thus, "Shalimar" was the first fragrance that was psychologically examined. Until then, only the usual sociodemographic data of the users such as age groups, income, etc. were known. However, what made the special fascination of this classic from the Amber-Oriental fragrance family was the psychology behind this choice of fragrance, which no one could explain so far. Therefore, it was unsure how to optimize and make the perfume, which was created in 1925, more attractive for existing, but also for new user groups. At first, I assumed that the specific psychological criteria of the Shalimar users

12

could be recognized with different personality tests such as the Freiburg Personality Inventory (FPI) or the Eysenck Personality Inventory (EPI). But that turned out to be difficult. So I could only detect slight statistical trends with my colleagues at the University of Freiburg. The use of "Shalimar" correlated slightly negatively with emotional instability or, expressed positively, with emotional strength and self-confidence. Some other fragrance preferences of the control group from the fresh-citrus area, which were determined in a blind test ("Eau de Lancôme"), correlated weakly with extraversion, that is, being open and active.

The correlations between the determined color and fragrance preferences were much stronger. Before we checked their stated fragrance preferences in a blind test, I had our fragrance users fill out a color test. This was a short form of the color pyramid test by Melcher. Subjects who, for example, preferred the stimulation colors red-green-yellow, showed the highest preferences for fragrances from the fresh-citrus fragrance family. The violet and dark blue color spectrum correlated highly with "Shalimar". This is how we came across the connection between fragrance, color and mood.

The Psychological Institute of the University of Freiburg was once a stronghold of color psychology (projective tests). This now paid off. What finally broke the breakthrough were the deep-psychologically oriented conversations after the tests that my colleagues and I had with our test persons and the perfume employees. The latter were helpful to us in recruiting the users of fragrance. We suspected that the choice of fragrance might be based on other factors that went beyond self-descriptions ("This is me: yes—no") in paper-and-pencil personality tests (or cognitive tests). We recognized that the choice of fragrance had to be linked to the ideal self, its experience desires ("This is how I would like to be") and mood needs ("This is how I would like to feel"). Desires and needs that are not always fully conscious. This left us with the following questions:

- What is the ideal self, how does it operate, according to which criteria does it make olfactory-aesthetic fragrance choices, and how can it be empirically captured as a personality instance?
- Which aspects of the ideal self, of experience desires and moods are relevant to the perfume industry, to the development and marketing of perfumes?

30 years ago, these were neither self-evident questions in psychology nor in the perfume industry or in the field of fragrance development or marketing. Although the connection between fragrance and mood has been described in poetry for centuries, it has only been systematically investigated in the last two decades. Although the importance of mood in fragrance choice has always been recognized in the perfume industry, it has had no significant impact on sales advice.

For sales training, customer types were described, for example "the sporty one". Most of the time, however, the advice was based on external features such as hair color and concluded from this on fragrance preferences. If this approach was too uncertain, it simply recommended the latest fragrances or, based on the fragrance family of the favorite fragrance, presented a similar perfume. Classics such as "Shalimar", for example, were therefore suggested less and less. Instead, one concentrated for the sake of simplicity on new products. It was therefore only logical to develop a psychological fragrance test for perfume advice, a kind of quick guide to ideal fragrances. This guide to women's notes became known as the Color Rosette Test© and was first published in the late 1980s (Mensing and Beck 1988). This was one of the first fragrance tests to be used in trade and industry.

The color rosette test is based on the connection between scent, color and self-experience. For the psychology of scent, as already mentioned, eight fragrance families/

fragrance directions of ladies' and six of men's perfumes are of particular interest. Each direction is characterized by its own psychology of emotions and desires. Four to five fragrance directions for men and women are recognized by the brain as such, since they can apparently be assigned to stimulation needs of specific brain regions. Further research in neuroperfumery will be based on or verify these findings in detail. It is difficult to assign individual perfumes to these fragrance families and to research their role in the connection between scent and brain. Because more and more perfumes carry so-called crossover characteristics from different fragrance families. Also, we still know too little about how the brain perceives individual ingredients in a total creation. So far, mainly individual aromas from fragrance families have been analyzed.

12.2 Psychological Scent Choice— The Slightly Different Experience of Perfumes

In practice, the Moodform-Test© has proven to be a reliable guide in the practice of perfumery. Of course, this test does not claim to be an exact tool in the strict scientific sense, even if not all of the criteria commonly used in psychology for test development were dispensed with. Nevertheless, color tests, to which the Moodform-Test also belongs, have been criticized by empirical psychology to some extent. As mentioned above, there are a number of factors in perfume selection, such as associations and memories of perfumes and their ingredients, which can influence and influence the test result. Given the complexity of the possible influencing factors, it is rather surprising that the Moodform-Test has proven to be a quick guide in the practice of perfumery. But that does not mean that it actually measures what it purports to measure. With psychological tests, in particular color tests, one can never exclude—even if they appear to work well—that other relationships decide on results and success than one assumes. More on the subject of color, psychology, fragrance and test in a moment, but three introductory articles on the subject should already be mentioned here: *Jue JS, Ha JH. (2022) Exploring the relationships between personality and color preferences. Front Psychol. 13: 1065372. Levitan C A et al. (2014) Cross-Cultural Color-Odor Associations. PLoS ONE 9(7): e101651. Sorokowski P, Wrembel M. (2014) Color studies in applied psychology and social sciences: An overview. Polish Journal of Applied Psychology vol. 12 (2).*

The Moodform-Test is based on the findings of Neuroperfumery and especially the color, form and fragrance psychology, whose relationships with current experience desires can be quickly checked by each individual. The Moodform-Test is a psychological scent and care test that consists of a short form for women from five color shapes with 15 solutions. In each case, three perfumes and three care products are recommended to match the experience desires.

The Moodform test is about answering the following questions:
- Which scent direction makes you feel particularly comfortable now?
- Which perfumes and care products match the respective experience wishes?
- Which brain area (hypothalamus, hippocampus, amygdala, orbitofrontal cortex or prefrontal cortex) is currently particularly open to olfactory stimulation? The answer can be indirectly derived from the above-mentioned findings of neuroperfumery with regard to the fragrance preferences of individual brain areas.

The Moodform test can therefore show in a fragrance and care consultation what attracts one olfactorily and what is sought as psychological experience. This can also be used to postulate which brain areas are open to fragrance stimulation. Of course, nothing works one hundred percent in psychology. This is not the point. Because people are too complex and the fragrance experience is too subjective. Nevertheless, the Moodform test offers a good basis for consultation in practice. The results can be discussed in more depth with the customer. This makes the Moodform test a good tool for the experience perfumery.

Here are the test questions, the solutions follow later:

» *Which of the five Moodforms (color forms) appeals to you most at the moment? Choose another color form that you also like. You can also choose the same Moodform twice* (❑ Fig. 12.1).

But first of all, back to the perfume tests. By now, several are used in consultation practice. With their help, an especially personal perfume experience is to be offered to the customer. For example, conclusions are drawn about perfume preferences from so-called mood boards (picture collages), i.e. preferences for certain mood or lifestyle pictures. Advertisements in particular visualize desired experiences and are already attractive as a kind of self-test due to the degree of attraction.

Mood boards that reflect different moods such as romantic, adventurous or urban experiences are also often used in the development of a new perfume. With their help, the desired olfactory impression or the aura of a fragrance is described or associated. The color of the pictures always has a great influence on the viewer and the associated perfume associations. I personally like to use colors and shapes as shown in the test above. They lead to surprising results in perfume advice, because the solution is not easy to guess in advance, unlike pictorial representations. There are interesting connections between color and fragrance, for example.

■ **Color, fragrance and self-experience**

The connection between color, fragrance and self-experience is well confirmed (Schifferstein and Tanudjaja 2004; Tamura et al. 2018). A fragrance direction can even be visualized with colors and the associated self-experience can be expressed. This makes color-fragrance tests possible. Although pictures or mood boards can also support the visualization of a fragrance direction well. In doing so, coordinated pictures and fragrances lead to clearer and more vivid inner images. This is also confirmed by studies of brand communication (Rempel 2006). However, pictures and mood boards in a fragrance test carry the risk of triggering additional associations due to personal memories, cultural influences or social desirability, and thus falsifying the test result. This would be the case, for example, with the depiction of a romantic couple at sunset.

In order to exclude further associations, many psychologists like to work with so-called projective tests on specific topics, such as the standardized ink blots of the Rorschach test. These include classic psychological color tests such as the Color Pyramid Test by Max Pfister, the Lüscher Color Test and the recently developed

❑ **Fig. 12.1**   The five Moodforms (color forms)

"Manchester Colour Wheel", where no correct or incorrect answers are suggested. Many projective tests also work with open questions or impulses.

There is certainly also legitimate criticism, especially of the classical projective test procedures, specifically of psychological color tests, for example due to the multiple meanings of individual colors (Sorokowski et al. 2014). The evaluation of classical projective tests is often seen more as an art than as a scientific method. Nevertheless: The close connection between scent and color makes them—provided that there are no color vision deficiencies, anosmia (loss of smell), hyperosmia (oversensitivity to odors) or hyposmia (reduced sense of smell)—interesting as a projective test for the perfume industry and especially for the determination of fragrance preferences. Because the test person can quickly check for themselves whether corresponding fragrance recommendations are actually suitable for them. The typical perfume lover is unlikely to be one of the 5% of the population who suffer from almost complete anosmia, or one of the 20% with reduced sense of smell (Hatt and Dee 2012).

The above-mentioned connection between color, scent and self-experience makes it possible to develop a psychological color-scent test. For my first scent test, the color rosette test©, I have already described in other publications how to proceed when developing such a color-scent test, how to validate a projective test in the best possible way according to scientific guidelines, so that it also measures what it claims, and also meets the requirements for reliability (reliability of the results) (Mensing and Beck 1988).

With psychological tests developed according to scientific criteria—including some projective tests—one can gain a good first impression of a topic and use the result as a basis for consultation. This also applies to color tests, especially if they are based on newer findings in color research. In the development and application of psychological tests, one is happy to work with statistical probabilities, the so-called significance level or the probability of error. They also show how likely it is that one will have to reject an assumption. It is considered very good if one is "wrong" in only 5% of the cases or deviates from an assumption. A probability of error of 10% is still considered acceptable in the social sciences.

However, with psychological color-scent tests, one must expect an even higher rate of non-hits. So you never know without exact testing whether the test person may be slightly colorblind. Of course there are also test procedures that can be applied before or after color tests. With men, it can be assumed that 8 to 10% have problems correctly identifying colors.

When it comes to fragrances, there is another point to consider: they are associated with fragrance socialization or corresponding memories. They ensure that perfumes can be experienced differently by individuals. For example, a particular floral fragrance or a perfume that smells like roses can evoke the memory of a person who was experienced as unpleasant. Older color tests also often suffer from significant methodological and image problems due to their application and statements. For example, one simply assigned the test person to a red type based on single color choices. This cannot work for the sole reason that a single color like red has too many different meanings. Color tests were therefore often only half-heartedly made or initially only tested.

Nevertheless, there are many good reasons for color-scent tests. They can be developed based on new findings in color psychology, fragrance psychology, and even neuroperfumery. In addition, perfume consulting is faced with the following problem: as a rule, the perception of fragrance already decreases after four to six fragrances have been smelled. Color-scent tests therefore offer a first good orientation and do not overload the nose with fragrances that it does not even want to smell.

However, the assignment of color and fragrance is not one-dimensional. You cannot assign a fragrance to just one color, even if you talk about green notes in the perfume shop. Perfumes are too complex to be represented by just one color. A color is also associated with several emotions (red is associated with love, but also with danger and thus with fear). In order to express perfumes and emotions, color groups are therefore better suited. They can map both fragrance and emotional directions more specifically.

Colors are like scents and emotions always in motion. They are electromagnetic radiation of the visible range of light, whose wavelengths move about between 750 nm (red) and 380 nm (violet). Even if each color corresponds to a specific wavelength, no color stands alone. The effect of each color is determined by its context, which can give it a different meaning. So, as mentioned, red can stand for danger, blood and love. Color groups, in which different wavelengths interact, are therefore less context-dependent and can better describe a direction of meaning and experience. So the color combination red-magenta-rosé will be interpreted more in the direction of love than to be associated with blood. Three to four individual colors in combination are often enough to better visualize a general scent direction. This is the basis for the Color Mood Grid©, another quick test for determining directions of experience and scent (❏ Fig. 12.2).

The choice of adjacent color groups or directions gives a first indication of associated experience and scent directions. ❏ Fig. 12.3 gives an example of eight scents for women's notes discussed in ▶ Chap. 5.

With the Color Mood Grid© a complex mapping can also be carried out. This shows how to delimit perfumes from each other. This supports perfume marketing from concept ideas to, for example, the

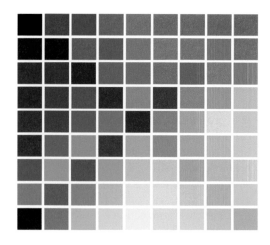

❏ **Fig. 12.2**   Color Mood Grid©

color scheme of the packaging for new perfumes and supplements the positioning of a brand. ❏ Fig. 12.4 shows an example of the Color Mood Grid©, to which 16 fragrance directions for men's and women's notes are assigned.

Mood-Grids are also a tool in psychological research. They are also used for the evaluation of the success of therapies and the observation of psychological changes (Parkinson et al. 1996). In this case, two Mood-Grids—a Double Mood Grid© (❏ Fig. 12.5)—are used.

In the Mood Grids, you are first asked to describe how you feel at the moment ("Current Self") and then how you would like to feel ("Ideal Self"):

1. "Please describe how you feel at this moment."
   You mark a white cell that best describes it. You can also describe the intensity on a four-point scale. Inside is the strength "1" (weakly expressed) and outside the strength "4" (very strongly expressed).
2. "Please describe how you would like to feel if you are looking for a change in your experience."
   As in 1., the corresponding cell is marked.

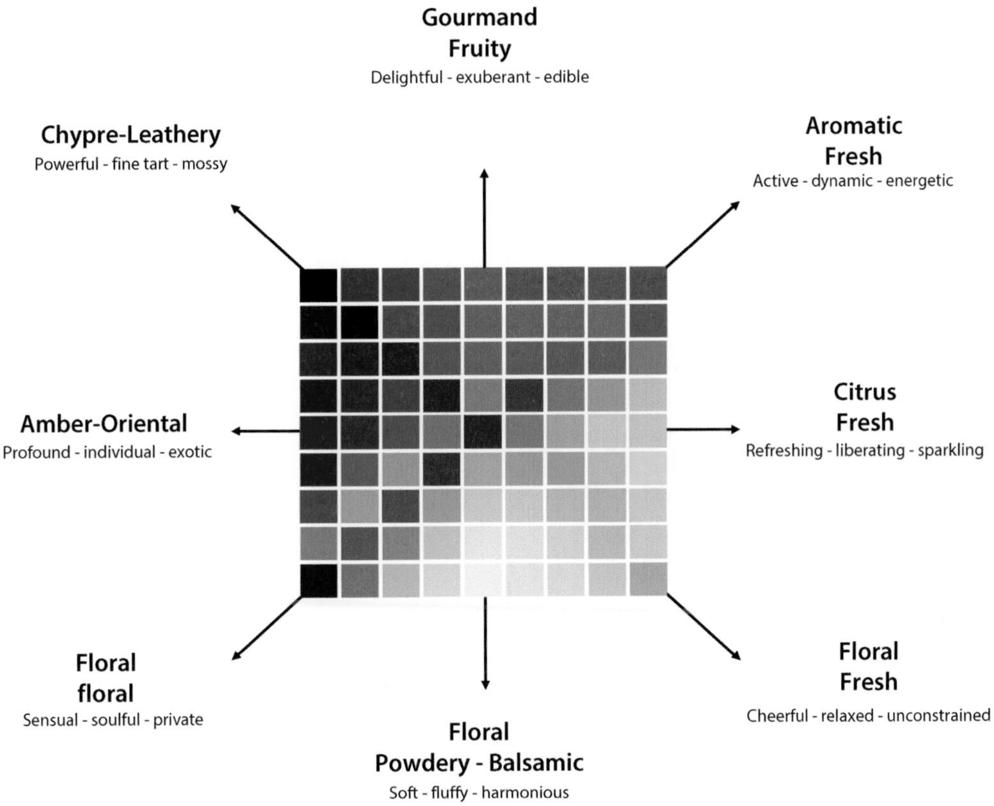

Gourmand
Fruity
Delightful - exuberant - edible

Chypre-Leathery
Powerful - fine tart - mossy

Aromatic
Fresh
Active - dynamic - energetic

Amber-Oriental
Profound - individual - exotic

Citrus
Fresh
Refreshing - liberating - sparkling

Floral
floral
Sensual - soulful - private

Floral
Fresh
Cheerful - relaxed - unconstrained

Floral
Powdery - Balsamic
Soft - fluffy - harmonious

◘ **Fig. 12.3**   Color Mood Grid© with 8 experience and scent directions for women's notes

There are a total of 8 experience dimensions to choose from for self-description (◘ Fig. 12.6).

Based on the comparison of the two Mood Grids, possible discrepancies between the "current self" and the "ideal self" can be visualized and addressed in psychological research (Higgins 1987).

■ **Color and shape**

You can further illustrate your experience and fragrance wishes in a fragrance test by supporting the effect of colors through shaping. The relationships between color and shape (form) were a favorite project of the painter Wassily Kandinsky. Color in shape was implemented in the Moodform-Test© which could be answered above.

It shows the connection between preferences for fragrance directions and feeling good. The short form of the test consists, as already mentioned, of five color symbols, the extended version of eight symbols. You should choose the most attractive and second most attractive symbol. The moment is in the foreground. After all, perfume is also an offer of transformation to the current mood, just like someone wants to experience the moment. Therefore, it is not asked for general favorite colors, but for colors or color symbols that attract you at the moment. This should be a spontaneous decision. Whether the colors suit you or whether they match is irrelevant for a color-fragrance test like the Moodform-Test.

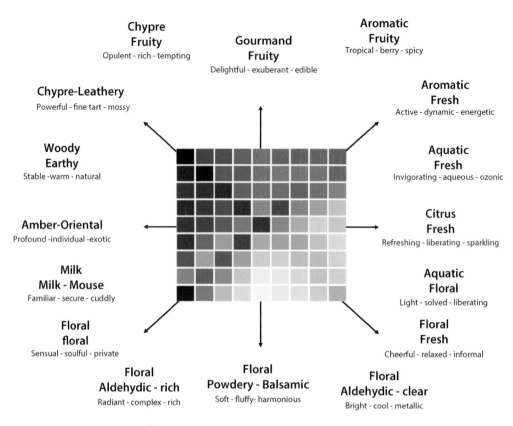

Fig. 12.4   Color Mood Grid© with 16 fragrance directions for men's and women's notes

12

12.3 Neuropsychological Fragrance Test: Experience Wishes and Perfume Preferences

12.3.1 Moodform-Test©—Test Instructions & Solutions

The questions of the Moodform-fragrance test are repeated here:

❯ Which of the five color symbols, i.e. Moodform color shapes, speaks to you at the moment and which as second? You can also choose the same color shape twice.

The Moodform-Test shows, as a so-called projective test, the current, also psychological experience wishes of a customer and what attracts him as a perfume, better said, as a fragrance direction. It reveals the stimulation needs of individual brain regions and their networks (such as the search of the orbitofrontal cortex for citrus freshness or the hippocampus for relaxing floral and flower notes) and thus serves as applied neuroperfumery in fragrance advice.

The individual color shapes were developed in such a way that they each reflect their own dynamics through form and color. The development of the Moodform-Test, which is a further development of the Color Rosette Test, included fra-

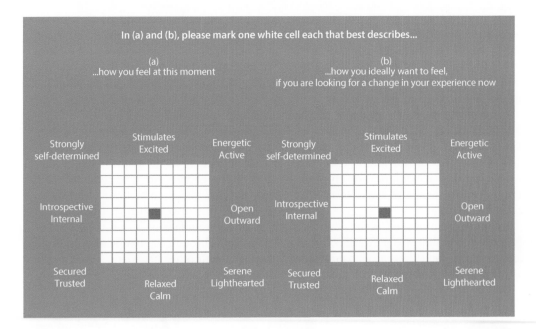

◘ Fig. 12.5 Double Mood Grid©

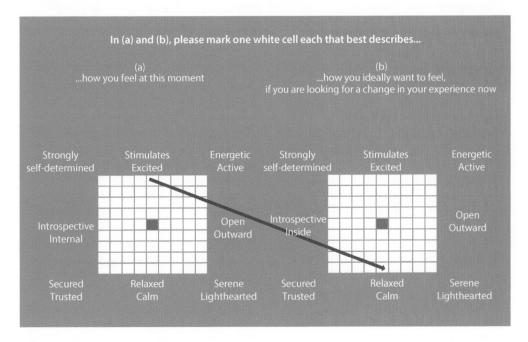

◘ Fig. 12.6 Double Mood Grid© with "Mismatches"

grance and color psychological findings. Also, the relationships between color and shape were taken into account, as described by Wassily Kandinsky and Johannes Itten. For example, violet is assigned to the ellipse, yellow and green to the triangle.

In color theory, it is repeatedly emphasized that colors in shapes connect people with energetic, but also with psychological contexts. Color vibrations have their own energy, they act invigorating or inhibiting on our organism and are thus also perceived by the brain.

In the psychology of scent, there are also proven relationships between color and scent choices, such as that the preference for violet hues is associated with greater ac-ceptance of amber-oriental or skin notes, or that green-yellow hues point to a greater inclination towards fresh-citrus notes. From these relationships, one can draw a certain, albeit cautious conclusion, which brain areas and networks are involved in the processing of color and form stimuli.

Certainly, the orbitofrontal cortex (OFC) establishes multisensory relationships, for example between olfactory and color impressions. But it is assumed that well 40% of the human brain is involved in seeing. It is also known that the perception of color and shape is only composed in the brain. Through the close connection of the OFC with other brain areas such as the amygdala, it can be assumed that certain

stimuli stimulate specific areas in the feedback. As I said, there are first assumptions for these relationships, the confirmation of which must still be provided by further comprehensive research. However, I do not want to question the Moodform test. Often tests work, even if the relationships have to be explained differently later.

In the following, I explain the emotional dynamics of the individual Moodforms arranged from left to right here:

— The **violet spiral** is the most closed form and is intended to symbolize deep inner well-being and self-confidence—a process we associate with the amygdala to calm the fear network. From the spiral, which can rotate downwards, but also upwards (outside), a dynamics with a range of experience wishes can arise:
To feel more individual, extravagant and reflective, to discover or rediscover one's own creativity and potential, to experience or feel more oneself. To feel more sensual and attractive. To collect new experiences and to dive into other worlds.
Scent direction: The scale ranges from floriental notes (warm-flowery) such as warm skin notes to full-bodied, amber-oriental perfumes.

In the system of the already mentioned Big-Five personality dimensions, the violet spiral visualizes the self-description "openness to (self-)experience" to the greatest extent. However, an empirical analysis of possible relationships is still pending.

— The **pink flower** is the most emotional form and is intended to visualize relaxation and inner peace—a process we associate with the hippocampus and the attempt to calm the stress network. From the flowery form certain experience wishes grow:

To feel more sheltered and harmonious, to be natural. To find more inner peace, comfort and trust while having less stress. To recharge the soul battery for a restart or simply to let the soul dangle.

Scent direction: The scale ranges from light floral notes such as white tea notes to relax and clean notes that remind of freshly washed.

In the system of the Big-Five personality dimensions, the pink flower visualizes the self-description "agreeableness", "friendliness or social compatibility" to the greatest extent.

- The **blue square** is the most pragmatic and self-confident form. It visualizes performance orientation and self-strength and can be associated with the prefrontal cortex. The colors and shapes used visualize specific desired experiences:

To experience oneself as even more successful, responsible and conscientious. To be more rewarded and recognized. To improve oneself even further, to enjoy what has been achieved and to be able to afford things.

Fragrance direction: The range for men extends from fresh aqua notes to aromatic, mossy and herbaceous fragrances, in particular fougère notes. For women, the range extends from radiant floral notes, in particular from the aldehyde spectrum, to rich and elegant, but also expressive floral chypre notes.

In the Big Five personality dimensions system, the blue square visualizes the self-description "conscientiousness" to the greatest extent.

- The **popping egg** is the happiest and most carefree form. It visualizes fun and joie de vivre and can be associated with the hypothalamus and the pleasure network. In its dynamics, color and shape refer to further desired experiences:

To be more pampered. To have more pleasure, fun and variety. To fall in love or to fall in love again. To do spontaneous things like travel. To avoid routine and rather try something new.

Fragrance direction: The range extends from fresh fruit notes to rich gourmand notes that remind of classic desserts and sweets.

In the Big Five personality dimensions system, the popping egg visualizes the self-description "extraversion" in combination with "spontaneousness" and "sensation seeking"—the search for variety and new experiences—to the greatest extent.

- The **flying triangle** is an expression of movement, of having free space and being active. This typically extroverted experience can be associated with the orbitofrontal cortex. In their dynamics, color and form are associated with specific experience desires:

Search for enthusiasm and gain. To be more self-responsible and unrestricted. To make decisions quickly and be decisive and risk-taking. To approach things resolutely and achieve them efficiently.

Scent direction: The scale ranges from fresh green leaf notes like mint to cool, refreshing citrus notes, inspired by bergamot, grapefruit, mandarin, lemon, orange or lime and other notes.

In the Big Five personality dimensions system, the flying triangle visualizes the self-description "extraversion" in combination with "openness to new experiences" the best.

12.3.2 The Moodform Test© as a Perfume and Care Guide for Women: Solutions from First and Second Choice

Now to the individual Moodform solutions, which consist of the first and second

12

choice. Here you can choose a color form twice. In the short form of the test it does not matter which color form you choose first. In total, there are 15 solutions for women in the short form of this perfume test. For each solution I have proposed perfume suggestions as well as suitable care products or the care experience corresponding to the recommended scents next to the perfume suggestions. They will certainly have to be supplemented and updated from time to time. This way, the Moodform Test, when used in perfume consulting, can also be used for targeted consulting and sampling of care products. Of course, it can also work the other way around.

Although skin needs and skin condition always play a role in care, product effect, expectation and application also depend on experience desires. In addition, there are more and more care brands and products that adjust to the skin's own care needs with intelligent technology. So you can also do a well-being consultation in care, which is more focused on psychological experience desires.

Under "What is important now and is being sought more", you will find short psychological keywords for the result, which you should communicate in your own words before showing the three perfumes mentioned to the customer and smelling them with him. Accordingly, one should proceed with a care test. Here is the solution matrix for the Moodforms with the individual solutions below. It does not matter which color form was chosen first in this short form of the test, as already mentioned:

First or second choice		Solution
Violet spiral	Violet spiral	1
Violet spiral	Pink flower	2
Violet spiral	Blue square	3
Violet Spiral	Popping Egg	4
Violet Spiral	Flying Triangle	5

First or second choice		Solution
Pink Flower	Pink Flower	6
Pink Flower	Blue Square	7
Pink flower	Popping egg	8
Pink flower	Flying triangle	9
Blue square	Blue square	10
Blue square	Popping egg	11
Blue square	Flying triangle	12
Popping egg	Popping egg	13
Popping egg	Flying triangle	14
Flying triangle	Flying triangle	15

Individual solutions

1. Solution:

"Violet Spiral" with **"violet spiral"**

What is now important and more sought after:

- To experience oneself more individually and deeply.
- To pursue more personal freedom and own interests.
- To express more creativity and artistic interests.

Amber-Oriental scents are ideal to bring you closer to this ideal experience:

- "Alien" by Thierry Mugler
- "Silver Rain" by La Prairie
- "Shalimar" by Guerlain
- The ideal care must adjust itself personally to the skin. Thereby, products have to work as multi-talents:
  - "Le Privilège Base Traitante" by Rivoli
  - "Oil Absolue" by Filorga
  - "Pep Start Exfoliating Cleanser" by Clinique

2. Solution:

"Violet Spiral" with **"pink flower"**

What is important now and is being sought more:

- To be pampered more.
- To discover an exotic world full of secrets.
- To experience more feelings and sensual moments.

The fragrance direction "**Floriental-oriental**" combines warm floral notes, rare fruits and vibrating wood notes:

- "My Name" by Trussardi
- "Angel Muse" by Thierry Mugler
- "Secret Obsession" by Calvin Klein
- The ideal care must pamper and relax with a unique texture and sensual fragrance:
  - "Oil Therapy Body Lotion" by Biotherm
  - "Aroma Care" by Clarins
  - "SOS Comfort Balm Mask" by Clarins

3. Solution:
"**Violet Spiral**" with "**blue square**"
What is now important and more sought after:

- To experience more extravagance and exclusivity.
- To enrich and show one's own personal style.
- To feel more inner strength within oneself.
- To discover something precious and high-quality where others overlook it.

The fragrance direction "**Chypre-oriental**" has a self-confident and exquisite radiance:

- "Coco Noir" by Chanel
- "Chypre Rouge" by Serge Lutens
- "Want" by Dsquared
- The ideal care must experience the skin as precious and of high quality and perfectly nourish it as an all-round high-performance care:

- "Deep Comfort Hand Cream" by Clarins
- "Absolue Precious Cells" by Lancôme
- "Cellular Swiss Ice Crystal Eye Care" by La Prairie

4. Solution:
"**Violet Spiral**" with "**popping egg**"
What is now important and more sought after:

- To have more joie de vivre and fun.
- To pamper the senses of the ideal partner and one's own in a sensual way.
- To live out more fantasies and luxurious dreams.

The fragrance direction "**Gourmand**" is extremely sensual and invites you to seduce the senses:

- "Poison Girl" by Dior
- "Si Lolita" by Lolita Lempicka
- "Classique" by Jean-Paul Gaultier
- The ideal care must pamper the skin and support it at the same time:
  - "Ibuki Gentle Cleanser" by Shiseido
  - "Émulsion Ré-équilibrante" by Rivoli
  - "Silky Peeling Powder" by Kanebo

5. Solution:
"**Violet Spiral**" with "**flying triangle**"
What is now important and more sought after:

- To be more open in all directions and unbound.
- To be active sometimes and to retreat into one's own world sometimes.
- To show more passion and to claim one's rights.

The fragrance direction "**Fresh-Oriental**" combines freshness and warmth in a fascinating fragrance direction:

- "Aromatic Elixir" by Clinique

- "Cologne" by Thierry Mugler
- "Olympéa Aqua" by Paco Rabanne
- The ideal care prepares the skin perfectly for the next day, cares for and regenerates it deeply:
  - "Sleep and Peel Cream" by Filorga
  - "L'Eau de Nuit" by Rivoli
  - "Botanical Detox Cure Night" by Sisley

6. Solution:
"Pink Flower" with **"pink flower"**
What is important now and is being sought more:
- To find more peace, harmony and balance.
- To experience more support and less pressure from others.
- To experience more attention and tenderness from people who are important to one.

The fragrance direction **"Floriental"** offers the most beautiful romantic floral notes, which do the soul more than good with fine fruit notes:
- "Donna" by Valentino
- "Jeanne Lanvin" by Lanvin
- "La Femme" by Prada
- The ideal care provides the skin with optimal care and at the same time gives a sensually relaxing experience:
  - "Pflegeölbad" by Kneipp
  - "Skin Meditation" by Declaré
  - "Hydra Zen" by Lancôme

7. Solution:
"Pink Flower" with **"blue square"**
What is now important and more sought after:
- To experience more sensitivity and style from others.
- To have more privacy and less hustle and bustle.
- To experience more elegance, luxury and beauty.

Delicate floral notes from the fragrance direction **"Floral-floral"** enchant with rich and powdery heart and base:
- "Delicate Rose" by Trussardi
- "Summer" by Kenzo
- "Coco Mademoiselle" by Chanel
- The ideal care must be of high quality, pure and free of mineral oils and parabens. It not only cares visibly, but also builds the skin from the inside:
  - "Le Regard Crème Lissante" by Rivoli
  - "Black Rose Mask" by Sisley
  - "Blue Therapy Serum" by Biotherm

8. Solution:
"Pink Flower" with **"popping egg"**
What is important now and more sought after:
- To be pampered spontaneously for a change.
- To flirt with someone in order to test one's own attractiveness.
- To have more joie de vivre and to be with people who make one laugh.

The fragrance family **"Fruity-Oriental"** smells like a summer flirtation—animating and lustful for life, but also sensual enjoyment:
- "Miss Dior Absolutely Blooming" by Dior
- "Trésor in Love" by Lancôme
- "Womanity Eau Pour Elles" by Thierry Mugler
- The ideal care acts immediately and makes you attractive for any situation and season:
  - "Mission Perfect Serum" by Clarins
  - "BB Perfect Cream" by Filorga
  - "Total Eye Revitalizer" by Biotherm

9. Solution:
"Pink Flower" with **"flying triangle"**

12

What is now important and more sought after:

- Increasing well-being and gaining new vitality are now the priority.
- To feel healthy and vital.
- To have new energy and more youthful vigor.

The fragrance family **"Fresh-flowery"** has many scents that increase well-being and have a positive effect on mind, body and soul:

- "Eau Vitaminée" by Biotherm
- "Olympéa Aqua" by Paco Rabanne
- "Donna Aqua" by Valentino
- The ideal care builds on for a restart, protects against harmful environmental influences and prepares the skin for everything that comes:
  - "The Renewal Oil" by La Mer
  - "Le Visage Mousse Nettoyante" by Rivoli
  - "Gesichtsspray Skin Meditation" by Declaré

10. Solution:
"Blue Square" with **"blue square"**
What is important now and is being sought more:

- To treat oneself more, to do something good simply to reward oneself for what has been achieved.
- To receive more respect and deserved recognition for one's own performance.
- To invest more in oneself and one's personal style and not just think of others.

The fragrance family **"Floral-aldehydic"** gets its name from the radiant and valuable top notes of noble perfumes that shine with a lot of flower purity and brilliance:

- "J'adore" by Dior
- "Cristalle" by Chanel

- "Calèche" by Hermès
- The ideal care is a small team of top products to individually care for the skin:
  - "Prime Renewing Pack" by Valmont
  - "Replenishing Body Cream" by Shiseido
  - "Skin Caviar Eye Lift Augenserum" by La Prairie

11. Solution:
"Blue Square" with **"popping egg"**
What is important now and more sought after:

- To break free from constraints and stagnation.
- To consider a new beginning and to be able to let go.
- To have more joy in life and to see things with humor sometimes.

The fragrance family **"Floral-fruity"** promotes a positive feeling of life and also demands to enjoy more:

- "Chance Eau Vive" by Chanel
- "Miss Dior Chérie" by Dior
- "Gabrielle" by Chanel
- The ideal care consists of intelligent anti-aging products that adapt to the individual needs of the skin:
  - "Masqué Réparateur" by Rivoli
  - "Wonder Mud Maske" by Biotherm
  - "Power Infusing Concentrate" by Shiseido

12. Solution:
"Blue Square" with **"flying triangle"**
What is important now and more sought after:

- Bringing more swing into one's own life.
- Gaining more energy for new projects.
- Motivating oneself to achieve goals and meet responsibilities.

The fragrance direction "**Chypre-Fresh**" comes with a lot of inner will, radiates power for new things and supports motivation:

- "Miss Dior" by Dior
- "Donna" by Trussardi
- "Calyx" by Clinique
- The ideal care are innovative and intelligent pick-me-ups that quickly provide the skin with everything it needs day and night—quickly and effectively:
  - "Hydro Energy" by Declaré
  - "Scrub & Mask" by Filorga
  - "Advanced Night Repair" by Estée Lauder

13. Solution:

"Popping Egg" with **"popping egg"**

What is now important and more sought after:

- To experience more variety, spontaneity and fun together and to enjoy it.
- To simply turn things upside down and see them differently.
- To fall in love or to fall in love again.
- To avoid routine and rather try something new.

The fragrance direction **"Gourmand-fruity"** offers spontaneous moments of happiness with many fantasies and surprises:

- "Lady Million" by Paco Rabanne
- "Fuel for Life Unlimited" by Diesel
- "Chance Eau Tendre" by Chanel
- The ideal care is pleasure and effect in one. Ideally, it comes as a To-go product:
  - "Lip Sugar Scrub and Glow" by Dior
  - "Bio Performance Glow Revival" by Shiseido
  - "Mini Glow Drops" by Dr. Barbara Sturm

14. Solution:

"Popping Egg" with **"flying triangle"**

What is now important and more sought after:

- To experiment with something new and also to provoke.
- To take on more and also to risk something.
- Just break out and gain new experiences.
- To be among people who spread good mood and no gloom.

The fragrance **"Fruity-fresh"** stimulates, inspires and provides a good mood feeling all day:

- "Aqua Allegoria Pamplelune" by Guerlain
- "Orange Tonic" by Azzaro
- "Un Jardin Sur le Nil" by Hermès
- The ideal care must not only provide a lot of moisture, it must also offer beauty from the inside and outside:
  - "Hydra-Hyal Ultra Plumping Serum" by Filorga
  - "Le Visage Sérum Lumière" by Rivoli
  - "Timeblock Vital Aging Nutrition" by Swiss Biologie

15. Solution:

"Flying Triangle" with **"flying triangle"**

What is important now and is being sought more:

- To live more sports and fitness and also to measure with others.
- To be active for oneself, to achieve one's own goals and not to have to wait for others.
- To be more self-responsible and free to breathe.

The fragrance direction **"Green-Citrus"** invigorates, stimulates, regenerates and sets a refreshing sign of life for itself:

- "Escale à Portofino" by Dior

- "Energizing Fragrance" by Shiseido
- "Chance Eau Fraîche" by Chanel

- The ideal care must be invigorating and already radiate new energy when applied. Ideal are products that can be combined:
  - "Pep Start" by Clinique
  - "Énergie de Vie" by Lancôme
  - "L'Eau de Jour" by Rivoli

12.4 Perfume Experience: Practice and Methods for more Perfume Enthusiasm

Of course, you can't offer a perfume experience to all customers and in all consultation situations, as I describe it below. As a fragrance consultant, you need empathy to decide whether and which methods of the perfume experience to use. A customer who obviously asks for a perfume under time pressure will find it inappropriate if he is confronted with—from his point of view—time-consuming methods of the perfume experience. It also depends on regional, mental and personality-specific peculiarities whether one can address a customer in this way.

Of course, with methods of experience perfumery, there is also the risk that some customers will initially feel a certain irritation. That is why I recommend that everyone in the perfume consulting practice rely on their gut feeling when it comes to the methods I have and will be presenting. It will signal to them what is possible and could work with a customer.

If you forego experience perfumery and its methods, you run the risk of not sufficiently distinguishing yourself from competitors in your advice and becoming interchangeable for customers. The use of experience perfumery makes a perfume selection unique. Once a customer has experienced

this service, they will be happy to come back—ideally even if they could buy the same scent cheaper elsewhere. Even more: the probability is even quite high that he will recommend this perfume to others and thus generate new customers.

12.4.1 Dancing and Smelling

- **Scented Dancing**

One method of experience perfumery is scented dancing. For a perfume shop this might be a somewhat unusual suggestion, but not for a surprise party or an evening or customer event with a personal perfume test.

The messenger substances serotonin and especially dopamine are considered happiness hormones. As already mentioned, dopamine release is controlled by the hypothalamus and its network, which is decisive for libido, for example, and communicates closely with the sense of smell. Dopamine conveys many positive emotional experiences that are perceived as rewarding and mood-enhancing for the psyche. Conversely, motivation, low mood and mood swings are associated with low dopamine levels.

Findings from neuroperfumery show that, with scents matched to desires in combination with other sensory stimuli, it is possible to naturally and healthily increase the dopamine level. Examples of such experiential desires are:

- To feel more active and dynamic.
- To be more relaxed and stress-free.
- To feel sensual, attractive and overall comfortable.
- To have more fun and joy in life.

You can achieve the optimal increase in dopamine levels most quickly if a specific desire for experience is literally fused with coordinated sensory impressions. For example, if the desire to experience more fun and joy of life is supported by matching color impressions, music, dance, and gustatory and olfactory selected fragrance notes. For this, perfumes from the "Gourmand" fragrance direction are ideal. An example of this fragrance direction is "New York Nights" by Bond No. 9 with a salted caramel note.

Dancing plays a special role in increasing the experience of fragrance, because dancing leads to an increased release of serotonin (Christensen and Chang 2018). Therefore, professional dancers have an average higher serotonin level, but even occasional dancers already show effects. Especially a very fast dance with hopping, jumping and laughing immediately releases a lot of serotonin. Do the self-test: Dance, for example, to the song "Cotton Eye Joe" by Rednex or to other versions (▶ Cotton Eye Joe).

■ **Those who dance often can smell better**

That dancing reduces stress has been known for a long time. Also that it is successfully used in therapies against depression and for more quality of life in Parkinson's. But that dancing also promotes smelling is relatively new. Dancing, like all forms of physical activity, not only supports the ability to smell, but also ensures that the sense of smell remains fit into old age.

Several studies point to a direct connection between physical activity and improved smell. Even if the effect does not always occur immediately, the studies show that the sense of smell improves with longer physical training. It can therefore be said: Those who dance often can smell better and preserve this ability longer. Why this is so, the neuroperfumery does not know exactly yet. But dancing also seems to stimulate—in addition to brain areas responsible for motor skills—the hippocampus, the olfactory memory, for example. This makes fragrance impressions more vivid and easier to recall.

But dancing also stimulates the orbitofrontal cortex, among other things. the seat of multisensory experience and important parts of the personality. Newer studies also show that synchronizing movements with beats addresses both the ear and the brain. Dance and music together form a real double pleasure. The resulting greater effect ensures a significant improvement in the ability to smell.

■ **Expressing emotions and moods of scents**

Scent can be expressed surprisingly well and creatively in dance. Dutch art historian and curator Caro Verbeek confirms this. In her article "Dancing Scent and Aromatizing Movement—Hackathon Conclusions" (the summary of a multi-day event on dance, movement and scent), she describes how something fascinating happened when she asked participants of the smell workshop to express individual scents through gestures, shapes and movements. The participants did not hesitate to make very expressive movements with their fingers, hands, arms and sometimes even their whole body. Verbeek (2018) writes: "Movement allows us to express emotions. In fact, a whole range of emotions. Because that's what perfumes do. They tell a whole story."

But Verbeek's attempt also shows that you don't need special kinesthetic intelligence to dance or express the emotions and moods of a perfume. There is no "right" or "wrong" in the experience of scent. And that also applies to the dance of scent. Of course, an experienced dancer with knowledge of perfumery will want to associate individual fragrance directions with dance styles. So you could visualize perfumes of the "Gourmand" fragrance direction and their typical ingredients with emotions and moods through figures of Latin American dances such as tango, bachata and merengue.

12.4.2 Goals, Steps, and Examples for the Experience Perfumery

The goal of the experience perfumery is to create anticipation and curiosity in the customer , then to create a pleasant, unexpectedly positive tension during the experience consultation, which can increase to fascination in the customer. Scent, but also care and make-up should trigger positive experiences, even moments of happiness in the customer, from which the consultant also benefits. This cannot be achieved solely by the presentation such as the form, color and material of the design of the products. Trend researchers, market insiders, but also more and more retailers have recognized that the customer does not primarily buy a product or service, but a positive experience. The psychological technique of surprising the customer positively with the unexpected plays a special role in the consultation situation.

It is known from sales psychology that customers are fascinated and bound to us above all with an unexpectedly positive experience. Whoever succeeds in doing this already has the upper hand in retail today. This is especially true for the offline perfumery, which can play its difference to the online sales through modern experience consulting. Below I present a general course of action for the experience increase in the perfumery, which can be transferred to other industries and product areas.

1st step of the perfume experience: Generate anticipation through curiosity and suspense

■ ■ "World Dance Day" as Cross-Promotion
To stay with the example of "dancing with perfume", for example, you could invite perfume customers and non-dancers to a party with perfume tests on "World Dance Day" in April in the perfume shop or another place, where they would also get something nice to smell and taste. This would be an ideal opportunity to win new customers, but also to remain attractive to regular customers. For this you could team up with a local dance school and even offer periodic dance evenings on a smaller scale. In the spirit of a Cross-Promotion, perfumery shop and dance school could promote each other. So every dance student could get a discount on a perfume and every customer who, for example, buys a perfume from the "Gourmand" range, could receive a voucher from the dance school to learn Latin American dances. Instead of only in the evening, dancing with perfume could also be invited during the day—for example, on the quiet Monday business days—to the perfume shop. Just imagine the curiosity of the passers-by who see customers dancing in the perfume shop. With good weather, the perfume dance could even be moved outside and lead to a gathering of spectators. For this I recommend the music video of the song "Brave" by Sara Bareilles (▶ Sara Bareilles - Brave).

A little courage belongs to the perfume experience already, but also a camera, so that it looks official and the video can be put online later.

2nd step: Offer surprises and unexpected things

■ ■ Self-tests as icebreakers
Ideally, everyone will learn something about themselves in this consultation phase, e.g. through the Moodform test. The goal of every fragrance test is to infer possible perfume preferences from desired experiences. So you could—to stay with the example of fragrance dancing—also offer a dance test with a matching perfume as a result. Ideas for this can be found under ▶ testedich…a German website, which dance fits you. Although fragrance tests do not have a hundred percent hit rate, they trigger interesting conversations and are ideally suited for advertising—by mailings,

in the window or a menu board at the entrance: "This week fragrance dancing—what your favorite dance says about your perfume preferences." Of course, no customer has to participate in a test. If someone comes into the store for a specific perfume, they can of course start smelling right away. Nevertheless: To build up further curiosity with personal expectations before smelling the perfume increases the tension.

■■ Positive irritation

Fragrance tests like consultation tools of all kinds and especially unexpected customer approaches not only generate curiosity, but also a moment of positive irritation. Artists often use them at exhibitions to prevent visitors from "passing through". These irritations cause one to be initially baffled, but then to increase attention.

When entering a store, one expects more or less consciously the usual questions and therefore often does not even listen properly—which is also true for the seller. He already believes to know what the customer wants. This "talking past each other" has fascinated the American sociologist Talcott Parsons (1902–1979). He liked to answer the usual greeting formula of the host with: "I'm sorry I'm late. I still had to kill my mother-in-law." In many cases then the answer was: "That doesn't matter, come right in." It always took a while until the host asked: "But you didn't really do that, did you?"

An irritation that is also a form of surprise therefore works in the head. It usually manages to make a reaction with a delay. That's exactly what you need when a customer has no time, is a casual or random customer, a new customer or not a customer yet. So you can win him for your own perfume store and also invite him to the next fragrance dancing.

■■ Amaze customers, especially new customers and non-customers

The best customers for surprises are new customers and especially non-customers who only come into the store to change coins for the parking meter, for example. An answer to customer questions or requests could look like this—after all, you have nothing to lose:

"I'll do that right away. By the way, on the occasion of the 'World Day of Dance', we are today presenting perfumes to various dance styles. Which dance style do you like the most? Do you love Latin American dances like tango or merengue (while making a dancing gesture)? Then I'll show you perfumes that match. Here are the coins for the parking meter." Even if the customer politely declines, he will keep the consultation offer in mind and realize that this perfumery is different. And maybe he will tell the experience, which in turn is a nice advertisement.

Under ▶ www.kleiner-kalender.de you will find many more occasions with which you can surprise in perfume advice as part of a promotion or event. In the following I will give two more examples with which I have had good experiences:

■■ World Turtle Day

This day is particularly suitable for an event on a weekend. After all, most people dream on Fridays of relaxing, regenerating and pampering their senses over the weekend. Psychologically, it is about the art of doing nothing, the art of being at home in oneself, withdrawing, gathering strength and maintaining peace. Since we have to learn this again and again, we are also happy to be advised about it.

To begin the fragrance consultation with a positive irritation, one could, for example, say: "Peace and serenity—these are the secrets of the turtle, which can be satisfied with its lifestyle up to 250 years old. In our fast-paced and hectic world, we humans should take this to heart! Welcome to World Turtle Day (-week) with the most beautiful perfumes to relax and find peace and serenity again."

If someone says this in his own words, 90% of the customers, but also all friends, will agree verbally or non-verbally. They will be grateful for any tips for weekend relaxation and also like to hear more about turtles while perfumes for World Turtle Day have already been set up on a display: most beautiful relax scents of the perfumery or from the own fragrance bar.

■■ **To the "Day of the Constitution"**

Even the "Day of the Constitution" is ideal for creating a positive irritation in the fragrance consultation and promoting perfumes that are not easy to sell:

For example in "1949 Article 5 of the German constitution proclaimed: Everyone has the right to express their opinion freely in words, writing and pictures, that also applies to perfumes. Today we show extravagant scents that have stimulated the senses since the introduction of the Law, but have also been discussed contrarily. Then as now it always takes some courage to show your style."

This is done by pointing to a display with four to a maximum of six unusual perfumes. More scents would put the customer's brain under stress when making a decision.

■■ **Experience display for the special**

You explain to customers or friends—if you want to animate them to smell or discuss at home or in the perfume shop—that there will always be new displays on different topics with a fragrance installation to experience perfumes . When choosing a topic, you should always make sure that the chosen topic also touches you yourself. You should not let displays from brands or companies stand in your perfume shop, so that

there would be no room left for your own experience display.

Ideal as a display are elegant floor tester stands in the field of vision—stable and large enough (about W 40 cm × H 154 cm × D 40 cm) to decorate the perfumes with some matching things. Suitable for the mentioned examples, there could be an issue of the constitution or a book about dancing or about turtles. This decoration can also be used for the shop window or as a night decoration behind the door. In front of the entrance you should point to the promotion with a blackboard. In the store you can also use an easel, which gives the display an additional creative touch. This is further increased if the invitations to the event are written by hand.

The display should be a permanent part of the perfume shop near the entrance. This way you can engage the customer in a conversation right when they enter the store and explain to them that new perfumes are decorated on the display for different occasions. The perfumes should be from different manufacturers. This shows expertise and competence in the perfume shop.

■■ **Jam Experiment**

A frequently asked question is: What sells better and is emotionally more appealing—a large or a targeted selection? Prof. Sheena Iyengar from Columbia Business School made a surprising discovery at the so-called Jam experiment. A delicatessen in her old home offered 300 different olive oil varieties and almost as many jams. In the experiment, customers could test 24 jams on certain days and only 6 varieties on other days. The result: Customers who were allowed to choose from 24 varieties bought almost 30% less than those who only tested 6 varieties.

Although the customers were initially attracted by the larger selection, they apparently experienced the decision-making as emotionally demanding. They appreciate the selection, but were ultimately not up to it. The success of a smaller range, however, requires additional information: The customer must know that his selection options are theoretically much greater. Therefore, stationary retail is well advised, for example, to indicate a large selection of stickers on the shop window: "We carry 1001 wonderful perfumes." However, only a pre-selection of four to six products should be offered on the "try-on table" in the store.

In order to interest and fascinate the brain, especially the amygdala, again and again, perfumes should be reassembled according to the visiting rhythm of the customers. The amygdala is oriented towards the present. For her, it is less interesting, for example, that Easter is in two weeks. In the foreground is her current experience. Of course, both aspects can be combined in the customer approach. This makes the day-to-day reference without forgetting the gift purchase.

■■ The Display as Treasure Island

Neuroperfumery gives another tip on how to make the display even more attractive for customers. The amygdala is happy to go on a treasure hunt for "mini-must-haves" (small things that one must have). She doesn't want to miss anything. With that I have already revealed how to present perfumes even more targeted and emotionally fascinating.

The Nordstrom department store group was one of the first in the USA to show how to address and fascinate customers' amygdalas even more. One is less focused on brand presence, but rather on the individual product or the product experience. Customers are helped by their own beauty concierges to find the best in each category—for example, with mascaras. In special trend zones or on labeled displays, the latest beauty trends of the season are presented by different brands, with the number of products exhibited per category again limited to four to six.

Perfumes are introduced in small sizes, so to speak as pocket-sized, in order to meet the needs of their target groups. For this reason, more and more brand manufacturers are producing their own editions in sizes between 15 and 25 ml. Certainly, the Corona pandemic has changed many things, including the service for customers and the department store groups themselves, as many doors had to close. Nevertheless, the smaller perfume sizes could be a long-term winner of the pandemic.

For years, drugstores have also been offering small sizes—not only to meet the needs of their target groups. The magic word is "mini-must-haves", as they are also known from the surprise shipments of beauty boxes.

The beauty concierge, who at least advised at Nordstrom before Corona, but also made sure that customers could explore the "mini-must-haves" on their own, so to speak, their own treasure hunt. In this way, a playful atmosphere with surprise possibilities was created, whereby the lucky treasure hunter is allowed to touch all the small treasures on the display and the whole thing takes place in as glamorous a launch atmosphere as possible. Of course, this type of tactile experience will be offered again after the pandemic. What has increased is the so-called unboxing (unpacking). To increase the anticipation of the treasure hunt, there is often first the unboxing. Then the social media shows how the new mini-must-have perfume of a brand arrives and is unpacked, and invites you to discover it on site or online.

This is then further increased by "visual storytelling", i.e. telling stories, supported by pictures, the anticipation so that the customer already knows roughly what to discover next in their own treasure hunt in the store.

This brings us to another fascinating phenomenon: if you like something, you want to know more about it, especially if you have already bought it. So to speak, as confirmation that the purchase was correct.

It is essential to confirm the customer's decision again by further storytelling, so that the consulting experience remains present. To come back to the turtle's anniversary, you could say on the doorstep when saying goodbye, if the customer showed a positive reaction to the keyword "turtle" at the beginning of the consultation:

"You have chosen a wonderful perfume for relaxing and you will feel really comfortable with it. By the way, turtles are among the most extraordinary things that the animal kingdom has ever produced! Which other animal, for example, manages to get so old and paddle calmly for 3000 km through the world's oceans? Turtles know very well how to protect themselves and when it is important to withdraw—after all, they have survived for millions of years."

With these afterthoughts that match the feeling of experience of the purchased perfume, you will achieve something very important: you communicate that you do not just see the customer as a buyer, but that he is also important to you as a person, because you take the time and give information that goes beyond the purchased perfume.

- **3. Step: Smelling**

Important: First, the seller should smell the sprayed perfume strip himself to check the perfume and its intensity. Many perfumers wave the perfume strip lightly and smell it themselves before handing it over, so that the alcohol also evaporates somewhat. This is also a gesture to prepare the perfume for the customer's smell, and conveys expertise, as we know it from a sommelier who first tastes it himself. How to smell together with the customer, I have already discussed in ▶ Chap. 11 and elsewhere. As I said, before and while smelling, do not name the individual ingredients, at least not in detailed

list of top, middle and base notes, this distracts and also carries the risk that the customer does not react positively to all ingredients. Better first describe the fragrance experience in general in images that everyone can relate to, for example "like a tropical waterfall in which the sunlight is reflected" or "like a wonderful summer evening with sunset". Before addressing ingredients in detail, the bottle with cap should be shown to convince the orbitofrontal cortex of its value. For this, the customer should be given the bottle immediately after the perfume strip. If it is held for longer, it is a sign that the scent is liked.

- **4. Step: Beginning of the interactive part after liking the fragrance in the initial smell and in the top note**

Customers or guests of an event can and should become active with the perfume shown themselves—provided it agrees. The creative work with a perfume, which we discuss below, makes the scent experience more personal and promotes identification ("my perfume").

If the first scent does not please, the other perfumes are shown. Even if the first perfume has already arrived very well, customers usually want to smell further, and they expect that at least three scents will be shown. In order to avoid tiring the nose, a maximum of six perfumes should be shown. Although smelling coffee beans clears the nose, it still costs the brain energy to smell. As I said, you get tired quickly when you smell and you get a little hungry.

There are various techniques and methods for the interactive part, which have a very positive effect on the further perception of scent. It is important that the customer has fun. Otherwise you should rather skip this part of the promotion or the event. Here are some examples of how the perception of scents that please can be further intensified.

▪▪ Scent painting as an event

Scent painting is particularly suitable as an event at the perfume counter or at a fragrance seminar. For this purpose, wax crayons or washable finger paints are distributed in about 16 different colors, each with a brush, which the customer can apply to sheets of an A4 block. With the following colors, scents can be described particularly well, with three to four colors to be selected: violet, lilac, pink, pink, dark and light blue, dark and light green, yellow, sand, white, black, gray, red, turquoise and orange. Customers are then asked to first paint or express the headnote in color and, if desired, the heart note that unfolds about ten minutes after application in color and—if desired—in shapes. If the scent pleases, it can also be applied to the skin. After the scent consultation and discussion to experience the perfume, the customer is given the painting. It is good if he signs it. This further increases the identification with the perfume. For this purpose, silver or gold metallic pens are suitable, which give the image a touch of additional artistic value. If the picture is not completely dry yet, the customer can pick it up later.

I personally like to offer customers a scent-color consultation. It takes about three to five minutes. I use my color psychological knowledge to explain the scent experience even further. To increase the customer's curiosity and surprise, I first ask him to choose three to four favorite colors from the 16 mentioned above and paint a picture with them. Then I give a color psychological analysis and recommend three to four perfumes of different brands that I place in front of the customer and let him smell them. Finally, I offer the customer to create his own personal perfume together with him from the two best scents by means of perfume layering. This is a special moment for the customer. He experiences the creation of his own perfume, which only he wears and which he can adapt to his needs and his sense of smell by applying the individual scents in different ways. By the way: Crayons and finger paints that can be applied with a brush remind of childhood and have a stress-relieving effect. They are ideal for customers who are looking for relaxation after a strenuous week.

In the context of my events on olfactory and color psychology ("Soul needs scent"), I have made the following experiences:

— Cross-brand promotions are more attractive, more successful and attract more customers because they are experienced by the consumer as neutral. Customers want advice that is tailored to their needs. This is possible with fragrance brands from different companies.

— Customers are looking for a very personal scent that not only smells good, but also works for them. When advising on fragrances, you have to distinguish between what the customer wants, for example to underline his personality, and what he needs to feel good and attractive. More and more customers are unconsciously looking for a fragrant, very personal effect perfume. They want a personal experience with an effect. The fragrance becomes the most beautiful of all drugs. It is increasingly decisive for the purchase to recognize at the beginning of the fragrance advice what the customer needs and what he is looking for as an experience.

— Customers love comprehensible psychological, in particular neuroscientific fragrance self-tests that give them insights into which perfume works best for them and what they need in terms of fragrance to feel more attractive and comfortable. Furthermore, the customer must experience himself actively in the tests, as is the case, for example, with the test with the favorite colors. Ideally, he can take the test results home with him.

— Customers want to be active themselves in the choice of fragrance and create a personal perfume for themselves.

They are therefore very open and curious with regard to perfume layering and often buy two perfumes instead of one. The possibility of receiving an individual perfume is very attractive to the customer. Experiencing oneself as a perfumer significantly enhances his shopping experience.

- Customers are happy to recommend an event to friends. Events that take place at the same location within a short time frame of three to six weeks or are announced as a double event at the beginning are usually well attended.
- There is an event flow that quickly arouses the interest of many customers and even moves them to wait in line until it is their turn. The waiting line even increases the attractiveness of the event.

Here is the procedure for a promotion that can be carried out as an event workshop with four people for up to 100 customers. The event took place on the occasion of the German "Duftstars 2019" (an event at which the best perfumes of the year are awarded) with the Breuninger department store in Düsseldorf.

The customers received the following information as an invitation to the event:

Neuroperfumery

A fragrance psychologist explains the effect and world of perfumes with the groundbreaking findings of brain research.

A lecture and workshop with a psychological fragrance test and the opportunity to create your own perfume.

Discover with a neuropsychological self-test the stimulation needs of individual brain regions such as the relationships between perfume, personality and moods.

Become your own perfumer and create scents that become well-smelling personal effect perfumes with which you can increase joie de vivre, self-confidence and attractiveness for yourself and for others, because soul needs fragrance.

The following took place at the event evening itself:

1. Introduction with lecture and self-test (20 min)

   Around 100 customers came to the evening event. After an introduction to neuroperfumery, all participants took a self-test (neuropsychological perfume color test) during the lecture. Afterwards, the individual solutions were presented to all participants in the lecture. Finally, the participants were explained the result of the self-test in the form of perfume layering, and tips were given on how to best mix one's own perfume.

2. Scent painting and perfume self-layering

   At various stations, participants were able to take part in a workshop themselves or wait for the lecturer for an individual analysis and consultation. Those who wanted to be active right away could paint a picture with their favorite colors, which was then olfactorily interpreted and assigned scent directions. Most people took advantage of this offer. In line with the individual solutions mentioned in the lecture, perfume testers were grouped and arranged by scent family on tables. The participants were able to carry out their own perfume layering. The lecturer was accompanied by three helpers.

■■ **Scent design, scent sculpture, perfume modulation**

For this event, clay is distributed in different colors and participants are asked to first design the top note—if time permits, also the heart note—of a layered perfume. This is especially suitable for modeling sets with 24 colors as well as with spatulas or other parts. It is exciting to observe how participants experience their new perfume and design it in color and form.

At another event—ideally with eight to twelve participants—one's new perfume can be represented as a sculpture. Alternatively to clay, modeling compound can be used, which dries without oven heat. Afterwards, the raw artwork can be further processed by drilling, grinding, sawing or painting—by the way also a fun for men. The resulting sculptures can be exhibited in one's own perfume shop or shared via social media. Through this kind of event - perfume as sculpture - , it is even possible that the local press will report.

▪▪ Scent of Rhythm

In this event form, participants are asked to put together the appropriate music for the perfume art work created by perfume layering using their own iPod or laptop. This can be developed into a perfume film, in which one or more key scenes describe the mood of the new perfume. Multisensory works of art arise from perfume modulation.

▪▪ Dancing Scent

In this event, the participants express their individual perfume experience while perfume-layering through gestures, movements and dance. In each individual performance, the spectators smell the relevant perfume. The individual dance performances should not be too short. A duration of about three minutes has proven to be effective. In this way, the release of the happiness hormones serotonin and dopamine is achieved through the joyful combination of physical activity with sensory stimuli.

12.4.3 Rediscovering the World of Perfumes Again and Again

The experience perfume store is particularly attractive when it is integrated into an extended artistic context. Exhibitions that offer a sensually creative journey of contemporary scents are usually well attended and attract target groups that one would like to interest in the stationary perfume store. An example of this was the exhibition "A Sensory Journey Through Contemporary Scent" with installations by ten perfumes as well as events and workshops, for example with the perfumer and sound artist Paul Schütze at the Somerset House in London in 2017. The aim was to be inspired by scents anew. Less known, but rather experimental creations of niche perfumes were in the foreground. The exhibition with these less known perfumes was a success. This shows that many perfume users feel the increasing need to experience perfume as a unique and experimental work of art—as with perfume layering.

» *Strategic luck-sampling*

Consumers are especially attracted by chances to win. Here are some tips on how to attract and inspire new customers with a promotion without incurring large marketing costs.

There are gift boxes and treasure chests in all colors and sizes. For this action you need a dice as well as gift boxes/treasure chests in violet-lilac, rose-toned, dark blue, pink-toned and red-yellow, the basic colors of the mood-forms (color forms) presented above, or boxes with a neutral white background, on which these moodforms are printed.

As an invitation to this promotion, you can write on a blackboard:

In each of the five different colored treasure chests, there are six different perfume samples , which are numbered from 1 to 6 and match the scent of the treasure chest (see Moodform Test). So with five treasure chests 30 perfumes can be advertised at the same time. Ideally, each perfume sample has a small lucky message, similar to lucky cookies. For inspiration for lucky messages, sentences from the Moodform solution texts can be used—"What is important now."

To save costs, customers only find a lucky message in the treasure chests. It comes with a fragrance suggestion that is printed on a perfume strip. For the fragrance consultation, only the respective perfume has to be sprayed on the perfume strip.

The lucky scent should be shown together with a second perfume, which is also suitable for layering for a personal scent.

» *Fascination of the not beautiful smell*

Perfumery can also be an invitation to smell differently. The enjoyment of fragrance is complex and not only limited to the beautiful smell, as we know it. This is also due to the trigeminal nerve, which runs through the face and is important for the sense of smell. It can cause very surprising, confusing effects and fragrance enjoyment, but also olfactory outrage.

The trigeminal nerve is also known as the sneeze nerve. For example, one reacts to sudden bright light with still unexplained, repeated sneezing, because it tickles or itches in the nose and one also has to rub one's eyes. But the same feeling in the nose also occurs when smelling. The trigeminal nerve conducts touch stimuli, but also pain to the brain. It can also measure temperatures, and you can even smell with it—and not necessarily unpleasantly. Because what the trigeminal nerve mostly feels, we also enjoy: a little cooling, warming, tingling, sharp, corrosive or biting. Not surprisingly, eucalyptus and menthol, for example as mint, as well as ginger or pepper also work in perfumes via this nerve. With it you can smell and feel things with additional effect, which goes beyond the usual "pleasant vs. unpleasant" and gives us pleasure despite or precisely because of the slight tingling. "Even if someone is completely anosmic, you can still stimulate the trigeminal nerve with appropriate fragrance substances" (Hatt and Dee 2012). This means: You can still experience a kind of olfactory pleasure, even if it is no longer in the normal range of smell and is experienced as unsatisfactory on its own.

The phenomenon known as trigeminal smell can be further intensified if the smell also has something stinging and disgusting. What we perceive as disgust was learned during olfactory socialization—for example during toilet training. If the smell also has something stinging, not only the olfactory memory, but also the trigeminal nerve is activated. This is known as a combination, for example with stale, disgusting-stinging urine smells. Here, learned disgust and innate aversion to stinging come together very typically.

Interestingly, it is precisely the most provoking smells that the history of perfumery knows. Their effect in scents is legendary and has always played a role in great perfumes like "Jicky" or "Shalimar"—even if it is somewhat hidden. The now forbidden secretion of the civet cat, which smells very animal, quite urinous and even fermented, is a classic example. In its pure form it actually smells disgusting-stinking, but in dilution it has a strange pleasure fascination, with cascades of associations in different facets running in the brain. Imaging studies show that when smelling of stinging-disgusting-smelling substances, different areas are activated immediately, because the brain obviously cannot calm down.

Sometimes art also likes to play with these scent impressions, because this type of smell quickly crosses boundaries and it is difficult to distance oneself from them. These smells can hardly be disciplined by the brain. You can get used to them, but they are intrusive, because short-term and at intervals always smelled again, it is difficult to withdraw. Subliminally, one knows that they have something to do with one's own self, with one's own humanization, with the arduous development to the aesthetic personality. Nevertheless, they are part of oneself. It is difficult to reject them. Sigmund Freud even went so far as to say that children in the anal phase (2–3 years of age) are proud of their excrement and experience it as a gift to their parents. The excretion is thus also an expression of adaptation, as the retention is an expression of defiance. The destructive of the analiticity is also reflected in vulgar expressions, e.g. as "fear", "fraud", "to shit on something", but also in the "reprimand" (Helle 2019).

Art that plays with these smells becomes very emotional and original. An ambivalent pleasure arises in the facets between attractive and disgusting, with which the French smell artist and perfumer Christophe Laudamiel likes to play. His works evoke olfactory associations, e.g. between sea, alcohol, wood, urine smells and skin, and present a very faceted pleasure, for which simple evaluations such as "pleasant" or "unpleasant" no longer want to fit, because stink becomes part of the pleasure experience.

» *Have you become interested in this chapter to express your favorite perfume, for example through dance or painting? I would be very pleased about that and, for that reason alone, this chapter would have been worth it.*

Summary

In this chapter we have looked at how to offer the customer, especially in the stationary perfume trade, an experience enhancement and in this context have discussed tools and methods of experience-oriented perfume retailing for more perfume enthusiasm in consulting practice.

The goal of experience-oriented perfume retailing is, first of all, to create curiosity and anticipation for getting to know perfumes, which then leads to a pleasant, unexpectedly positive, but also exciting consultation that can increase to fascination. As tools, we have learned psychological fragrance consulting methods such as the Moodform-Test©, with which the reader can test himself. The test is based on findings from neuroperfumery, color and fragrance psychology and connections with experience wishes. A method for experiencing perfumes more intensively and originally is fragrance dancing, but there are also special occasions such as World Turtle Day(-Week), to which one can present the customer with the most beautiful perfumes to relax and experience inner peace and calm again on an experience display. Furthermore, we discussed methods of experience-oriented perfume retailing such as the treasure island, fragrance painting, the fragrance-color consultation as well as strategic luck-sampling.

References

Christensen JF, Chang DS (2018) Tanzen ist die beste Medizin: warum es uns gesünder, klüger und glücklicher macht. Rowohlt, Reinbek

Hatt H, Dee R (2012) Das kleine Buch vom Riechen und Schmecken. Albrecht Klaus, München

Helle M (2019) Psychotherapie. Springer, Berlin

Higgins ET (1987) Self-discrepancy: a theory relating self and affect. Psychol Rev 94:319–340

Mensing J, Beck C (1988) The psychology of fragrance selection. In: Van Toller S, Dodd GH (Eds) Perfumery: the psychology and biology of fragrance. Chapman & Hall, London, pp 185–204

Parkinson B et al (1996) Changing moods: the psychology of mood and mood regulation. Addison Wesley Longman, New York

Rempel J (2006) Olfaktorische Reize in der Markenkommunikation: Theoretische Grundlagen und empirische Erkenntnisse zum Einsatz von Düften. Springer, Wiesbaden

Schifferstein HNJ, Tanudjaja I (2004) Visualising fragrances through colours: The mediating role of emotions. Perception 33(10):1249–1266

Sorokowski P et al (2014) Color studies in applied psychology and social sciences: an overview. Polish J Appl Psychol 12(2):9

Tamura K et al (2018) Olfactory modulation of colour working memory: how does citrus-like smell influence the memory of orange colour? PLoS ONE 13(9):e0203876

Verbeek C (2018) Dancing scent and aromatizing movement. Hackathon conclusions. Mediamatic Art centre, Amsterdam

Stationary Perfumery in Change

Welcome to the Race for the Stationary Perfumery of Tomorrow

Contents

© The Author(s), under exclusive license to Springer-Verlag GmbH, DE, part of Springer Nature 2023
J. Mensing, *Beautiful SCENT*,
https://doi.org/10.1007/978-3-662-67259-4_13

13

Trailer

There is a lot of fear for the future in stationary perfume shops.

Many retailers are pessimistic when asked if they can continue to exist as a specialty store. There are certainly many competitors today for the classic perfume shop— not just online—trying to break into the premium perfume market, but one must also ask: "Do consumers have new needs that the classic stationary perfume shop does not or not yet cover?" As we will see in this chapter, consumer behavior is changing rapidly in the perfume industry. For the future of perfume retailing, it is therefore more important than ever to understand and respond to how customers want to experience themselves today and which experience desires are increasing. This also brings us to the topic of store design, how a stationary perfume shop can also be designed.

13.1 Elephant Races: The Competition for the "Perfume Sales Location" of Tomorrow

The future of stationary, owner-operated perfume shops is a mostly pessimistically discussed topic not only in Germany. The focus is on the effects of E-Commerce and in particular the perfume online shops on the development of sales of stationary perfume shops with the main areas of fragrance, body care and decorative cosmetics (make-up). But worries are also caused by the growing importance of drugstores, pharmacies, food retailers, but also discounters. They have become increasingly closer to the core range and service of the perfume shop in recent years. That is why strategies for positioning are now urgently needed. This is especially true for medium-sized perfume specialty stores that live from prestigious mid- and high-priced cosmetics.

The total market for all beauty products sold in Germany reached just over 14.3 billion euros in 2022. That is 5.4 percent more than in the previous "Corona year". This includes women's and men's fragrances, skin and facial care products, decorative cosmetics, oral and dental care products, bath and shower additives, deodorants, soaps, shaving care products and other beauty care products. The stationary, mostly owner-operated perfume shops only benefit from this total sales and growth (with forecast of 2.5 percent for 2023) to a small extent, if at all. They have core competencies, but these are also offered by competitors from other industries who are constantly catching up. Let's discuss this using Germany as an example, because classic perfume retail has already changed drastically in many other countries.

13.1.1 Core Competencies of Stationary Perfume Shops

Core competence 1 of the perfume shop is **beauty**. That used to be its unique selling proposition—with the exception of hairdressers, spas, medical spas, cosmetic surgeons and other beauty providers. Increasingly, more and more consumers associate beauty above all with drugstores, which have been recording much higher growth rates for years and have taken a dominant position in the perfume industry in terms of beauty products. Even the food trade and pharmacies have gained ground in this core competence—although to a lesser extent. The competence of caring cosmetics is of course determined to a large extent by the beauty image. In addition, decorative cosmetics and perfumes aim at **attractiveness**, which plays a role above all when choosing a fragrance.

Core competence 2 of the perfume shop is **competentadvice**. However, pharmacies score much better in this area due to the combination of health and trust. Perfume stores are now only trusted to the same extent or less than discounters. Drugstores and the food trade are even enjoying greater trust.

Core competence 3 of the perfume shop is **well-being**, with drugstores catching up here. Well-being is closely linked to the experience of advice. Nevertheless, it is surprising how little well-being is recognized by perfume stores as a major image carrier. So this area is hardly integrated into the training of perfume salespeople in terms of customer approach. Its implementation is also only marginally visible in store design. An exception are the "first in beauty" perfume stores in Germany, which have taken up this topic in their customer magazine (Pilatus 2020).

Core competence 4 of the perfume shop are **good brands**. Also here the drugstore catches up from the perspective of the consumer, while the pharmacies lag behind in this image assessment. The winner in this core competence is the food trade, with which corresponding brands are particularly often associated. From the perspective of the perfume industry and the food trade, this means that in terms of a good brand image, an overlap of beauty and food like Beauty Food is not a problem.

Core competence 5 of the perfume shop is **quality**. Here, the perfume shop still has the upper hand over drugstores and pharmacies, but can learn something from the better image of the food trade. It is still more strongly associated with good quality, especially in combination with pleasure, bio, freshness and authenticity.

Discounters are currently still known for their low prices and special offers, but are becoming more and more similar to the food trade with its supermarkets. In fact, the boundaries of the individual distribution channels are beginning to blur.

Now the following questions arise: Will the stationary perfume shop, as we know it, survive as a specialist shop? Do consumers have new needs that the classic perfume shop no longer covers?

The takeover of stationary, especially owner-operated perfumeries by large perfume groups has not only started in Germany for some time now for reasons of age of the owners. Customers are changing in their wishes and needs. In some countries, such as the USA, classical perfumeries have long been insignificant in terms of market share. There the trend started early towards multi-store operators. Not only for the perfume retail trade, but also for consumer research in Germany, there is an increasing trend: Customers who are offered parallel offers by several distribution channels are not only more satisfied and more willing to spend, but also go shopping more often.

In fact, consumer behavior is currently changing rapidly. For the future of perfume retailing, it is therefore more important than ever to understand how customers want to experience themselves and which needs, experience wishes and habits are increasing—but above all, what fascinates them, how they can be bound to themselves and how they can be positioned accordingly. Taking these points into account, the following simple conclusion can be drawn for the trade: In the race for the perfume of tomorrow, it is not about off- or online, but about how to fascinate customers and—no matter how, when and where they shop—how the business does not lose sales.

Greg Wasson, former president of Walgreens USA, a drugstore chain and one of the world's largest drugstores, puts it in a nutshell: "The company doesn't care where the customer shops, as long as it's at Walgreens—in the store, via the app, online, whenever, wherever he wants to shop." With its about 9000 stores and its share in Walgreens Boots Alliance, which also operates mainly in England, Walgreens has been doing a lot for years to not lose any potential customers. For this purpose, advice, service and range have been expanded in the stores for years. The same applies to the competitor CVS with close to 10,000 stores, which has also primarily invested in customer advice.

The large number of stationary, easily accessible stores is an advantage. But in the age of the Internet, something else is added. The two large perfume and cosmetics retailers have been addressing the "omnichannel consumer", that is, the customer with a preference for cross-channel shopping, for some time now. Accordingly, in addition to perfume and drugstore articles, there are also pharmacy products and a food area at Walgreens. This attracts consumers for whom products, services and information from various product categories are available with a mouse click. In some

Walgreen-Boots stores, deliveries are even made twice a day. Multi-store operators not only reach customers more quickly, but also provide them with an increasingly exciting and stimulating shopping experience than the traditional retailer.

There will certainly continue to be a potential customer base for the perfume retailer in Germany and the neighboring countries, but the average customer base will be increasingly older. In particular, the European consumer appreciates specialist shops in his inner cities, which—as individual cities have had to experience—help significantly against the desolation or an interchangeable shopping experience. Nevertheless, there is something that can be learned from the strategies of American drugstores, for example, which led to the "perfume taxi", which was introduced by some stationary perfume shops with the Corona pandemic.

Drugstores in this country have also intensified their face-to-face customer advice in recent years. This applies to the drugstore company Müller, the drugstore market leader dm, Rossmann as the second largest drugstore chain in Germany and the perfumery group Douglas. All of them have put the stationary owner-operated perfume shops under a lot of pressure—also because they can address a younger clientele.

13.1.2 The Role of Drugstores, Pharmacies, and Grocery Stores

▪ **Drugstores are positioning themselves more and more as perfumeries**

In recent years, drugstores in Germany have strategically expanded their range in the direction of perfumeries and health food stores (typically known as a Reformhaus). At the same time, they are approaching the pharmacy market with online pharmacies. This not only did good for their image, they developed into attractive multi-store operations as a result—like the market leader dm. dm has been cooperating with the Swiss online pharmacy "Zur Rose" since 2013 at least in some countries. As is well known, the drugstore chain dm has been planning its own range of medicines for a long time. According to its own statement, it is or wants to be the European market leader in pharmaceuticals and experiments with a pick-up service or pick-up service for medications in over 2000 branches in Germany (over 3,800 stores in 12 European countries and 66000 employees (2021)). With sales of around 33 billion euros, Germany is the largest pharmaceutical market in Europe—ahead of France and Italy."Zur Rose" seems to have plans of its own. The company is active in numerous European countries; its most important market is Germany, where it operates the online pharmacy DocMorris. "Zur Rose" apparently now sees the online success of the perfumery retailer Douglas, since it's not a big step from Douglas' beauty business to "Zur Rose's" core business – health.

In the competition for the German perfumery of tomorrow, the health image and the greater trust associated with pharmacies are decisive. With the aging of the population, businesses associated with health in general are gaining in importance. Pharmacies have an almost insurmountable lead in terms of health image, even if the number of stationary pharmacies is declining. They are trusted much more than perfumeries.

Whoever can offer health and trust in combination with beauty, attractiveness and well-being in the eyes of consumers and manages to advise competently on new, innovative fragrances and care products has the perfumery of tomorrow almost in the palm of their hand.

Drugstores are already well on their way, because for them the affiliation with pharmacies has already paid off for image reasons alone. So the health image of the drugstores is much better than that of the perfumery. Apart from health food stores/organic stores , only drugstores and pharmacies are associated with health in Germany, based on large shopping areas. Traditional food retailers or supermarkets and discounters have the greatest need to catch up in this area, but also are well on their way.

- **Pharmacies and the run to natural cosmetics**

Only insiders know what role pharmacies with their image of health and trust could play in the perfumery of tomorrow in Germany. The two companies Walgreens and CVS, which belong to the world's largest cosmetics retailers, are primarily pharmacies. Good sales can be generated especially with non-prescription drugs such as cold products. So already only cold products lead customers into the stores, which then buy drugstore goods and perfume articles. Therefore one must ask: Why are not pharmacies in Germany the perfumeries of tomorrow?

In order to protect the health of the German population, the legislature is particularly vigilant over pharmacies. Thus, a pharmacist can only own four stationary stores, can only be represented in his pharmacy for a certain period of time, and even his profit is fixed. This will not change any time soon. So far, all federal governments have held the view that an owner-operated pharmacy with freelance pharmacists working on the basis of high professional standards is best able to ensure the proper provision of medicinal products to the population.

While it is possible to build and maintain pharmacy chains via cooperations, licenses, consulting fees and other services, such as rental contracts, to which perfume shops can also be attached, the individual pharmacists must still be officially independent according to the legislature. Pharmacists who, for example, join together in a cooperation like Pluspunkt (a pharmacy cooperation), are quickly put under pressure as a franchise concept.

Various aspects make the situation even more difficult. For example, a whole range of lucrative additional services, such as vaccinations, which are common in the USA or in Great Britain, are traditionally prohibited for German pharmacists. But Corona brought a change of heart. Since 2020, pharmacists have been allowed to vaccinate customers in pharmacies under strict technical, spatial and organizational conditions. Another hot topic is the sale of advertising space or the support or reimbursement of advertising, decoration and placement costs by pharmaceutical manufacturers, as they are comparable to those provided by perfume and cosmetics manufacturers for perfume shops. They could be considered as influencing patients.

It is no wonder that the number of stationary pharmacies in Germany has declined in recent years and that there were already over 3000 (internet) mail-order pharmacies in Germany in 2020. However, they do not belong to pharmacies licensed in Germany, but operate directly and indirectly from a foreign location, such as the Netherlands or Switzerland. This makes the entrepreneurial situation of German pharmacists increasingly difficult. In the coming years, pharmacies and perfume shops as multi-store businesses are unlikely to play a significant role in the market.

If German pharmacists were less restricted by the legislature and had a different attitude to perfume and cosmetics, the premium brands associated with them and the associated service offerings, which they often see as an image-enhancing, but less as a sales potential, the future of the perfume industry in Germany would have a different dynamics. With the next generation of care, beauty from the outside and inside—for example, beauty from the inside through beauty food dietary supplements—the pharmacy has the chance to position itself even more strongly in the healthcare future

market with the image of health, beauty, attractiveness and innovation.

In one cosmetics segment, the overlap of health and beauty is already very clearly reflected: in natural cosmetics. Pharmacies and drugstores are just beginning to bind this area more to themselves.

In 2018 alone, sales here increased by up to 7% annually (according to the forecast with an annual sales growth of over 5% until 2027), and over a million new customers were addressed. This makes natural cosmetics in Germany a special phenomenon in recent years. It rose to the pearl of trade. Not only does it attract younger target groups, but it also offers different positioning options: from inexpensive to premium/luxury. Another plus: new discoveries in the field of ingredients made natural cosmetics even more effective and special. This opens up new cosmetic segments, for example vegan high-tech cosmetics or a new generation of - active scent cosmetics. They could provide needle-free alternatives to injections in the field of cosmetic surgery—possibly even in combination with well-being and relaxation, as known from aromatherapy.

- ■ **The food trade and the race for the perfumeries of tomorrow**

The food trade is closely monitoring the drugstores. Above all, they see how they are infiltrating the state-protected pharmacy market in Germany and are also getting closer to the perfumeries. For the two market leaders Edeka and Rewe, the pharmacy market is not new. Rewe had already experimented with the mail-order pharmacy DocMorris. However, when the pressure from the pharmacy lobby became too great, they withdrew for the time being. The discounter Netto, which belongs to the Edeka Group, cooperated with the Dutch mail-order pharmacy Almedica until it had to file for bankruptcy. In recent years, it has become clear that the food trade also wants to position itself more strongly in the healthcare future market with the image of health, beauty, attractiveness and innovation.

The food trade with its supermarkets has undergone a profound change. Supermarkets have been winning designer prizes for the "art of shopping" for a few years. Because the importance of the ambience for the shopping pleasure was recognized. As a consequence, supermarkets began to present themselves more and more like luxury brands for product effect and experience.

As a result, there was a lot of redecoration. In order to increase the visual well-being of the customers, shelves were rearranged and adjusted, narrow check-out tables replaced massive cash registers; instead of standing in long queues, customers are politely sent to a free colleague by employees, or even checked out at self-service checkouts.

Despite these innovations, the question arises: Would a customer also be willing to buy luxury perfume brands such as Creed and Bond No. 9 or luxury care from La Prairie at Edeka and Co.? Or vegan fragrance-effective cosmetics from the freshness shelf? Possibly not yet. However, there are more and more supermarkets that look like designer temples inside and out and have greatly upgraded their drugstore department.

13.2 Stationary Perfumery in Transition: Developments & Trends

13.2.1 Outstanding Food Markets as Role Models

Two food stores have won designer prizes for some time now—a signal of the direction in which this industry is developing as a whole. But they also show how a perfumery in the 21st century could present its product groups and increase the fun of shopping and the shopping experience. These are the Coop store in Novoli (Florence) and "Hiebers Frische Center" in Bad Krozingen near Freiburg /Germany. They were awarded the "Euroshop Retail Design Prize" as supermarkets of the year in New York and have found many imitators in their industry.

The 2500 square metre Coop store surprises with the impression of a market hall—but without a colourful jumble, but with an intelligently minimalist pleasure-oriented shopping experience that looks fresh, well-groomed and clean. This is achieved above all by the impressive product presentation at the fresh islands. The dominant colours here are beige and grey. Large glass surfaces, combined with materials from a comfortable Italian kitchen, create a warm and modern atmosphere at the same time. This is underlined by fresh, regional products of the highest quality. In addition, there is plenty of luxury space. Nobody has to squeeze past crammed shelves—similar to the set of a fictional supermarket that Chanel chose for the presentation of its winter fashion in 2014.

In this Coop store, everything is logically arranged. You almost create your next menu while strolling by and get information on how to refine your own kitchen. The employees are dressed in fresh white and remind one of cooks or culinary experts. Individual stops provide inspiration for new culinary combinations and discoveries in a funny and witty way. The neighbourhood of the university campus fits this type of relaxed finding, discovering and exploring. Not surprisingly, a café and a restaurant are part of this shopping paradise. Comfortable parking is a matter of course. Everything is oriented towards customer comfort. They even go one step further: Inviting areas offer space for tastings, seminars and customer events.

With over 2000 square metres, "Hiebers Frische Center" in the spa town of

Bad Krozingen also offers further inspiration for a perfumery of the future. As the first Edeka market, the center was already awarded the Euroshop Retail Design Prize as supermarket of the year in 2013. Even from the outside, the store impresses with a new design concept and unusually elegant architecture such as a 15-degree inclined matt black lacquered exterior wall.

Even the interior has surprises in store. Everything revolves around the creation of product worlds in which products should appear as a harmonious whole and be set in scene by special materials, lighting and floor coverings. Minimalist line drawings and black and white lettering on the back walls identify each department.

"Learn first, then buy" is the motto. Accordingly, "instructional boards", which also serve as design elements, provide information on the individual product areas. Since the demand for fresh, chilled products is constantly increasing, Hieber has equipped the refrigerated shelves with elegant wooden frames in which, for example, smoothies immediately catch the eye. The drugstore department in the middle of the store glows blue-violet-silver like a spaceship.

However, Hieber and Coop in Novoli also want to be a social meeting place for their customers. Business-related events such as cooking courses, wine tastings or market tours are announced via till receipt and the small screens of the self-service checkouts.

In order to involve customers and bind them even more to the business, a customer council was set up which gives suggestions for expanding the range and redesigning the store, but also keeps an eye on environmental guidelines.

Finally, there is an own "quality seal" which communicates the self-claim. Hieber attaches particular importance to the standards of its own employees. They may not be among the best-paid in the industry, but they benefit from flexible work scheduling, lounge with internet access and terrace, continuous training as well as spacious, pop-papered changing rooms with chandeliers. All this is intended to promote the motivation of the employees.

13.2.2 New Beauty Worlds: More and More Perfumeries are Upgrading

In the meantime, a number of stationary perfumeries—including the Douglas PRO Store in Hamburg-Eppendorf and the Douglas store in downtown Düsseldorf (Joachim-Erwin-Platz 1 - opened 2023) with a futuristic look and feel or the flagship store on Düsseldorf's Königsallee (opened 2022), which is actually no longer a perfumery, but rather a beauty and well-being paradise—have already shown how they could position themselves better in general in the future as well as which services and products would have to be offered. The focus is still mainly on skin care and decorative cosmetics. However, it has been recognized that the expert knowledge-dependent advisory level with regard to innovative products and the associated service would have to be given greater importance.

In this context, upgrading has been taking place in the field of care cabins for some time now. They can score points above all in contrast to internet shops with special pampering experiences. But one has also come to an extremely important realization: perfume shops must also carry well-known products from pharmacy cosmetics, such as the well-known brand La Roche-Posay. With special products against skin irritation and problems, the perfume shop increases its health image and gains trust. This should be supported by professionals with pharmaceutical knowledge. Often an experience perfume shop is created that offers services such as those previously only associated with plastic surgery practices and aesthetic dermatology. As early as 2018, perfume shop Douglas examined whether it would be possible to offer treatments, e.g. with Botox, in branches. This was not easy to realize, among other things, because a Botox injection can only be carried out in Germany (unlike in the USA) by a doctor (and not by a dentist). With hyaluron injections, this can still be administered by a naturopath. Meanwhile, some perfume shops work together with doctors and naturopaths. The trend is towards specialist doctors from the fields of plastic surgery or aesthetic dermatology, who, like Just & Peek in Frankfurt am Main, advise and perform injections themselves in stationary perfume shops with expert knowledge and comprehensive information for an intervention. In other words, many perfume shops have now realized that their customers want additional and new service and experience offerings. Fulfilling them now gives perfume shops the chance to expand their core competencies: the experience of "Medical Beautytainment", in order to realize their desired look above all for more joie de vivre and happiness of customers. This is accompanied by serious and specific offers of medical aesthetics and is forcing itself as a trend because more and more customers feel significantly younger than their external appearance suggests and therefore demand solutions. As a result, as already mentioned, an experience perfume shop is currently being created that also offers services such as those previously only associated with plastic surgery practices and aesthetic dermatology.

The range of many perfume shops is also constantly being expanded by special care brands, especially for skin over 50, many of which are not yet represented on the German market or were not previously. In this context, customers can have a personalized facial cream made live on site using artificial intelligence.

Perfume shops can position themselves against the potential supremacy of the food trade by—as already the pharmacies—increasingly relying on beauty food. These are dietary supplements that are supposed to ensure more beautiful and attractive skin, hair and nails. They are offered as a drink, powder or even as a chewing gum, for example in the form of delicious gummy bears. The latter should protect against thin or dull hair and even hair loss with their high vitamin content.

■ **The stationary perfumery of tomorrow: A center of freshness**

The promise and experience of freshness triggers a fascinating psychological association chain in everyone. Because freshness is first of all more or less consciously associated with purity, cleanliness and hygiene, but also with psychological cleansing and personal restart, good air and ultimately with health. Freshness is therefore not only personally, but also socially desirable.

Whoever promises freshness also promises honesty, reason, reliability, quality, untouchedness, spotlessness, honor, conscience, decency, morality, even youthfulness. In addition, freshness is also associated with new, well-being, satisfaction, sympathy and above all with recognition. It is perceived as invigorating, refreshing, sparkling, encouraging, strengthening, stimulating, refreshing, vitalizing, as clarity, brilliance, cleanliness and above all as fragrance. In this context, the color warm-gray, dryness, dusty, sticky-slimy, greasy and of course stench, but also musty and stuffy do not look fresh. The latter is especially a problem of cosmetic departments not only of old department stores. In particular, insufficient ventilation possibilities reduce a fresh experience during fragrance advice.

The food trade has set the perfect staging of freshness. The customer experience associated with it stimulates all senses—tactile, visual, olfactory and gustatory. In the following I show some examples of what the stationary perfumery could take over from the freshness trends of the food trade:

- The staging of healthy freshness with products that are e.g. offered from a noble freshness shelf. Transferred to the perfumery, these could be special smoothies for the beauty from within, but also cooled eau de toilettes, for example from the fragrance direction citrus notes. The perfumery could become a real temple for this growing target group that regularly consumes juices, smoothies and superfood shots. Especially green drinks of this kind promise beauty from within through detoxification. Fat burning and strengthening the immune system. New trendy and valuable vital substances are used: algae such as the detoxifying chlorella or the metabolism-stimulating spirulina, environmental toxins-degrading goji berries or the anti-inflammatory ground substances from the plant lucuma, also called "gold of the Incas". With beauty and health from the freshness shelf of a wellness perfumery, one could also address younger target groups.
- The association of pure, pure and healthy enjoyment by presenting beauty-food dietary supplements in front of a white tiled background in connection with cooking recipes that promise especially also fragrant.
- Multi-sensory installations on presentation islands where perfume ingredients such as kumquats or mint can be served on ice and then smelled or eaten.
- Green walls as nature-related, elegant presentations, for example with different moisture-providing orchids. Small labeled boards indicate in which perfumes the plants can be found.
- Freshness stands, shelves and tables with newly introduced or newly listed perfumes, care products and decorative cosmetics.

- Creative areas for customers to mix different perfume ingredients and fragrance components with the help of, for example, a perfume organ that perfumers work with. Here the customer could create a new perfume under the guidance of an expert with perfume layering. Alternatively, instructions can be given on teaching boards.
- Constantly changing visualized information on individual product areas and topics. Example: Roses, which are used as perfume ingredients, are picked early in the morning to gain them fresh for perfume production in the best quality.
- Social meeting point with café and seminar or event options where, for example, perfume lovers can make new, invigorating contacts.
- Offer that as a member of the customer council you can influence the range or the listing of new brands.

Own, regional and indie brands (independent, often still young perfume, make-up and cosmetic brands) are also gaining importance in the perfume shop.

The trend towards the upscale supermarket is now also evident at discounters—and not just at Aldi in London, where it is trying to go new ways with Aldi local. A good example of this is Lidl in Germany. The discounter is making great efforts to more closely match the image of a supermarket.

Both branches, supermarkets and discounters, have become much more self-confident. The sales of own brands have increased steadily over the past years, while those of branded products have decreased. This fits the international trend of consuming products without brand names. In Switzerland and Spain, the share of own brands of all products sold amounted to 53 and 51% respectively in recent years.

The great majority of Germans now believe that there is no big qualitative difference between brands and own brands. This was not the case a few years ago. The old divisions "brand = valuable" and "own brand = cheap" no longer apply. So far it has been assumed that customers would reach for own brands for reasons of thrift. However, luxury food, fresh organic products and specialties are now also sold in supermarkets under the name of the retailer. For example, Rewe offers the own brand "Rewe frei von" as a specialty with different gluten- and lactose-free products.

Even if this is not yet a big business, it still binds the customer, especially families, to the retailer. It is enough if only one family member is allergic. The probability is quite high that all other purchases will be made in this supermarket. Often, such own brands even develop into a certain type of lifestyle and thus into image carriers, as they are considered healthier.

Also, regional products from local manufacturers such as Black Forest Milk (Schwarzwaldmilch) or eggs from a certain farmer are the winners. They are rated as fresher and above all as authentic. We trust them more because we also associate them with being limited and of high quality.

In the end, it all comes down to the fact that own brands and the local range will be seen as qualitatively equal in the first step and, in the second step, possibly even as qualitatively better than known brands by the consumer. This would then be the final triumph of premium own brands of various kinds and various branches in the supermarket, but also the starting signal for the rapid deflation of large, established brands e.g. from the perfume industry.

- **Indie brands**

It is no coincidence that the Douglas perfumery is experimenting more and more with indie brands. These are usually young, owner- or founder-run cosmetic brands without ties to large corporations, as already mentioned. In 2019, Tina Müller, the former CEO of the Douglas Group, announced at a conference organized by the Perfume Association that the sales of Indie Brands had grown three times faster since 2012 than the turnover of established brands. Meanwhile, large brands such as Estée Lauder have reacted to the new movements in the market and are buying more and more indie brands, including, for example, "too faced" with the best-selling mascara "Better Than Sex". Originally, the good-mood brand was only available online at Amazon. Then Douglas and Sephora grabbed it and discovered how much fun customers had with this brand's personal makeup advice. It remains to be seen whether indie brands will continue to develop over the years or whether they are just a short-lived trend. Nevertheless: With indie brands, a perfumery is always on the winning side, because customers always need something new. For trade, it is important to weigh up how to divide investments between large brands and small indie brands wisely.

The food trade could also become attractive in the next few years for perfumery customers who buy medium and high prices – especially in the area of care products. For some time now, this industry has been unrivalled in the image areas of pleasure, freshness, bio and authenticity. In the image areas of trust and security, it has also moved closer to pharmacies.

- **Premium natural cosmetics straight from the fridge**

In addition, the food trade is also catching up in the field of health with mail-order pharmacies and natural cosmetics. Because consumers have been rewarding the large selection, good quality and low price of branded products for some time. However, this industry is still not associated with the areas of beauty and advice—especially in the medium and high-priced cosmetic segment. But this can change quickly if food retailers offer premium natural cosmetics as their own brand straight from the fridge for extra freshness without preservatives and customers are advised by specialists for inner and outer beauty—of course in combination with matching culinary recipes. The luxury Edekas of the future will therefore have everything it takes to attract medium and high-priced perfumery customers. But do they also offer well-being?

13.2.3 The Trend Towards the "Face-to-Face" Well-being Meeting Point

Currently, the upscale food market is developing into a social "face-to-face" meeting point. In addition, the trend towards the experience restaurant can be observed here. The dishes are not only created to be taken away in front of the guest's eyes, but also to be eaten on site. The elegant combination of supermarket and gastronomy is intended to inspire and bring together customers. Although the development of the food market into a social meeting point started a little bumpily in recent years with freshly prepared pizzas, pasta dishes, sandwiches and desserts, this is the best move that supermarkets can currently make: the offer of places where people feel comfortable is becoming increasingly important.

- **Strategic opportunities for stationary perfumeries: What role does "face-to-face" advice play?**

The desire for personal fragrance and cosmetic advice—also with regard to cosmetic treatments—is increasing. Perfumeries with beauty cabins can confirm this. "Face to face" is considered to be an increasingly strong trend towards well-being and wellness. Why? Anyone who is active in life coaching knows that with many people the mood slowly sinks with increasing age. This is not only true for temporary workers and job seekers, but above all for full-time employees in the private sector. At the same time, the feeling of being alone and with-

out real friends grows. Studies show that the Internet and its social media intensify this feeling of isolation. The simulated networking quickly develops into a personal exile, as the following poem by R. Chandra (2013) describes:

» *I am all over the internet, but nowhere to be found.*
Cyberian exile is complete and profound.
"Connected" to everyone, but alone at my screen -
I'm dissolving into a Silicon dream.

Of course, the Internet and social media such as Instagram, Pinterest, Facebook or Twitter also have many positive aspects and, like the entire E-Commerce, will continue to increase in importance. Mobile commerce in particular is becoming increasingly important. 7.26 billion people currently use smartphones (2023).This not only offers stationary perfume retailers disadvantages, but also unanticipated advantages—not only with regard to perfume lover blogs, with which one can inform oneself about perfumery and cosmetic products. More and more people are taking advantage of the opportunity to pick up, exchange or return their online orders in stores. The aspect of sustainability also plays a role here."Two-thirds of millennials are paying increasing attention to sustainability factors - both in terms of products, services and companies themselves. In the following generations, it should be even more. For example, it is becoming more important how returns or packaging are handled or which companies are actively involved in environmental protection"(Jessica Krahl - Online Handel 2023). Also, more and more people want to have the opportunity to try products physically, but still want to read online reviews and find out about possible alternatives. Even if it seems incredible at first glance, social networks even promote the social importance of stationary stores. However, this is hardly aware of the trade.

Until approximately 2016/17, it was assumed that the internet would destroy the livelihoods of many brick-and-mortar businesses, which has partially come to pass. The future will also show, however, that the two need each other. In the medium term, the internet will actually need brick-and-mortar businesses more, because the convenience of the internet poses a danger. You can lose yourself and isolate yourself relatively quickly, as the Corona pandemic has shown, even if you didn't want to. In isolation, you quickly forget how to be a friend in real life. Psychology has known this for a long time and therefore always recommends direct, personal contact. Because a friend you meet in real life is emotionally much more valuable than one you find online.

It is particularly ideal to spend time in small, real groups (Nef 2008). Already, three good friends in physical contact are much more significant for your own well-being than many in the virtual world. Furthermore, studies show that real friends—compared to more anonymous ones from the web—usually have higher moral and ethical values and contribute more to happiness and satisfaction. Internet relationships also have a number of negative side effects. For example, on Facebook, TikTok and Twitter, hidden envy or feelings of inferiority quickly arises. Or you pretend to be someone else in order to gain recognition and recommendations. This phenomenon is especially common among younger bloggers.

So, to lift your own mood, to feel better, it is better to make plans with friends for a relaxed evening, a joint shopping trip, or a smell tour of a perfume shop. In between, you can have a coffee together and then let your hair down. You know what to expect, and you can look forward to it for a long time—and then enjoy it when the time comes.

This type of small happiness is contagious. Because a happy partner increases the probability of being happy yourself by almost 10%. The probability that your mood will benefit from the good moods of friends, colleagues, and neighbors is even more than 30% (Fowler and Christakis 2009). Conversely, the complaints of colleagues, for example about burnout, can also bring you down. Personal beauty, attractiveness, and wellbeing consulting are the great social contribution and strategic opportunity of brick-and-mortar perfume shops. These functions will become even more important in the future.

If we transfer the observations about mood-enhancing to the perfume retail trade, its customers should benefit twice:

— First, through the product range itself, as perfumes and cosmetics directly target the user's personality, the enhancement of his mood, personal attractiveness and beauty, as well as the experience of individual pleasure and well-being.

— Second, through the professional, whose contagious good mood the customer benefits from in the personal consultation. Because this positive mood, as I said, rubs off on the customer to around 30%.

All this makes it possible that the stationary well-being perfume shop with its personal advice will still be in demand in the future. The social function as a place of personal encounter, where one can look each other in the eye with trust, and the professional competence of the stationary perfume shop will even increase in value—especially the more the Internet is experienced as a personal exile with its many social networks. Middle and higher-priced customers may be quite willing to pay for individual well-being advice.

13.3 Sales Location Perfume Shop: Methods and Strategies for Winning (New) Customers

13.3.1 Target Groups of the Perfume Shop of Tomorrow

Let me draw your attention to some target groups at the end of this topic and ask whether we are hitting the taste of these groups, how a perfume shop should look and what it offers, in the right way. Maybe shopfitters, who are mostly male, are overlooking something that certain - especially female - customers actually want.

First, I would like to talk about the youngest target group of the perfume shop and give you some tips on how to win them over to the perfume shop, but also what they can offer a perfume shop and how to stay in touch with them.

— Teens today are the style consultants for their parents and often even their grandparents. They have a great influence on the choice of fragrances, make-up and care. Even if the parents are already regular customers, the children contribute significantly to maintaining the image of a perfume shop. Therefore, a stationary perfume shop should make a special effort to contact 15- to 17-year-olds and offer them periodic teen workshops on fragrances, make-up and care, and then also win them over for internships.

— Even 12-year-olds are socially networked today. With the apps on their smartphones, they have mobile masterpieces of communication. This is especially true for spontaneous group and opinion formation. So for every occasion group chats can be formed in which quickly 30 people can come together. As a perfume shop, you can offer scent sampling courses at events, where you can test your own sense of smell—

for example, as in entrance examinations for training as a perfumer, where you have to find the similarities of three scents or simply recognize scents in a blind test. Also suitable for getting to know the perfume shop are some tools and methods of the experience perfumery, for example, to design scents through dough as a sculpture, as I have presented in ▶ Chap. 12. Of course, everyone can gain something for their "nasal performance". The works of art are also exhibited or spread through the social networks by the perfume shop.

— When hiring new employees, it should be checked how active the possible future employee is in social networks such as Facebook or Instagram. In addition, you should look at how this person communicates, how they represent themselves online, whether they upload videos of themselves and how many followers they have.

— The younger target group should be given a lot of time to look around the store themselves. Teens don't like it when someone tries to force something on them and they get the impression that someone much older is trying to sell them something.

— The currently dominant life feeling, or the new experience hunger caused by Corona, with the desire for new variety, spontaneity and enjoyment is also reflected in the fragrance trend "Gourmand-fruity". These pleasure fragrances should therefore be offered outside or directly at the entrance of the store for self-testing. In combination with music videos, they have a particularly magnetic effect on younger target groups. The fragrance direction "Gourmand-fruity" is also particularly suitable for the fragrance mailings discussed in ▶ Chap. 7 or for the digital marketing of perfumes, especially if the fragrance preferences of potential new customers are not yet known.

- As a food lifestyle trend, conscious nutrition has crystallized out with regional typical, but also exotic specialties in combination with health and fitness. This is also reflected in the numerous juices that can be found in the refrigerated shelves. The trend towards juice could also be implemented in the perfume shop with a freshness or cooling shelf. On certain days, then regular and new customers could be invited to a "Vegan Happy Beauty Hour", where a connection to selected fragrance substances could then be made.

- In the future market Health & Beauty there is a trend towards special products that address the specific health need "free from". Already one in five consumers is interested in products that are free of certain ingredients and therefore, in his opinion, better suited for sensitive skin. In your own perfume shop you could also set up a special "free from" zone.

In general, it is assumed that the location of a store is decisive for its success. But wouldn't it also be possible that a stationary perfume shop in a 2B or three-location, for example in a barn outside the fringe area and despite low footfall, be successful? Starbucks has tried this successfully with a flagship store in Seattle. Not only do customers save on the relatively high parking fees, but a gourmet café world has also been created that is worth experiencing. It is a roastery in the perfectly renovated building of a car dealer from the 1920s.

The new guideline can be summarized as follows: If people no longer come to Starbucks because of the pedestrian zone, they will have to come to Starbucks in the future. This can also be transferred to some abandoned German inner cities. For example, the Starbucks branch mentioned should have developed into an important meeting place for customers since it opened in December 2014. This may also be a 2B location that has enough parking and is well served by public transport. The size of this café not only allowed for a market hall similar to the aforementioned Coop branch in Italian Novoli, but also the installation of a complete roasting plant as a display object.

In a comparatively large perfume shop, one or more illuminated, gigantic perfume bottles could serve as an eye-catcher. There would also be the possibility to offer very special personal additional services. So something handmade could be created in front of the customers' eyes, for example a Bespoke cosmetics formulated for special skin needs (e.g. as personalized skin care). Or a workstation for a perfumer is set up, who creates their own perfume in front of the customers.

13.3.2 Shopfitting

This brings us to the topic of shopfitting, how a stationary perfume shop could also be designed.

How to set up a world of fragrances and cosmetics is a much-discussed topic among shopfitters. The discussion also depends on whether the shopfitter is a man or a woman. Many men tend to clear, uniform design language—minimally decorated, with a renunciation of color in favor of the colors black, white, chrome, which rather carry the spirit of the Bauhaus style. This will certainly please many women. But if it comes to a shop, i.e. a room in which one feels comfortable, they have additional wishes that often do not quite fit with a minimalist decoration. There is the desire to browse, to decorate, to style, to dream, to marvel, to trust, to discover beauty for all senses. This appeals to women of all ages. There are magazines like *Landlust*, which are read by over 3 million, mostly women. The focus of *Landlust* is on feeling good, relaxation, deceleration, mindfulness, searching for the special in the hidden. The address is consistently a lovingly emotional-valuing attitude.

In the last five years, *Landlust* has been copied by at least 20 similar magazines like *Landhaus*. One might think that the British-inspired lifestyle magazines for the home and garden primarily address a classic target group—the well-situated and affluent, "nature-loving, value-conservative reader" with a high orientation towards brands and quality. Not only. As LOHAS, a consumer group has been described for years that is committed to "green" values for environmental issues, but also health and politics. It is certainly not wrong to say that this was often the mothers, whose children are now campaigning for "Fridays for Future" or for the Last Generation a national student-led climate research organization. With these movements, a design trend has grown more and more over the last three decades, which is based on sustainability, a real lust for decoration with an ecological balance in mind and an authentic feel-good atmosphere: Shabby Chic.

The design trend is said to have been invented by Rachel Ashwell, who showed a creative flea market look in her store in California with Shabby Chic. From this, a store concept has developed that is characterized by imperfect elegance, inviting you to browse and discover. It encourages women in particular to be creative and design their own. With Shabby Chic, every woman is immediately a decoration expert—she can decorate beautifully with simple elements and above all create a lovingly-healed, nostalgically-valuable and relaxing-authentic feel-good atmosphere. Shabby Chic & Dekolust means living and gardening to dream and do it yourself. You can rediscover things, decorate them differently indoors and outdoors. Nothing is lost, everything has its value.

Shabby Chic & Dekolust is simply, but as mentioned also elegant and offers endless individual expression possibilities which can be supplemented, set in scene and collected. Shabby Chic can be an old and modern

store or everything in one. Dekolust needs space, open rooms with surprise and uniqueness in the assortment. An authentic setting is ideal – old warehouses, barns, factories, garden houses, etc. As the perfume store of tomorrow, it could be a "Garden Perfume Café & Juice Bar", a place where friends can meet. This Garden Perfume Café & Juice Bar also has the perfume and beauty tools of tomorrow's perfume store. This is not a contradiction because two currently rapidly growing and increasingly merging target groups (digital creatives and urban trendsetters) combine health awareness and sustainability with technology and nature relatedness, individuality and community spirit and celebrate everything together as pleasure. In this way, the latest unusual high-tech consultation tools can be available in the Garden Perfume Café & Juice Bar – for example, the already mentioned SniffPhone, but also ModiFace, a tool that allows you to test and compare make-up products from different brands, but also to see how a care product can work in the next few weeks, for example, by answering the question of how, for example, with 40% fewer wrinkles around the eyes. Or you can undergo plastic surgery simulation and experience what you might look like after cosmetic surgery, whether you actually want it and whether you better rely on certified natural vegan active cosmetics as an effective alternative to cosmetic surgery. Furthermore, interactive programs on the topics of beauty, nutrition and health are available as well as cooking courses. But you will also be able to create your own active perfumes and make smell scans with them to see how a new favorite scent works on you and your partner in the brain.

» *This chapter was certainly something for insiders of the trade, but you now know some of the possible reasons why your local perfume store may have to close. If you read this, a lot will have changed for the better in our fast-paced time – hopefully. But remember: it is up to our shopping habits whether the cultural asset perfume will continue to be available in stationary, owner-operated perfume stores.*

Summary

The chapter began with the fundamental question: Do target groups that want to address the stationary perfume trade have other, additional, new needs that are not or only partially covered? It is therefore important to pursue this, because a competition for the "sales location perfume" has broken out. Not only perfume on-line shops worry the stationary perfume trade, but also the growing importance of drugstores, pharmacies, even food retailers. They have all approached the core range and service of the perfume trade more strongly in recent years.

Against this background, this chapter has given inspiration for a perfume shop of the future, combined with concrete tips on how to position yourself better, that is, more attractively, and also win new customers. For example, it was suggested to expand the range of the perfume shop of tomorrow and, for example, offer it as a center of fresh pampering experiences for the beauty from within and outside—with premium natural cosmetics and beauty drinks from the refrigerator. Furthermore, the topic of shopfitting was addressed or how to set up a perfume and cosmetics world so that it inspires as a meeting place and place of contagious good mood.

References

Chandra R (2013) I have met the internet and it is not us. Hashtag heroes and internet convos. Psychology Today, Published on December 26, 2013 in The Pacific Heart

Fowler JH, Christakis NA (2009) The dynamic spread of happiness in a large social network. Br Med J 337(768):a2338

Nef—New Economics Foundation (2008) Five ways to well-being. The foresight project on mental capital and wellbeing. Government Office for Science, UK

Pilatus C (2020) Bewegung, Flow, im Fluss sein. Essence 2020—first in beauty. Kundenmagazin

Perfume Trends, International Fragrance Preferences and Mentalities

Fragrance Preferences of Individual Markets in Comparison

Contents

© The Author(s), under exclusive license to Springer-Verlag GmbH, DE, part of Springer Nature 2023
J. Mensing, *Beautiful SCENT*,
https://doi.org/10.1007/978-3-662-67259-4_14

14

Trailer

In this chapter, I would like to introduce you to perfume trends, as well as, one could almost say, long-term fragrance trends of individual fragrance markets, i.e. countries, and show how they differ from the German women's fragrance market. In fact, there are great differences between the markets. Individual fragrance directions, but also ingredients have a different meaning and are smelled in a different context. What fascinates a fragrance psychologist in particular is how strongly fragrance preferences are rooted in the mentalities of individual countries and how they show themselves again and again over generations of fragrance lovers.

Of course, this can be explained by the fragrance socialization. But somehow one always has the feeling that there must be other reasons that explain long-term national fragrance preferences. Should the explanatory models of the Swiss psychoanalyst Carl Gustav Jung and the French philosopher and sociologist Maurice Halbwachs, which

postulated a collective unconscious, also be relevant for the national choice of fragrance and, for example, the preference of the English women's fragrance market, which has loved floral fragrances and, above all, roses for generations, explain? According to this theory of psychoanalysis, historical situations and memories of national importance of individual social groups and societies are stored in the unconscious collective memory. An example would be England's victory under Queen Elizabeth I. against the Spanish Armada in 1588, one of the greatest military victories in English history. The legend tells that Queen Elizabeth I., the virgin English queen, fought her battles in this olfactory aura of a cloud of lavish rose fragrance. The royal nature of the rose was revived by the English royal family again and again in the following. So Elizabeth II. is said to have been enveloped in rose fragrance after the victorious Second World War and at the wedding with Prince Philip (1947). Success, victory and virginity have been attributes of the English rose ever since. I am sure you will read this chapter with interest as a perfume lover who also likes to "smell" beyond the horizon.

14.1 Evaluating Fragrance Preferences

Different market research firms such as Euromonitor international (► www.euromonitor.com/fragrances), Hitlists of large perfume groups such as Douglas as well as economic data from publishers, e.g. from market intern, offer regular insight into the best-selling perfumes of individual retailers, segments and countries. As a rule, the current top 10 for men's and women's fragrances are given, but there are also hitlists with the top 100 best-selling perfumes over the year in a country.

If statistics from different consecutive months and years are available, it is interesting to see which specific perfumes dominate over a period of time in individual fragrance markets. Personally, I like to go back three and more years from the current year. For example, you can then see which women's perfumes have become or are real cult classics in the top charts in individual countries, and to what extent a cult classic of a country like "Dolce & Gabbana light blue" dominates in Germany as well as in other countries or is rather a national, country-specific success phenomenon. Based on these comparative studies, the following questions can then be asked:

- How does the German fragrance market differ from other markets?
- Are there trends in fragrances that are similar to those in fashion?
- Are markets increasingly determined by national preferences?

The answer to these questions will then, as I will show you in an example in a moment, reveal a lot about the DNA of a fragrance market and, from a fragrance psychological point of view, say a lot about its consumers.

14.1.1 Characteristics of the German Women's Fragrance Market

As mentioned before, some statistics also offer a monthly comparison, and one can deduce emerging seasonal perfume trends from them over various years. Thus, the "Jil Sander Sun" perfume, which was introduced to the market in 1989, has become a perennial hit in the German summer women's perfume market—yes, in recent years even a winter hit. In February 2020, the perfume was ranked 4th among Douglas's current best-selling perfumes. The magazine "stern" names 2023 in an article - timeless fragrance bestseller - Jil Sander Sun among the top 10". Jil Sander Sun" is available in other countries too, but it has become a real cult classic in Germany in particular, because the perfume from the "Floral-Ori-

ental" fragrance family has repeatedly been found in the German top charts. If you smell the facets of "Jil Sander Sun", you may also come to a different conclusion about the fragrance impression. The perfume presents itself rather as a crossover of two fragrance families: Gourmand-fruity accents play with a floraloriental sensuality. It is a mixture that appeals to German fragrance users olfactorily above all.

Let me first go into the floral -oriental part, i.e. the Floral-Oriental impression of "Jil Sander Sun". With the fragrance preference "Floriental", the focus is on the search for security, tenderness and harmony. You want to feel more comfortable in your own skin. This fragrance direction with its subtle warmth that it radiates is almost made to merge sensually and gently with the self-smell of the skin.

In fact, the fragrance users of "Jil Sander Sun" appreciate that the perfume smells of summer and sun-drenched skin on the skin and triggers corresponding memories. Indirectly, you are also longing for people who are warm-hearted, who take into account feelings and one's own well-being. This gives rise to the desire to share feelings and thoughts, to feel cared for at the same time and to live out dreams and romantic fantasies unpretentiously. In other words: This fragrance mood pervades much sensuality, but also speaks of withdrawal from a hectic and rational world. The whole thing can intensify in the experience and then be paired with a touch of nostalgia and gentle melancholy, where there is also a hint of world-weariness in the air.

I would not call this mood typically German, because other cultures such as the Portuguese also know it with "Saudade". Nevertheless, the emotional withdrawal into one's own world full of fantasy and romance, especially into the realm of nature, is not unknown to Germans.

Heinrich von Ofterdingen (1772–1801) dreams, as reported above, in Novalis' novel of a blue flower, the quintessential symbol of German Romanticism, and then sets out in search of it:

» *"What attracted him with full force was a tall, light blue flower, which initially stood at the source and touched him with its wide, shiny leaves. Countless flowers of all colors surrounded it, and the most delicious scent filled the air."*

Romantic fantasy flower notes have remained a popular theme in German perfumery to this day. They usually fall into the fragrance category "Floriental"—for years the second favorite fragrance category of Germany's women.

The note created by the French perfumer Pierre Bourdon obviously meets summer and winter experience wishes of many German female and certainly also male fragrance fans with its fruitiness in the scent (from African orange blossoms, bergamot, Amalfi lemon and black currant). Also the heart note, which smells of ylang-ylang, heliotrope and rose, in combination with the base note (ambra, vanilla and tonka bean and the legendary Pierre-Bourdon tobacco sweetness), which has many elements of the fragrance category "Gourmand", delights German women. From the consumers who use "Jil Sander Sun" all year round, one often hears that it is a fragrance that has captured tanned skin in a bottle in the sun. Obviously, the perfume triggers the call for the south over the Alps in Germany, one of the countries with the greatest travel and wanderlust, and certainly also in winter for distant places. The longing for the sunny south, but also the dream of tropical and exotic regions and thus the search for places with more joie de vivre, where one can live happy and sensual fantasies, is deeply rooted in the German psyche. In ► Chap. 8 I have already followed the influence of German Romanticism from the end of the 18th century to the 19th century on German olfactory experience. Artists, thinkers and poets such as Wolfgang von Goethe (1749–1832) have en-

thusiastically described the magical attraction of the south in travel reports and poems and have described it longingly again and again. Goethe's much-quoted opening in Mignon (around 1795/96) is a call for this epoch dreaming of the south, which still has its effect today.

» *"Do you know the land where the lemons bloom, In the dark foliage the gold oranges glow, A gentle wind from the blue sky blows, The myrtle is still and high the laurel stands? Do you know it well? There! There I would like to go with you, O my beloved."* („Kennst du das Land, wo die Zitronen blühn, Im dunkeln Laub die Goldorangen glühn, Ein sanfter Wind vom blauen Him mel weht, Die Myrte still und hoch der Lor beer steht? Kennst du es wohl? Dahin! Dahin möcht' ich mit dir, O mein Geliebter, ziehn.")

To anticipate it at this point: None of the great perfume markets in the world loves "gourmand-fruity" as much as the German, especially not in combination with the pampering feeling of the florientals, which interpret the fruity romantic-tender, almost nostalgic. There were years in the German women's perfume market, including mass market, in which over 30% of the fragrances launched belonged to the "gourmand-fruity" fragrance family and more than 25% of the new introductions could be attributed to the florientals. Currently one can say: Over 50% of the German perfume users love the perfume spectrum "gourmand-fruity-floriental". Therefore, one must above all ask oneself as a perfume psychologist what makes this large fragrance family so attractive to perfume users. Are perfumes of this direction mood enhancers and at the same time soul comfort for perfume users who long for the sunny south, for places with more joie de vivre?

As already discussed in ► Sect. 5.4, the search for more fun, variety, spontaneity and above all enjoyment is in the foreground with the fragrance preference "gourmand-fruity". It is the desire for more joie de vivre, lightness of being, optimism and humor. One wants to be free and playful, to be pampered, to live out fantasies, to implement spontaneous ideas and to provoke a little. They are primarily holiday wishes and -fantasies, in which one wants to experience everything together and at once, to break out of the everyday life, because one has a chronic psychological need for this type of self-experience.

Large and small celebrities (celebrities) primarily lead this ideal experience in front of younger target groups as self-evident, almost as a normal daily claim. Therefore, sociologists have been observing an astonishing phenomenon for years: The ideal experience—to have fun, to be happy, spontaneous, carefree and to provoke a little, while enjoying more and being pampered—has become a latent permanent self-claim, with the motto: That's how it has to be, I need that!

This self-claim is communicated above all in the fragrance preference "Gourmand-fruity", but also in music, e.g. by Katy Perry in her music videos like "This Is How We Do" and also in her album *Witness* with songs like "Bon Appétit" it is about enjoyable seduction. Even with her own perfumes, which Katy Perry brought to the market, she remained true to her motto: "I'm just having a lot of fun." With her fragrance "Spring Reign" (2015), to name just one of her numerous perfumes, "Fruity" is interpreted as cheeky-sexy-enjoyable. Based on her music video for "This is how we do", one could say that the fragrance is like a "twerking ice cream", which dances enjoyably with provocative hip movements. Katy Perry's newer perfumes like "Indi Visible" (2018) also continue the trend of enjoyment with a lot of vanilla, coconut and rum.

No other fragrance family like "Gourmand-fruity" has been launched in Germany so often in recent years. Sometimes new fragrances are based on the smells of

Mediterranean fruits, sometimes on exotic tropical fantasy pleasures or a mixture of both. More and more perfume brands offer sun-hungry people north of the Alps a cheerful, southern-smelling garden of delicious fruits. The "Dolce & Gabbana Fruit Collection", which came on the market in 2020, is an example. Lancôme's "La Vie Est Belle Iris Absolu" - a floral-fruity gourmand fragrance for women (2023) - with orange blossom and fig is another example.

Even the men's fragrance market in Germany knows the fragrance preference "Gourmand-fruity". However, in comparison to the women's perfumes of this fragrance family, it is a small market, although it is growing. After all, there were already over 200 pure Gourmand-fruity men's fragrances on the market in 2017, and their number has increased significantly since then. One example, already mentioned, is "Pirates' Grand Reserve" by Atkinsons, which surprises with a rum accord and reminds one of the ring-shaped, French yeast pastry "Baba au Rhum". A real trend has developed in Gourmand-fruity unisex notes in recent years. Examples are "Lychee & White Mint" from the "4711 Acqua Colonia" series or "Pomegranate & Eucalyptus" from the same line; both fragrances are created for women and men alike.

Currently over 1500 fragrances are available in the German fragrance market from the "Gourmand-fruity Unisex" fragrance family, and the trend is rising. For better overview, this fragrance family is now divided into numerous subgroups for women's, men's and unisex notes:

- Amber-Oriental-Gourmand/Vanilla
- Citrus-Gourmand
- Chypre-fruity
- Aromatic-fruity
- Floral-fruity
- Floral-fruity-Gourmand
- Fruity-woody
- Fruity-fresh
- Sparkling-Gourmand
- Fruity-floriental

It can be said that in recent years, especially in Germany but also in the USA, the number of perfumes from the gourmand-fruity direction has grown enormously.

This brings us to the topic of "fragrance trends" and the question of which fragrance direction will "gourmand-fruity" replace or which new facets of "gourmand-fruity" will gain importance in the coming years.

14.1.2 Perfume Trends or: How a Fragrance and Trend Coach Thinks

The days when the perfume industry only knew *one* big fragrance trend are long gone. Today there are many fragrance trends at the same time and even real anti-trends. What has not changed is the influence of the zeitgeist on the perfume industry and thus the psychosocial factors, social events and developments. So with the beginning of the Covid-19 pandemic in 2020, three fragrance directions in particular became more accepted:

- Light, naturally-soothing, floral and plant notes that stimulate the imagination and relax the especially stress-prone hippocampus and its network during the pandemic.
- Citrusy-aromatic notes that maintain or even increase concentration during home office work and give the person's mood an extraverted "freshness-kick".
- Gourmand notes that remind one of desserts and sweets. They especially address the hypothalamus and its network in the brain. This area is responsible for well-being, reward and small moments of happiness. Gourmand notes also help against "snacking attacks".

But often it was pure chance that enriched the perfume industry with new fragrance ingredients or impressions, which then resulted in a trend. That these fragrance ingredients were recognized for their innovation in the perfume industry and then celebrated a triumph, always owed to the influence of the zeitgeist. Of course, commercial aspects also played a not insignificant role, for example when a certain perfume or a smell impression could be produced more cheaply.

Nevertheless, many new smell impressions have only arisen through long-term targeted research. Some of them were developed for other industries, but then also found application in the perfume industry. So the smell impression "cool" is also of interest in the flavor industry. Just as in the fragrance industry, here the zeitgeist, social events and innovations have an influence on the intensity of research activities.

In 1963, Walentina Terschlowa became the first woman in space. This stimulated the perfume industry; they increasingly sought unusual, modern fragrance notes for women. Accordingly, the smell of metal, steel, ice, snow and coolness inspired the perfume industry of the future from the 1960s and 1970s.

With certain aldehydes (e.g. C-10) such as rose oxide (a molecule from rose oil) and lavender, one came closer and closer to this fragrance impression from the 1960s and 1970s. Fragrance lovers of the fragrance direction "metal" are already looking forward to the olfactory journey into space. Here are some of the perfumes available on the market for this topic with the year of release:

- 1991: "Dreams" by O Boticário
- 1998: "Odeur 53" by Comme des Garçons
- 2017: "Methaldone" by Aether
- 2019: "Metallique" by Tom Ford

But back to the trends. Most of the time, changes in style can be seen first in fashion and music before they are reflected in fragrance trends. As a trend coach for perfumers, I pay special attention to change, i.e. the increasing or decreasing acceptance of music genres. It is interesting to see, for example, how the number of "Views" on Youtube changes. Here is a current example:

The music style "Sexy Female Rap" has been gaining more and more "views" in the last two or three years. For example, in October 2020, Cardi B's—WAP feat. Megan Thee Stallion's video on Youtube had 268 million "views". In May 2021, the video reached almost 400 million "views". As a fragrance coach who has advised and continues to advise many stars (such as Janet Jackson or Naomi Campbell) in fragrance development, I automatically ask myself how a perfume of the rappers Cardi. B or Doja Cat must smell for the target groups or which fragrance direction "Sexy Female Rap" reflects. In discussions with perfumers and in fragrance tests with target groups, one could come to the conclusion that a further development of the fragrance direction "Gourmand-fruity" is needed, namely with elements of the fragrance direction "Amber-Oriental-leathery". The following observations are also based on this: "Gourmand-fruity" stands for the search for fun, variety, spontaneity and above all for pleasure. However, the fragrance direction lacks something of the "I-strength". In addition, it is often too bravely interpreted for the target group, because in many female rap songs it is about "female empowerment", which is controversially and sexy. You could also say: Many fragrances of the fragrance direction "Gourmand-fruity", especially with the tendency "flowery", seem too old and too innocent to today's teenagers. Pop, which typically found its counterpart in "Gourmand-fruity", was omnipresent until just over a year ago, just like Katy Perry's sound, but with Corona and social political events, teenagers are now taking a break from pure pop.

Music and fragrance trends are in constant flux. Of course, individual fragrance brands love specific fragrance directions as a permanent trend, as we will see later. However, individual fragrance directions change or are reinterpreted. I therefore expect the great fragrance trend of recent years, "Gourmand-fruity" with a "flowery" orientation, which is associated with perfumes of pop stars such as Rihanna or Katy Perry, to get a rapping sister. Pop music will also develop further and become more socially relevant—and one can predict that Pop-Rap will gain even more importance.

As far as the olfactory implementation of rap is concerned, there have been fragrance creations since 2019/20 that come quite close to the life feeling expressed by the music. Perfumers began to combine "Gourmand-fruity" very innovatively with "Amber-Oriental-leathery" and thus to implement more edible eroticism in their fragrances. Probably the first perfumes of this kind were not aimed at the core target groups of rap, but they serve as good templates for future developments, for example "Tobacco Mandarin" by Byredo, which came onto the market in 2020. The perfume for women and men can be described as "amber-oriental-woody with fruity Gourmand accents". A leather note, which always vibrates a little, and tobacco leaves, which give the whole thing a certain sexy "I-strength", are well integrated. It is quite possible that a further development of "Tobacco Mandarin" with more leather, some smoky impressions and playful fruits will inspire a new fragrance direction: "Gourmand-fruity-leathery". This rich creations automatically lead to fragrances with a more intense charisma, which has now influenced the entire fragrance industry. Since 2021/22 we have seen an increased trend towards perfumes that are very expressive and that certainly existed before. They are referred to as "Beast-Mode" perfumes and are also particularly long-lasting with up to 18 hours' wear time. They are steadily gaining popularity as fragrances for men. Relatively new are now increasingly "Beast Mode" perfumes for women, such as those launched by Dior in 1985 with the classic Poison. As well as "Beast-Mode" perfumes for women and men, e.g. Baccarat Rouge 540 Extrait Limited Edition 2021 Maison Francis Kurkdjian, with the first version of the perfume being launched in 2015. Even beyond 2023, a special sillage - the perfume trail left in the air when someone leaves the room - will play a major role in new perfumes.

14.2 How the German Women's Fragrance Market Differs from Other Markets

What fragrance preferences does each country have? I would like to follow up on the question of women's notes. For a better international overview, the eight fragrance families presented in ▶ Chap. 5 were reduced to six and renamed as follows for the international comparison:

1. **Chypre** (Chypre-leathery)
2. **Citrus-Green** (Fresh-green-citrus/Aqua- & Ozon-notes)
3. **Fruity** (Gourmand-fruity)
4. **Floral-Aldeydic** (Floral-aldehydic)
5. **Floriental**
6. **Amber-Oriental**

This is certainly a very rough classification. For the comparison of specific countries, it is usually made more detailed in marketing and also analyzes perfumes that fall into two or more fragrance families (see ▶ Sect. 5.3). Nevertheless, this solution already gives a very good insight into the individual fragrance markets for ladies.

14.2.1 Scent Preferences in Spain

In the Spanish fragrance market two fragrance families dominate for women, which are different from Germany: fresh-green and floral notes. Many perfumes are crossovers with elements of both fragrance directions.

From a psychological point of view, the combination describes a buoyant-lively-natural femininity that is more oriented towards the classical female ideal. Traditionally, light, fresh-clean to energizing scents that forego olfactory provocations with their naturalness play a big role in the Span-

ish fragrance market. Examples are "Aqua Lavanda", "Agua Fresca", but also "Aire" and "Aqua de Loewe". If you assign the individual perfumes in the Spanish market from the top charts (top 100) to fragrance families over several years (2016–2019), this typical characteristic becomes apparent:

With a market share of around 27% and 26%, the fragrance families "Fresh-green-citrus / Aqua- & Ozonnotes" and "Floral" predominate, mainly from the aldehydic range. The fragrance families "Floriental" and "Gourmand-fruity" each account for about 15% of the market share. Amber-Oriental (8%) and Chypre notes (5%) have less market importance. Of course, if you only analyze perfumes from the top 25 for a specific year, other weights will result. Analyzed over several years on the basis of top 100, the market importance of "Fresh-green Aqua" and "Floral" is already very characteristic of the DNA of the Spanish fragrance market.

❏ Figures 14.1 and 14.2 show the German and Spanish women's fragrance market in comparison.

On a world map of scents, the Spanish ladies fragrance market is closer to South America than to Germany. In comparison, the German scent market is more to the east towards Poland and the Baltic states (see ❏ Fig. 14.7). The olfactory proximity to South America can certainly also be explained for the Spanish scent market, especially for the Aqua notes, with the climatic conditions. Fresh-green-citrusy Aqua or water fragrance notes are the ideal refreshment on hot days. Olfactory psychology, as discussed in ▶ Chap. 5 in the psychology of fragrance families, however, offers this unisex fragrance direction even more. It stands for new beginnings, the need for freedom and unboundness and of course the desire to experience oneself refreshed, alive, active and open. Behind it is a way of life that goes back to the discovery and conquest of the New World and further back in time and is closely linked to their cultural history. So the departure into the new world, when the colonial conquerors crossed the seas ex-

14

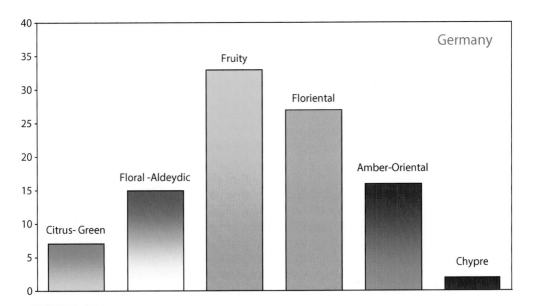

Fig. 14.1   Female notes: market share of fragrance families in Germany

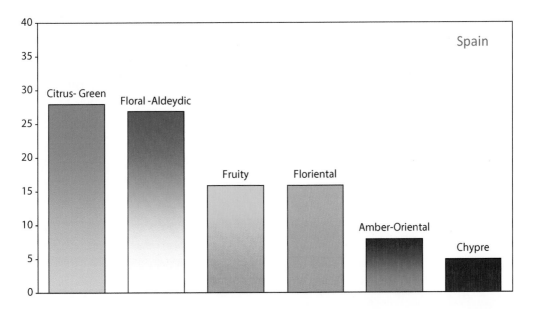

Fig. 14.2   Female notes: market share of fragrance families in Spain

pectantly in the early 16th century to search for the fabulous gold land "Eldorado", was formative for the society. But the Spaniards were also looking for a fountain of youth, a water with magically regenerative powers.

"Aqua" has always been something special for the Spaniards. It was a fragrant life and beauty elixir that the perfumers delivered to the Spanish court by the liter.

14.2.2 **Scent Preferences in Italy**

Of the large scent markets the Italian is the most balanced and liberal when it comes to the importance of the individual female fragrance families. On the world map of scents, Italy is therefore in the middle of Europe (see ◘ Fig. 14.7). If you assign the top 100 Italian perfumes over several years to fragrance families, almost a homogeneous distribution results.

"Gourmand-fruity" is just ahead with a share of around 23%, followed by "Floriental" with about 20%. "Floral", mainly from the aldehydic range, and the fragrance family "Amber-Oriental" each account for about 18%. "Fresh-green-citrus / Aqua- & Ozon-notes" are at about 16%. Only the fragrance family "Chypre" is slightly behind with a market share of just under 5% for women's notes (see ◘ Fig. 14.3).

I will come back to the reasons for the fragrance preferences of Italian women shortly, but in the next section we will first look at the fragrance preferences of French women.

14.2.3 **Scent Preferences in France**

Unlike in Germany, fruit-gourmand notes for women are the second-smallest segment in the France. Interestingly, many French perfumers do not see "Fruity" as its own fragrance family, but rather split it into "Fruity-Floral" or "Fruity-Floriental". "Gourmand" is often also not seen as its own fragrance direction, but rather assigned to other olfactory compounds such as Amber-Oriental notes. In fact, the Amber-Oriental fragrance direction is one of the great characteristics of the French market. Over the years, its share has been about 27% of the total fragrance market. The second most important fragrance family are the Florientals, which over the years have made up about 23% of the fragrance market, followed by the floral-aldehydic with almost the same market importance (◘ Fig. 14.4).

On the world map of female scents, France with the fragrance preference "Amber-Oriental" is more in the Indian Ocean, or Paul Gauguin greets from Tahiti in the Pacific with a flower arrangement from "Floral-Oriental/Floriental". Could it be

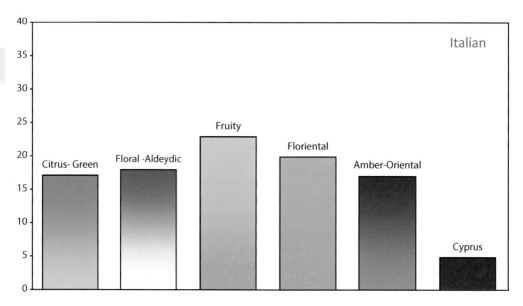

◘ **Fig. 14.3** Female notes: market share of fragrance families in Italy

that in the great fragrance preferences of the French, old colonial claims or today's overseas territories such as Martinique, Réunion or Bora Bora with their magical-exotic beauty are reflected? For the Germans, as already mentioned, it is the longing for the south beyond the Alps and for distant countries that expresses itself in their love for "Gourmand-fruity". French noses go one step further and want to discover fascinatingly extravagant and above all exotic-mysterious-magical depth in their perfumes. This is offered by floral-oriental perfumes, which often—and not by chance—also advertise with the name, such as "Magie Noire".

But back again to Italy. In the Italian perfume market for women's fragrances, which is characterized by the balance of different fragrance directions from different regions of the world map of scents, the former power is reflected again, more precisely: that of Roman noses, which in their best days could afford to procure only the best from all over the world and thus enjoy a sense of smell. One must seriously ask whether the fragrance preferences of individual cultures can be traced back to their glorious history and whether the then life feeling even today's

fragrance preferences or the way of using fragrance still influence. This thought is certainly very speculative. But let's follow it up: Since our sense of smell reacts not only to what has been learned, but also conservative, unconscious and genetically programmed, it could be that it is also linked to emotional moments of the respective national history. This means that impressions, longings and great events of a culture are more or less consciously also on an olfactory level, so to speak as a fragrance mentality, passed on to later generations with certain preferences. In this context, one could postulate that in national olfactory socialization not only learned what smells bad, but also find in the brain dreams and motives from the respective cultural history, which are so emotionally laden that they are also olfactorically passed on as a predisposition.

■ **Is there a collective olfactory subconscious?**

If that were the case, the concept of the collective unconscious of the psychiatrist Carl Gustav Jung (1875–1961) would also be of interest to the perfume industry. Perhaps the concept of the French philosopher and soci-

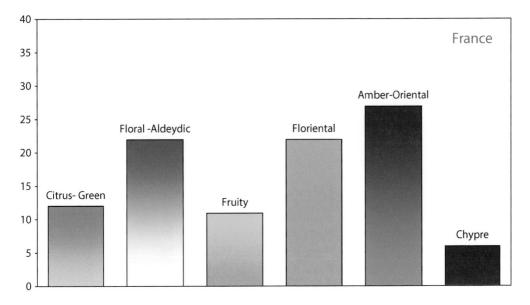

□ **Fig. 14.4** Female notes: Market share of fragrance families in France

ologist Maurice Halbwachs (1877–1945) is even better suited to the perfume industry. He speaks of unconscious collective memories and is rather oriented towards concrete historical situations and memories of individual social groups and societies. Transferred to the perfume industry, one could then ask whether one can also imagine a collective olfactory subconscious or a collective olfactory memory that is fed by the cultural history of a nation. However, international perfume markets know typical national market mentalities that are so strong and lasting that they should not be explained solely by national olfactory socialization.

The possible existence of a collective olfactory subconscious can also be concluded in other perfume markets, as I will show in the following section.

14.2.4 Fragrance Preferences in England and the USA

From a German perspective, the **English ladies' fragrance market** is very special. In it, innocent flowers from the floral-aldehydic fragrance family dominate. Over the years, this fragrance family has had a market share of around 27%. "Fruity" appears especially in combination with lightly "Gourmand-fruity" as well as "Floral-fruity". If one adds fragrances that can be classified as both light and full-bodied Gourmand-fruity as well as Floral-fruity, Fruity with 25% is the second largest group. In third place among the fragrance families with a market share in the English ladies' fragrance market of usually 17% are warm floral notes that fall into the "Floriental" fragrance direction. "Floral" thus appears in various guises in park and gardenland England and dominates the market. On average over the years with less than 14%, "Fresh-green-citrus / Aqua- & Ozon" is followed by oriental notes with around 12% and the extravagant Chypre notes with around 5% share (◘ Fig. 14.5).

The same is almost true for the **North American ladies' fragrance market**, where the floral, i.e. floral theme even dominates with around 29% over the years, slightly more than in England (◘ Fig. 14.6).

England and the USA are a flowery sister market, and their close connection and shared history is also reflected in the olfactory (◘ Fig. 14.7).

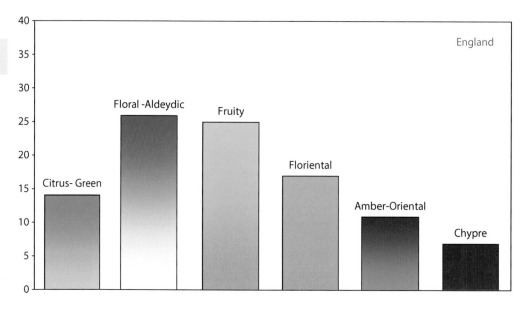

◘ **Fig. 14.5**   Female notes: market share of fragrance families in England

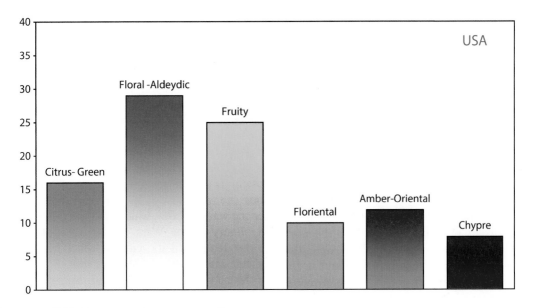

Fig. 14.6 Female notes: market share of fragrance families in the USA

Fig. 14.7 Female notes: Small world map of fragrance markets

How can the dominance of the flowery be explained in both markets? Certainly, you are not wrong if you include the English women's ideal that has grown over centuries in the analysis.

14.3 Scent Mentality: The Rose in the Anglo-Saxon Cultural Area

There have been repeated attempts in the history of perfume to damn it as the work of the devil and immoral. There were also such attempts in Victorian England in the 19th century. Often prudery was given as the reason by historians. But what is overlooked is that the Victorians had an extensive cleanliness mania. That was more than necessary because, above all, the big cities like London were dirty. Foreigners wondered how a wealthy person could live in a city covered with dirt, excrement, soot and dust. Flower notes, especially the smell of rose, must have smelled almost paradisiacal against this background, also because rose in particular rose combines well with almost all other smells and makes unpleasant smells more bearable. No wonder that with "Fleurs de Bulgarie" by Creed in 1845 a fresh smelling rose fragrance came onto the English market, which was probably created for the needs of Queen Victoria at first. The perfume house Creed was founded by a tailor in London in 1760. A better documented example of the importance of rose at that time is "White Rose" by Floris, which was probably brought onto the market even earlier, around 1800. The perfume house Floris was also founded in London in 1730 and is still in family ownership today.

Of course, other flowers and plants such as lavender were also used to cover up unpleasant smells or to perfume, but rose was already part of national identity in Victorian times.

The rose is said to have already been a symbol of love in early high cultures. The rose's triumphal march in the West began when the Virgin Mary was associated with the white rose. The rose not only became an eye-catcher in Gothic cathedrals, but was also used by the nobility—as in the English House of Lancaster, which has the red rose in its coat of arms. The House of York took the white rose as its coat of arms. After the wars of both, which also went down in history as the Wars of the Roses, the Tudor Rose with its red and white petals emerged from the red and white rose.

Much has been written about roses, their history, their scent and their aromatherapeutic effect (Kremp 2013), but this would go beyond the scope of this book. I would like to limit myself here to the meaning of the rose in England. The Tudor Rose has remained the symbol of England to this day. It has been immortalized in the coat of arms of the United Kingdom, but also in the coat of arms of Canada. The United States Army Institute of Heraldry also shows the Tudor Rose in its coat of arms. The institute supports the United States Army and dates back to an executive order by President Woodrow Wilson, the 28th President of the United States (1913–1921). The rose from the Tudor period still connects the two nations so closely today. The rose is therefore not simply seen as a flower by the English, but it embodies strength, albeit subtle, behind which stands a successful royal dynasty with tradition and aura. This suggests that the rose is perceived differently by the English than, for example, by the Germans. It also smells specifically for the English of strong, self-confident and successful femininity. This has reasons that go back to the 16th century and to a woman from the House of Tudor. The legend tells that Queen Elizabeth I. (1533–1603), the virgin English queen, is said to have fought her battles in a personal olfactory aura of fragrant roses. Elizabeth's victory over the Spanish Armada in 1588 was one of the greatest military successes in English history. Elizabeth was described as very intelligent, verbally gifted and above all clever as a child. Men who desired her probably made her wait in an attractive rose-scented cloud. One victim was the French Prince François-Hercule de Valois, whom she called "Frog". She danced with him and he smelled success, but she made him wait until he finally left demotivated. Elizabeth was not a beauty in the common sense as a young queen, but she could fascinate many men with humor, charm and wit. She was also an enthusiastic musician who enjoyed pleasure in every respect. It was Elizabeth who brought vanilla to the European continent as one of the first and enriched her dishes with it. She discovered her love for perfumes and beautiful scents at an early age. So she loved rose water steam baths, combined with other perfumes and aromatic herbs. Elizabeth was not only a rose lover—it is reported that she also appreciated other flowers like many other perfume ingredients. For her personal perfume creations, she specially brought Venetian perfumers to England.

In particular, Elisabeth gave the rose its own female imprint, both as a symbol and olfactorically. Through her, the English have associated the rose, whose aura then also transferred to other flowers, as royal and female, but also as royal strong, successful and intelligent. One can now ask whether these associations, if I understand them correctly, are still subtly passed on by English olfactory socialization today or whether parts of the "rose experience" are not stored in the collective olfactory memory. However, the rose is obviously more than a flower in England. She is royal and female and very complex. This also applies to the much-cited "English Rose", which is more than a naturally attractive girl or a woman with very pale skin. She also has the power and ability to hold her family empire together intellectually, witty and successfully and to achieve great things in her living space. The royal aspect of the rose is also regularly revived on the English women's perfume market at larger intervals. For example, it is said that Elizabeth II wore the "White Rose" by Floris at her wedding to Prince Philip in 1947.

14.4 Global vs. National Perfume Trends

Back to the two unanswered questions:

1. Are there trends in perfumes that show themselves similarly to fashion globally?
2. Are the markets increasingly determined by national preferences?

To put it in advance: Of course there are also trends in perfumes that show themselves globally. Just think of the above-mentioned current triumph of the fragrance family "Gourmand-fruity" over the fragrance direction "Floral", which is more or less strong in many fragrance markets, even if individual countries and regions, as we saw, have typical own fragrance preferences. There are also trends in ingredients. If you look at the markets more closely, you come to the following surprising finding:

Globality is increasing, but also national fragrance individuality!

How is that possible?

We have a divided perfume market in many European countries, in which two target groups in particular dominate, which often also obtain perfumes from different sales channels or sales locations:

— First, from the **classically stationary, owner-operated perfume shop** with customers "45+" who want to distinguish themselves from mass-produced perfumes with niche perfumes and premium classic fragrances. This customer group buys perfumes both online and offline, but is traditionally very interested in personal advice. The dealer tends to suggest scents that are only available from him or that are only difficult to obtain online. This often results in the presentation of scents in perfume advice that are not globally or nationally, or even regionally, adapted to the preferences of the customer. Even with the topic "Gourmand-fruity" there are, for example, slight regional differences in Germany: in northern Germany "fresh-fruity-Gourmand" dominates, in the Berlin area rather fruity-oriental notes. In the western part of the country, the combination "Floriental-Gourmand" is particularly popular and in Bavaria "Fruity-Gourmand". With regional preferences for fragrances, of course, in addition to mentality and olfactory socialization, dietary habits or taste preferences, but also the climate play a role. Just think of the enthusiasm for sweet mustard in or for the often relatively sweeter, therefore less tart beer of Bavaria. The influence of climate and air is particularly evident in the different latitudes of Germany. Chanel No 5 smells different on the streets of Hamburg than in Munich. How often it happened to me that I brought perfumes from Paris, which is still reached by the Atlantic sea air, wonderfully smelled, but did not develop the same class in my southern Baden. In southern Baden we have less often Föhn, a warm, dry downwind, as in Bavaria, but often an inversion weather, in which the upper air layers are warmer than the lower—with the effect that the same fragrance smells slightly different in the sunny highlands of the Black Forest, for example more fruity, as down in the foggy valley.

- Second, from **drugstores and perfume shops or perfume chains** with relatively younger customers, often between the ages of 16 and 34, who more often have the desire to be part of a group and to share experiences and events openly or to recommend products and brands. Often there is almost only self-service in these stores. The dealer then often relies on fragrances that are "pre-sold", that is, that are already known from advertising. And these are, if not always, often international perfumes with high brand awareness. Of course, more drugstores and perfume shops or chains rely on the above-mentioned indie brands and also carry nationally known fragrance brands in their range. Nevertheless, the place of sale plays an important role in whether the range of fragrances or brands is oriented towards global or national fragrance individuality.

As already mentioned, a change took place between 1999 and 2003, which finally contributed to the fact that the two sales locations—privately owned perfumery and perfume chains—developed silently. Of course, as we have already discussed, there have always been national preferences for fragrances, but around the turn of the millennium, many global perfumes dominated the market. The privately owned perfumery also had to take this into account and carry it in the range. There were at least four to five global fragrances per country, often seven to eight fragrances in Germany. Examples are "Allure", "Trésor " or "CK one", which almost dominated the respective lists of the top 10 perfumes worldwide. So you had to have these perfumes in stock as a specialist shop if you wanted to play a role in your place as a premium perfumery. From 2003, national fragrance individuality gradually increased in almost all European fragrance markets. At first it was fragrances of local designers as well as fragrances of national celebrities who were increasingly launched on their own markets. Of course, there have always been perfumes by national celebrities. Just think of the perfumes by Jil Sander and Wolfgang Joop, which came onto the German market as early as the 1980s. Then, around 2004/2005, noble privately owned perfumeries discovered the "niche fragrances" for themselves, the name was not yet born. They could of course also have global reach and be designer fragrances or fragrances of small and large celebrities, but most of the time they were collector's fragrances from small productions that did not claim global reach and were limited to less than 50 sales points, for example in Germany. National niche perfumes were then the logical further development in the niche fragrance market. One who recognized the trend early on was the perfumer and Berliner by choice Geza Schoen, who almost single-handedly brought "Molecule 01" by escentric molecules onto the German fragrance market in 2006.

To stand out with a fragrance as a signature is and was particularly popular with target groups that address the privately owned perfumery. Niche fragrances from small local manufacturers are and were ideal for this. The wearer wants private olfactory experience, to smell beautiful for himself and not be perceived by everyone in the environment. The trend towards niche perfumes has since undergone a further development, but this new trend is not limited to global or national fragrances: perfume layering, the currently most personal way to create new and unusual olfactory active perfume as a very own one, in order to smell and experience more beautiful.

» *You have now read the penultimate chapter of the book, and I hope you have enjoyed it so far. At the end of the book in Chap. 5 I would like to explain to you how and where modern perfumery originated and what role the plague played in it—here you will learn something surprising again.*

Summary

Using the example of preferred fragrance directions for women's notes, we have shown that there are sometimes quite large differences between the individual perfume markets. The German and French perfume markets are quite different, while Spain has many similarities with the South American market. Italy is located on a world map of scents or perfume markets in the middle of Europe; England and the USA, on the other hand, are a sister market in which the floral fragrance dominates.

With regard to the differences in fragrance directions and preferences for individual perfume ingredients, a lot can be explained by a fragrance socialization. But psycho-analytical explanation models could also play a role. According to C.G. Jung, historical situations and memories of the national greatness of individual opinion-makers, for example, are stored in an unconscious collective memory, which could also be coupled to olfactory experiences or memories.

Are perfume markets increasingly determined by national preferences? No, there is a global and a national fragrance trend.

Reference

Kremp D (2013) Majestät Rose. Rosenduft—der Liebe Lust—die Geheimnisse der Königin der Blume. Engelsdorfer, Leipzig

The Emergence of the Modern Perfumer

How and Where the Modern Perfumer Emerged, How Individual Perfumers Could Become Luxury Brands, and What Development Opportunities Female Perfumers have—An Epilogue

Contents

For a number of years, the perfumer has become a research subject in his own right. This has led to an artist's social history of creative noses that enriches and refreshes perfumery, because one is otherwise inclined to concentrate on questions of olfactory perception and the effect of scents. Investigating the perfumer in more detail, as I will show with a few examples, contributes to a better overall understanding of perfumery and, inter alia, makes clear its development into the modern perfumery of today. When researching the perfumer, what is of interest above all is his changing professional and self-image, but also his role and position in society. Of particular interest is the period between the Renaissance and the present day. The presentation of the results takes place as part of symposia, such as in October 2021 at the Palace of Versailles Research Center, to which the perfume house Guerlain and the University of Lyon invited experts (Camus and Wicky 2020). Those not involved will be

somewhat surprised by the newly awakened research interest—but not if they know how great the respective political, economic, and social situation can influence perfumers and their work.

15.1 How and Where the Modern Perfumer Came About

As we discussed in ▶ Chap. 1 some names of perfumers are already known from ancient times and antiquity. Due to their abilities, they must have played a significant role in society at that time, as can be seen from the first perfumer we have Tapputi Tapputi-Belatekallim created her perfumes around 1200 BC in Babylon and was originally a chemist. In the first place, the ruler, his harem, but also higher-ranking and influential people who lived at and around the palace, enjoyed her creations that took into account the individual status and role. Much more is known to perfume historians about the role of perfumers since the beginning of the Renaissance in Italy, in particular about perfumers in the cities Venice and Florence and their importance for the development of modern alcoholic perfumery in the Western world. They have contributed greatly to its construction after the fall of the Roman Empire and after the rediscovery of the great world of scents of the Orient, Africa and Asia by crusaders and by the trade in aromas and spices. Of course, traditions of Roman and Greek practices were also very helpful here. I have already discussed in Sec. 7.3 the great impact that alcoholic perfumery had on the experience of scent.

However, the Renaissance in Italy also brought about a gradual change in the professional and self-image of the creative noses, especially in Venice (Messinis 2017). More and more, perfumers were seen as independent artists and merchants from this epoch onwards, who not only created their perfumes, but also manufactured and sold them to wider population groups (Gobet u. Le Gall 2011). If you trace the production of scent back to the Bronze Age, it was quite different. The production of perfume was linked to the palace and controlled there. Perfume itself was a prestigious product, the enjoyment of which was mostly limited to those who, as already mentioned, lived in and around the palace. It is even assumed today that the use of perfume was limited to a very small circle of people, namely the royal family and the highest ranking. Accordingly, perfume was an olfactory expression of power, wealth, and luxury. An example of this is the use of perfume at the Mycenaean palace of Pylos (Greece) around 1200 BC (Murphy 2012).

One can ask why Venice played the largest role in the development of modern perfumery and why the lagoon city became the first center of European perfume in the 14th century. Since there are different opinions among perfume historians about which Italian city played the greater role in modern perfumery, I would like to go into the background of the question of modern perfumery and the situation of its then-perfumers in order to answer the question.

Florence was very open to artists and new artistic and fashionable developments through the patronage of the Medici noble family, which initially made its money in the textile industry. In particular, Catherine de' Medici was interested in perfumery. Her perfumers were very successful in the early 16th century in the then-modern art of glove perfumery, for which they created very innovative and attractive fragrance notes. In ▶ Sect. 5.10 I have given background information on this. Catherine herself or her perfumers such as René Bianchi (Renato Bianco, also known as Maître René or René le Florentin), who were dependent on her favor—even if they already had their own perfume businesses—, significantly influenced the French functional perfumery.

Nevertheless, I would like to focus on the development of perfumery in Venice, which came into contact with this in Florence long before, through returning crusaders and trade relations with the world of scents of the Orient, Africa and Asia. Venetian traders and perfumers were the first to have personal insights into the perfume production of the Orient and direct contact with the then important trading places for fragrances in the Middle East and North Africa.

While it was Dominican monks who produced mixtures such as the popular rose water since the 13th century in Florence, the perfumery originated in Venice from trade. This was often based on free trade agreements, from which the bourgeois perfumers also benefited greatly because fragrance materials could be purchased more cheaply for perfume production. This also had the effect that one was less dependent on the nobility and the church, both financially and as a client or protégé.

With the beginning of the Renaissance, but also earlier, the trade structures changed, from which the Republic of Venice benefited from free trade in many places. In this context, it became especially possible for the Venetians to borrow money cheaply for their own enterprises or to find one or more silent partners for their own business. Also, the success of the maritime power Venice has gradually improved the living and working conditions in this epoch, although not everywhere.

Between 1347 and 1575, the lagoon city was visited more than 20 times by the plague. As terrible as it was for the residents (only 70,000 of the 120,000 Venetians survived the great plague wave of 1347), the plague had a significant influence on and challenged perfume-makers during this period. This led to progress, but also to regression in the social role and scope of activity of perfume-makers. I would like to explain why this was the case below.

In Europe, especially during times of plague, smoking was used to disinfect, above all, the air in sickrooms. Because there were no real means of combating the Black Death other than hygiene and isolation, perfume-makers were in demand during the Renaissance and even in antiquity to develop and offer hygienic products and applications. The population was looking for health-promising, protective things that should also offer a pleasant smell.

The development of modern Western perfumery and thus also of alcoholic perfumery is therefore to be seen against the background of the plague-ridden Renaissance. Especially Venice had to experience how important hygiene or cleaning was not only for the enjoyment of life, but also for the survival of an entire society. No wonder then that the plague waves contributed to the fact that Venetian perfume-makers were the first in Europe to experiment with disinfecting alcohol in perfumes. Arab perfume-makers had been doing this since the 10th century AD, but the ancient Egyptians had already known alcoholic perfumery since about 400 BC, but had not really used it. With vinegar, which has an anti-inflammatory and antibacterial effect and contains small amounts of alcohol, but smells pungent, the healthy newcomers were washed naked in Venice before they came into 14-day quarantine. What could be more natural than to experiment with scented alcohol, which not only disinfected, but also produced a much nicer smell of plants, above all, than the scented creams and oils used in perfumery since the beginning. With this, one can say: From the need of the plague-ridden Renaissance and its perfume-makers it came now also in Europe to the progress—to the olfactory jump in the modern perfumery.

As in our time through Corona also the use of soap increased during the plague, in the production of which the Venetians of the Renaissance were already true masters.

Also, disinfectant incense was very popular again, which the ancient cultures had already used for air purification. Certainly, it did not serve the progress of perfumery if, during the plague, perfumers descended to "smokers" or merely supplied incense. What further hindered progress and quickly brought perfumers into dubious light was that, as was customary in antiquity, they again assumed the role of a physician or pharmacist to a greater extent during the plague. This had already been warned by Roman and Greek doctors in antiquity, because perfumers could only help with fragrance treatments in case of less serious illnesses, if at all. During the plague, as we know, the plight of the population was exploited from all sides—ecclesiastical as well as secular. Alleged doctors did a brisk trade in miraculous waters, herbal mixtures, fragrance prescriptions, smelling powders, and magical potions that promised to heal the sick and protect the healthy. Because there was a great demand for this, perfumers were also involved, who, like pharmacists, had the supplies for many substances that were needed for the mixtures. There has never been an oath of Hippocrates, the original Greek-language oath of medical ethics, with a corresponding oath of perfumers in the history of perfumery. There would have been many occasions for this. For example, the already mentioned René Bianchi is said to have murdered with perfumed, but also poisoned, gloves at the French court.

Venice developed a sense of hygiene at an early stage. The Magistrature of Venice issued strict disinfection rules and erected a plague hospital in 1423, in which, among other things, rosemary was systematically smoked for the masses. In the following years, the first quarantine station in the Western world was created. In 1490, a separate health authority was founded in Venice. At that time, the interest in myrrh, which was already highly valued by Greeks and Romans in antiquity and antiquity, increased sharply. It was introduced to Europe via Venice, Florence, Genoa, and finally Southern France.

As early as the 11th century, people in Venice had received luxurious perfumes from the Orient, from which they could learn how fragrances could smell in their glory. The perfumes that one learned to know in Venice came from Byzantium (Constantinople). This is evidenced by the move of Maria Argyropoulina (Maria Argyre) to Venice. She was a perfume lover from the ruling family of Byzantium, who in 1004 married the son of the Doge Orseolo in Venice. It is said that she impressed the Venetians more than with her heavy seductive perfumes. The marriage also had a very positive effect on trade between Byzantium and Venice. The Venetians learned from where their trading partner obtained the raw materials for the wonderful perfumes. One place was Trebizond (Trabzon) on the Black Sea coast of today's Turkey, where the Silk Road also ended. Other trading places were located on Cyprus, in Asia Minor and in Egyptian Alexandria. But Trebizond must have been a very important trading place not only for the perfume of these days. After some disagreements, i.e. wars, of which Genoa initially profited, a trade agreement was concluded between Byzantium and Venice in 1268, which now also allowed the Venetians free access to the Black Sea.

The trading skills of the Venetians led to valuable knowledge, which was the opportunity for the local perfumers. They began to experiment more, which led to the breakthrough of alcoholic perfumery. We can therefore say that the development towards modern Western perfumery was forced by plague waves and good trade agreements. This development took place with much pain. So Maria Argyropoulina died with her children a few years after her arrival in Venice of the plague, and many perfumers will have suffered the same fate.

15.2 How Individual Perfumers can Become Luxury Brands

The success of the Venetian perfumers in the Renaissance, however, is based on other factors as well. Written sources such as invoices, diaries, and correspondence will certainly confirm the overall better working conditions for perfumers that began in the Renaissance in further research—just as that in this epoch the costs for most of the fragrance ingredients needed for the production of a perfume were now possible for the majority of perfumers a lucrative work. There were also tax incentives that Venice offered its citizens and merchants. So there were times in its history when taxes and duties were levied on land ownership and absence from military service, but direct taxes were waived. Yes, even duties were completely waived if one exported as much in value as one imported. This also applied to perfume products, which, like spices, silk, pigments and precious stones, were among the most expensive goods.

Tax adjustments also helped to keep Venice attractive as a place to live despite epidemics and wars. This also ensured the influx of specialists and workers, including a sufficient number of apprentices and journeymen who had to be available to ensure the general organization of a perfume workshop and its development. Even in times of economic or health crisis, specialists were often kept against their will in the lagoon city. The glassmakers, who are still working on the lagoon island of Murano today, were even threatened with the death penalty if they were caught moving out of the city or if they passed on their knowledge. Venice, the cradle of Central European glass production, therefore offered its perfumers the environment for their work very early on. They were given containers by the Venetian glass factories to present their creations in an artistic way or to export them nationally and internationally. Today, the perfume brand "The Merchant of Venice" in particular advertises with the perfumes inspired by the local glass art, even though the centres of the perfume industry are now in cities such as Paris and New York. This environment, which developed mainly in Venice between the 14th and 15th centuries, thus also offered aspiring perfumers good prospects for their own careers. They could move from craftsman to investor in their own companies, or even to their own luxury brand with multinational distribution and to an influential figure in public life (Briot 2015).

It was mainly Italian and French perfumers who first used these new market and development opportunities and settled successfully in other countries. So in the 16th century all perfumers in England were either French or Italian (Dugan 2011). The first perfume still sold on the German market, "Farina", was, as mentioned in ▶ Sect. 7.1, created by an Italian, the perfumer Gian Paolo Feminis, from Venice to Cologne. There he and his family opened their own business and sold their scents to national and international customers. By the way, "Farina" was originally named "Aqua Admirabilis" by its Venetian creator.

Of course, the international distribution of fragrance products was not an entrepreneurial achievement of modern times. As we have seen in ▶ Chap. 1, there were already perfumers in antiquity who distributed their products internationally and made a name for themselves with their perfumes or already represented their own perfume brand. But it is no coincidence that the trading power of Venice shaped the history of perfume in Europe in the Renaissance and that the Italian perfumer in particular became a model for the modern, self-marketing cosmopolitan perfumer.

15.3 What Development Opportunities Do Perfumers Have?

Were there also the above-mentioned professional or development opportunities for female perfumers?

Female perfumers like Tapputi certainly played a big role since the beginning of perfumery, even though it was mostly dominated by men (rulers, nobles, priests). Being a perfumer was a profession that women have practiced since antiquity, as well as in modern times, and still do today with a lot of talent. So in 1551 there were eight female perfumers in Lisbon alone (Kennett 1975). In ▶ Sect. 5.9 we discussed ten of today's female "super noses". They alone have created over 700 global perfume hits by 2020. In addition, there are women like Helena Rubinstein (Fitoussi 2019) or Estée Lauder (Epstein 2000) who have very successfully built whole perfume and cosmetics empires or luxury brands. They had the passion for perfume that has been handed down to us by the Egyptian Pharaoh Hatshepsut (around 1495–1459 BC). She even undertook her own expedition to the gold land of Punt, south of the Horn of Africa, in order to get to her favorite scents. By ship and caravan, the Pharaoh brought back numerous frankincense trees from the trip, which she had planted around her temple (see ▶ Sect. 7.2).

I also reported on Elizabeth of Poland, Queen of Hungary (1326–1361 AD), who in the 14th century already brought her own perfume ("Aqua Reginae Hungariae") to the international market through early influencer marketing (see ▶ Sect. 7.4). We also talked about Catherine de' Medici, who, together with her perfumers, founded functional perfumery in France and thus indirectly helped her adopted homeland become an olfactory superpower.

So one can say: It was and is precisely in the field of perfumery that women were able to "pre-live" the modern woman who is also successful in her profession, because the perfume industry gave them (as well as men) the development opportunities that they then used. For female perfumers, there are still many setbacks in our time, as I will show below using the example of Patricia de Nicolaï.

The perfume industry has already reacted relatively early to the deepest and most far-reaching social change of the 20th century with its own perfume creations and fragrance directions—see the discussed Chypre notes (▶ Sect. 5.4)—the fight of women for equal rights and opportunities. This was, as we discussed in ▶ Sect. 8.6 , the trigger that the perfume "Jicky", originally intended as a men's fragrance (1889), became the fragrance of the early women's movement of the 19th century through the influence of the first feminist mass organization during the Paris Commune (from 1870). However, this does not mean that there was already a consciousness of equal opportunities in the perfume industry. In the perfume house Guerlain, which is often considered progressive in perfume history, the gender discrimination within the family was only too well known. Patricia de Nicolaï was born into the Guerlain perfume family; she is the great-granddaughter of Pierre Guerlain and the niece of Jean-Paul Guerlain. But she is also an example of what women can achieve if they do not give up their passion. In the 1980s, Patricia de Nicolaï founded her own luxury fragrance brand—Parfums de Nicolaï—after it became clear that her family would deny her the career of "Guerlain Master Perfumeur". The reason: In the famous perfume house, this title was traditionally passed on from father to son.

Today, more women are being trained at the famous French perfume school ISIPCA than men (see ▶ Sect. 6.1). The big perfume manufacturers see the same trend. More and more female perfumers are following in the footsteps of, for example, Germaine Cellier, who created perfumes such as Balmain's "Vent Vert" (1945/47) in the 1940s, or Josephine Catapano, who created the Estée Lauder classic "Youth Dew" (1952), or Sophia Grojsman, who created "White Linen" (Estée Lauder) in 1978 and "Trésor" (Lancôme) in 1990. The future of perfume, the professional and self-image of a perfumer or a female perfumer, as well as their role and position in society will therefore be shaped by the increasing number of female noses.

Christine Nagel, the first in-house perfumer of the fashion house Hermès, already recognized the trend in 2018 when she noticed that women now make up 80% of perfume students. For Christine Nagel there are good reasons for this. Perfume houses are increasingly relying on a master perfumer who is not only creative, but also decides on the latest perfume with female instinct. At Cartier, the manufacturer of luxury goods, it is her colleague Mathilde Laurent. There are still mostly male noses that have developed into their own luxury perfume brand, like Alberto Morillas. But also in this area, female noses are likely to catch up quickly and inspire perfume lovers with perfumes from their own luxury brands in the future. Today it can already be said: The noses of tomorrow will certainly be mostly female. It is the women who inspire and further develop perfume and thus "smelling better". The trend was already evident in the development of the perfume "Idôle" by Lancôme (2019), for which three ambitious female perfumers from different continents contributed their knowledge of perfume. More and more female perfumers are also winning renowned French FIFI Awards. For example, 2023 Daphné Bugey with 33 Abyssae by L'Artisan Parfumeur for men and women. But perhaps the best news: there are more perfumer teams consisting of female and male perfumers who receive valuable awards. Like Nadège Le Garlantezec, Shyamala Maisondieu, and Antoine Maisondieu with Paradoxe by Prada - the 2023 winner in the Best Women's Launch category.

Now we have arrived at the end of our joint journey into the world of scents, and I hope you enjoyed it as much as I did. Maybe I even inspired one or others of you to pursue a professional path in one of the areas of the scent universe in the future. Others may use the findings of neuroperfumery and scent psychology for self-therapy. However: I very much hope that you all enjoyed reading my book! Maybe we'll meet again at one of my seminars and workshops (▶ Online Perfume Academy).

Summary

We have first looked in this chapter at how and where the modern perfumer emerged in the Western world. The Renaissance in Italy is particularly interesting for the history of perfume. After the fall of the Roman Empire and the rediscovery of the great world of fragrances of the Orient, Africa and Asia by the Crusaders as well as by the trade in aromas and spices, it was above all the trading power of Venice that offered its perfumers development opportunities and inspiration through insider contacts and "favourable" import duties on fragrance raw materials. From this epoch onwards, perfumers were increasingly seen as independent—albeit not independent of rulers—artists and traders who could not only create their perfumes, but also produce and sell them to wider population groups. If you trace the production of perfume back to the Bronze Age, it was quite different. The production of perfume was linked to the palace and controlled from there.

Between 1347 and 1575, Venice was visited more than 20 times by the plague. This was of course terrible for the population, but it stimulated perfumers to further develop their art. Disinfecting alcohol was increasingly used as a fragrance carrier, which the ancient Egyptians already knew in 400 BC, but did not really use.

We have finally looked at the question of whether and, if so, how female perfumers can also benefit from modern perfumery. It turned out that they suffered from open gender discrimination for a long time. But there are also encouraging signs today. Women now make up 80% of perfume students, and luxury brands such as Cartier or Hermès are increasingly relying on a master perfumer for the creation of their perfumes.

References

Briot E (2015) La Fabrique des parfums: naissance d'une industrie de luxe. Vendemiaire, Paris

Camus A, Wicky E (2020) Le parfumeur: évolution d'une figure depuis la Renaissance. Appel à communication/Call for Papers. Centre de recherche du château de Versailles

Dugan H (2011) The ephemeral history of perfume: scent and sense in early modern England. The Johns Hopkins University Press, Baltimore

Epstein R (2000) Estée Lauder: Beauty business success. Franklin Watts, New York

Fitoussi M (2019) Helena Rubinstein: l'aventure de la beauté, catalogue d'exposition. Flammarion, Paris

Gobet M, Le Gall E (2011) Le parfum. H. Champion, Paris

Kennett F (1975) History of perfume. Harrap, Londres

Messinis A (2017) Storia del profumo a Venezia. Lineadacqua, Venise

Murphy JMA (2012) The scent of status: Prestige and perfume at the bronze age palace at Pylos. Southern Illinois University, Greece

Printed by Printforce, the Netherlands